Practical Data Communications Instrumentation and Control

Practical Data Communications for Instrumentation and Control

John Park ASD, IDC Technologies, Perth, Australia

Steve Mackay CPEng, BSc(ElecEng), BSc(Hons), MBA, IDC Technologies, Perth, Australia

Edwin Wright MIPENZ, BSc(Hons), BSc(Elec Eng), IDC Technologies, Perth, Australia

AMSTERDAM • BOSTON • HEIDELBERG • LONDON • NEW YORK • OXFORD
PARIS • SAN DIEGO • SAN FRANCISCO • SINGAPORE • SYDNEY • TOKYO

Newnes is an imprint of Elsevier

Newnes

Newnes
An imprint of Elsevier
Linacre House, Jordan Hill, Oxford OX2 8DP
200 Wheeler Road, Burlington, MA 01803

First published 2003

British Library Cataloguing in Publication Data
A catalogue record for this book is available from the British Library

ISBN   07506 57979

For information on all Newnes publications, visit
our website at www.newnespress.com

Typeset and Edited by Vivek Mehra, Mumbai, India
(vivekmehra@tatanova.com)

Printed and bound in Great Britain

Contents

7   Modems and multiplexers   145

8   Introduction to protocols   186

9   Open systems interconnection model   199

10   Industrial protocols   205

11   HART protocol   239

12   Open industrial Fieldbus and DeviceNet systems   248

13   Local area networks (LANs)   291

Preface

The challenge for the engineer and technician today is to make effective use of modern instrumentation and control systems and 'smart' instruments. This is achieved by linking equipment such as PCs, programmable logic controllers (PLCs), SCADA and distributed control systems, and simple instruments together with data communications systems that are correctly designed and implemented. In other words: *to fully utilize available technology*.

Practical Data Communications for Instrumentation and Control is a comprehensive book covering industrial data communications including RS-232, RS-422, RS-485, industrial protocols, industrial networks, and communication requirements for 'smart' instrumentation.

Once you have studied this book, you will be able to analyze, specify, and debug data communications systems in the instrumentation and control environment, with much of the material presented being derived from many years of experience of the authors. It is especially suited to those who work in an industrial environment and who have little previous experience in data communications and networking.

Typical people who will find this book useful include:

- Instrumentation and control engineers and technicians
- Process control engineers and technicians
- Electrical engineers
- Consulting engineers
- Process development engineers
- Design engineers
- Control systems sales engineers
- Maintenance supervisors

We would hope that you will gain the following from this book:

- The fundamentals of industrial data communications
- How to troubleshoot RS-232 and RS-485 links
- How to install communications cables
- The essentials of industrial Ethernet and local area networks
- How to troubleshoot industrial protocols such as Modbus
- The essentials of Fieldbus and DeviceNet standards

You should have a modicum of electrical knowledge and some exposure to industrial automation systems to derive maximum benefit from this book.

Why do we use RS-232, RS-422, RS-485 ?

One is often criticized for using these terms of reference, since in reality they are obsolete. However, if we briefly examine the history of the organization that defined these standards, it is not difficult to see why they are still in use today, and will probably continue as such.

The common serial interface RS-232 was defined by the Electronics Industry Association (EIA) of America. 'RS' stands for Recommended Standards, and the number (suffix -232) refers to the interface specification of the physical device. The EIA has since established many standards and amassed a library of white papers on various implementations of them. So to keep track of them all it made sense to change the prefix to EIA. (You might find it interesting to know that most of the white papers are NOT free).

The Telecommunications Industry Association (TIA) was formed in 1988, by merging the telecom arms of the EIA and the United States Telecommunications Suppliers Association. The prefix changed again to EIA/TIA-232, (along with all the other serial implementations of course). So now we have TIA-232, TIA-485 etc.

We should also point out that the TIA is a member of the Electronics Industries Alliance (EIA). The alliance is made up of several trade organizations (including the CEA, ECA, GEIA...) that represent the interests of manufacturers of electronics-related products. When someone refers to 'EIA' they are talking about the Alliance, not the Association!

If we still use the terms EIA-232, EIA-422 etc, then they are just as equally obsolete as the 'RS' equivalents. However, when they are referred to as TIA standards some people might give you a quizzical look and ask you to explain yourself... So to cut a long story short, one says 'RS-xxx' and the penny drops.

In the book you are about to read, the authors have painstakingly altered all references for serial interfaces to 'RS-xxx', after being told to change them BACK from 'EIA-xxx'! So from now on, we will continue to use the former terminology. This is a sensible idea, and we trust we are all in agreement!

Why do we use DB-25, DB-9, DB-xx ?

Originally developed by Cannon for military use, the D-sub(miniature) connectors are so-called because the shape of the housing's mating face is like a 'D'. The connectors have 9-, 15-, 25-, 37- and 50-pin configurations, designated DE-9, DA-15, DB-25, DC-37 and DD-50, respectively. Probably the most common connector in the early days was the 25-pin configuration (which has been around for about 40 years), because it permitted use of all available wiring options for the RS-232 interface.

It was expected that RS-232 might be used for synchronous data communications, requiring a timing signal, and thus the extra pin-outs. However this is rarely used in practice, so the smaller 9-position connectors have taken its place as the dominant configuration (for asynchronous serial communications).

Also available in the standard D-sub configurations are a series of high density options with 15-, 26-, 44-, and 62-pin positions. (Possibly there are more, and are usually variations on the original A,B,C,D, or E connector sizes). It is common practice for electronics manufacturers to denote all D-sub connectors with the DB- prefix... particularly for producers of components or board-level products and cables. This has spawned generations of electronics enthusiasts and corporations alike, who refer to the humble D-sub or 'D Connector' in this fashion. It is for this reason alone that we continue the trend for the benefit of the majority who are so familiar with the 'DB' terminology.

The structure of the book is as follows.

Chapter 1: Overview. This chapter gives a brief overview of what is covered in the book with an outline of the essentials and a historical background to industrial data communications.

Chapter 2: Basic principles. The aim of this chapter is to lay the groundwork for the more detailed information presented in the following chapters.

Chapter 3: Serial communication standards. This chapter discusses the main physical interface standards associated with data communications for instrumentation and control systems.

Chapter 4: Error detection. This chapter looks at how errors are produced and the types of error detection, control, and correction available.

Chapter 5: Cabling basics. This chapter discusses the issues in obtaining the best performance from a communication cable by selecting the correct type and size.

Chapter 6: Electrical noise and interference. This chapter examines the various categories of electrical noise and where each of the various noise reduction techniques applies.

Chapter 7: Modems and multiplexers. This chapter reviews the concepts of modems and multiplexers, their practical use, position and importance in the operation of a data communication system.

Chapter 8: Introduction to protocols. This chapter discusses the concept of a protocol which is defined as a set of rules governing the exchange of data between a transmitter and receiver over a communications link or network.

Chapter 9: Open systems interconnection model. The purpose of the Open Systems Interconnection reference model is to provide a common basis for the development of systems interconnection standards. An open system is a system that conforms to specifications and guidelines, which are 'open' to all.

Chapter 10: Industrial protocols. This chapter focusses on the software aspects of protocols (as opposed to the physical aspects which are covered in earlier chapters).

Chapter 11: HART protocol. The Highway Addressable Remote Transducer (HART) protocol is one of a number of smart instrumentation protocols designed for collecting data from instruments, sensors and actuators by digital communication techniques. This chapter examines this in some depth.

Chapter 12: Open industrial Fieldbus and DeviceNet systems. This chapter examines the different Fieldbus and DeviceNet systems on the market with an emphasis on ASI Bus, CanBus and DeviceNet, Interbus-S, Profibus and Foundation Fieldbus.

Chapter 13: Local area networks (LANs). This chapter focuses on networks generally used in industrial data communications with an emphasis on Ethernet.

1

Overview

This chapter introduces data communications, and provides a historical background. It discusses the need for standards in the data communications industry in terms of the physical transfer of information and the way in which data is handled. Finally, it takes a brief look at data communications as they apply to instrumentation and control systems.

Objectives

When you have completed studying this chapter you will be able to:
- Describe the basic principles of all communication systems
- Describe the historical background and evolution of data communications
- Explain the role of standards and protocols
- Describe the OSI model of communication layers
- Describe four important physical standards
- Explain the purpose of instrumentation and control system
- Describe the four most important control devices:
    - DCS
    - PLCs
    - Smart instruments
    - PCs

1.1 Introduction

Data communications is the transfer of information from one point to another. In this book, we are specifically concerned with digital data communication. In this context, 'data' refers to information that is represented by a sequence of zeros and ones; the same sort of data that is handled by computers. Many communications systems handle analog data; examples are the telephone system, radio, and television. Modern instrumentation is almost wholly concerned with the transfer of digital data.

Any communications system requires a transmitter to send information, a receiver to accept it and a link between the two. Types of link include copper wire, optical fiber, radio, and microwave.

Some short distance links use parallel connections; meaning that several wires are required to carry a signal. This sort of connection is confined to devices such as local printers. Virtually all modern data communication use serial links, in which the data is transmitted in sequence over a single circuit.

The digital data is sometimes transferred using a system that is primarily designed for analog communication. A modem, for example, works by using a digital data stream to modulate an analog signal that is sent over a telephone line. At the receiving end, another modem demodulates the signal to reproduce the original digital data. The word 'modem' comes from modulator and demodulator.

There must be mutual agreement on how data is to be encoded, that is, the receiver must be able to understand what the transmitter is sending. The structure in which devices communicate is known as a protocol.

In the past decade many standards and protocols have been established which allow data communications technology to be used more effectively in industry. Designers and users are beginning to realize the tremendous economic and productivity gains possible with the integration of discrete systems that are already in operation.

1.2   Historical background

Although there were many early systems (such as the French chain of semaphore stations) data communications in its modern electronic form started with the invention of the telegraph. The first systems used several parallel wires, but it soon became obvious that for long distances a serial method, over a single pair of wires, was the most economical.

The first practical telegraph system is generally attributed to Samuel Morse. At each end of a link, there was an operator with a sending key and sounder. A message was sent as an encoded series of 'dots' (short pulses) and 'dashes' (longer pulses). This became known as the Morse code and comprised of about 40 characters including the complete alphabet, numbers, and some punctuation. In operation, a sender would first transmit a starting sequence, which would be acknowledged by a receiver. The sender would then transmit the message and wait for a final acknowledgment. Signals could only be transmitted in one direction at a time.

Manual encoding and decoding limited transmission speeds and attempts were soon made to automate the process. The first development was 'teleprinting' in which the dots and dashes were recorded directly onto a rotating drum and could be decoded later by the operator.

The next stage was a machine that could decode the signal and print the actual characters by means of a wheel carrying the typefaces. Although this system persisted for many years, it suffered from synchronization problems.

Perhaps the most severe limitation of Morse code is its use of a variable number of elements to represent the different characters. This can vary from a single dot or dash, to up to six dots and/or dashes, and made it unsuitable for an automated system. An alternative 'code' was invented, in the late 1800s, by the French telegraphic engineer Maurice Emile Baudot. The Baudot code was the first uniform-length binary code. Each character was represented by a standard 5-bit character size. It encoded 32 (2^5) characters, which included all the letters of the alphabet, but no numerals.

The International Telecommunications Union (ITU) later adopted the code as the standard for telegraph communications and incorporated a 'shift' function to accommodate a further set of 32 characters. The term 'baud' was coined in Baudot's honor and used to indicate the rate at which a signal changes state. For example, 100 baud means 100 possible signal changes per second.

The telegraph system used electromechanical devices at each end of a link to encode and decode a message. Later machines allowed a user to encode a message off-line onto punched paper tape, and then transmit the message automatically via a tape reader. At the receiving end, an electric typewriter mechanism printed the text. Facsimile transmission using computer technology, more sophisticated encoding and communications systems, has almost replaced telegraph transmissions.

The steady evolution of data communications has led to the modern era of very high speed systems, built on the sound theoretical and practical foundations established by the early pioneers.

1.3 Standards

Protocols are the structures used within a communications system so that, for example, a computer can talk to a printer. Traditionally, developers of software and hardware platforms have developed protocols, which only their products can use. In order to develop more integrated instrumentation and control systems, standardization of these communication protocols is required.

Standards may evolve from the wide use of one manufacturer's protocol (a *de facto* standard) or may be specifically developed by bodies that represent an industry. Standards allow manufacturers to develop products that will communicate with equipment already in use, which for the customer simplifies the integration of products from different sources.

1.4 Open systems interconnection (OSI) model

The OSI model, developed by the International Standards Organization (ISO), is rapidly gaining industry support. The OSI model reduces every design and communication problem into a number of layers as shown in Figure 1.1. A physical interface standard such as RS-232 would fit into the 'physical layer', while the other layers relate to various other protocols.

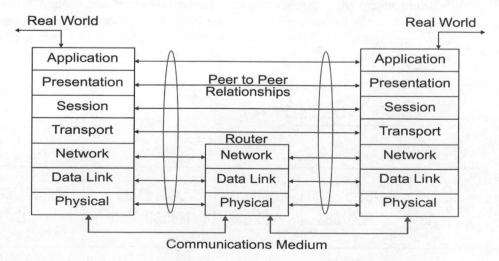

Figure 1.1
Representation of the OSI model

Messages or data are generally sent in packets, which are simply a sequence of bytes. The protocol defines the length of the packet, which is usually fixed. Each packet requires a source address and a destination address so that the system knows where to send it, and the receiver knows where it came from. A packet starts at the top of the protocol stack, the application layer, and passes down through the other software layers until it reaches the physical layer. It is then sent over the link. When traveling down the stack, the packet acquires additional header information at each layer. This tells the next layer down what to do with the packet. At the receiver end, the packet travels up the stack with each piece of header information being stripped off on the way. The application layer only receives the data sent by the application layer at the transmitter.

The arrows between layers in Figure 1.1 indicate that each layer reads the packet as coming from, or going to, the corresponding layer at the opposite end. This is known as peer-to-peer communication, although the actual packet is transported via the physical link. The middle stack in this particular case (representing a router) has only the three lower layers, which is all that is required for the correct transmission of a packet between two devices.

The OSI model is useful in providing a universal framework for all communication systems. However, it does not define the actual protocol to be used at each layer. It is anticipated that groups of manufacturers in different areas of industry will collaborate to define software and hardware standards appropriate to their particular industry. Those seeking an overall framework for their specific communications requirements have enthusiastically embraced the OSI model and used it as a basis for their industry specific standards, such as Fieldbus and HART.

Full market acceptance of these standards has been slow due to uncertainty about widespread acceptance of a particular standard, additional upfront cost to implement the standard, and concern about adequate support and training to maintain the systems.

1.5   Protocols

As previously mentioned, the OSI model provides a framework within which a specific protocol may be defined. A frame (packet) might consist of the following. The first byte can be a string of 1s and 0s to synchronize the receiver or flags to indicate the start of the frame (for use by the receiver). The second byte could contain the destination address detailing where the message is going. The third byte could contain the source address noting where the message originated. The bytes in the middle of the message could be the actual data that has to be sent from transmitter to receiver. The final byte(s) are end-of-frame indicators, which can be error detection codes and/or ending flags.

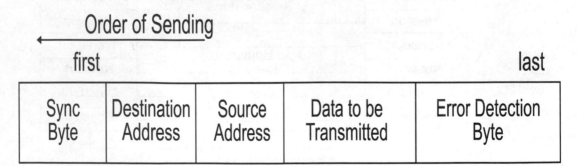

Figure 1.2
Basic structure of an information frame defined by a protocol

Protocols vary from the very simple (such as ASCII based protocols) to the very sophisticated, which operate at high speeds transferring megabits of data per second. There is no right or wrong protocol; the choice depends on the particular application.

1.6   Physical standards

RS-232 interface standard

The RS-232C interface standard was issued in the USA in 1969 to define the electrical and mechanical details of the interface between data terminal equipment (DTE) and data communications equipment (DCE) which employ serial binary data interchange.

In serial Data Communications the communications system might consist of:

- The DTE, a data sending terminal such as a computer, which is the source of the data (usually a series of characters coded into a suitable digital form)
- The DCE, which acts as a data converter (such as a modem) to convert the signal into a form suitable for the communications link e.g. analog signals for the telephone system
- The communications link itself, for example, a telephone system
- A suitable receiver, such as a modem, also a DCE, which converts the analog signal back to a form suitable for the receiving terminal
- A data receiving terminal, such as a printer, also a DTE, which receives the digital pulses for decoding back into a series of characters

Figure 1.3 illustrates the signal flows across a simple serial data communications link.

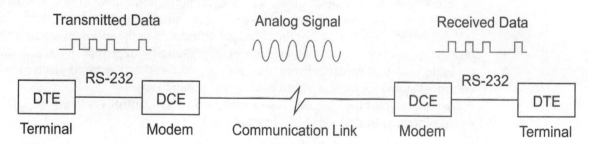

Figure 1.3
A typical serial data communications link

The RS-232C interface standard describes the interface between a terminal (DTE) and a modem (DCE) specifically for the transfer of serial binary digits. It leaves a lot of flexibility to the designers of the hardware and software protocols. With the passage of time, this interface standard has been adapted for use with numerous other types of equipment such as personal computers (PCs), printers, programmable controllers, programmable logic controllers (PLCs), instruments and so on. To recognize these additional applications, the latest version of the standard, RS-232E has expanded the meaning of the acronym DCE from 'data communications equipment' to the more general 'data circuit-terminating equipment".

RS-232 has a number of inherent weaknesses that make it unsuitable for data communications for instrumentation and control in an industrial environment. Consequently, other RS interface standards have been developed to overcome some of these limitations. The most commonly used among them for instrumentation and control systems are RS-423, RS-422 and RS-485. These will be described in more detail in Chapter 3.

RS-423 interface standard

The RS-423 interface standard is an unbalanced system similar to RS-232 with increased range and data transfer rates and up to 10 line receivers per line driver.

RS-422 interface standard

The RS-422 interface system is a balanced system with the same range as RS-423, with increased data rates and up to 10 line receivers per line driver.

RS-485 interface standard

The RS-485 is a balanced system with the same range as RS-422, but with increased data rates and up to 32 transmitters and receivers possible per line.

The RS-485 interface standard is very useful for instrumentation and control systems where several instruments or controllers may be connected together on the same multi-point network.

1.7   Modern instrumentation and control systems

In an instrumentation and control system, data is acquired by measuring instruments and is transmitted to a controller – typically a computer. The controller then transmits data (or control signals) to control devices, which act upon a given process.

Integration of a system enables data to be transferred quickly and effectively between different systems in a plant along a data communications link. This eliminates the need for expensive and unwieldy wiring looms and termination points.

Productivity and quality are the principal objectives in the efficient management of any production activity. Management can be substantially improved by the availability of accurate and timely data. From this we can surmise that a good instrumentation and control system can facilitate both quality and productivity.

The main purpose of an instrumentation and control system, in an industrial environment, is to provide the following:

- **Control of the processes and alarms**
  Traditionally, control of processes, such as temperature and flow, was provided by analog controllers operating on standard 4–20 mA loops. The 4–20 mA standard is utilized by equipment from a wide variety of suppliers. It is common for equipment from various sources to be mixed in the same control system. Stand-alone controllers and instruments have largely been replaced by integrated systems such as distributed control systems (DCS), described below.

- **Control of sequencing, interlocking and alarms**
  Typically, this was provided by relays, timers and other components hardwired into control panels and motor control centers. The sequence control, interlocking and alarm requirements have largely been replaced by PLCs, described in section 1.9.

- **An operator interface for display and control**

Traditionally, process and manufacturing plants were operated from local control panels by several operators, each responsible for a portion of the overall process. Modern control systems tend to use a central control room to monitor the entire plant. The control room is equipped with computer based operator workstations which gather data from the field instrumentation and use it for graphical display, to control processes, to monitor alarms, to control sequencing and for interlocking.

- **Management information**
  Management information was traditionally provided by taking readings from meters, chart recorders, counters, and transducers and from samples taken from the production process. This data is required to monitor the overall performance of a plant or process and to provide the data necessary to manage the process. Data acquisition is now integrated into the overall control system. This eliminates the gathering of information and reduces the time required to correlate and use the information to remove bottlenecks. Good management can achieve substantial productivity gains.

The ability of control equipment to fulfill these requirements has depended on the major advances that have taken place in the fields of integrated electronics, microprocessors and data communications.

The four devices that have made the most significant impact on how plants are controlled are:

- Distributed control system (DCS)
- Programmable logic controllers (PLCs)
- Smart instruments (SIs)
- PCs

1.8   Distributed control systems (DCSs)

A DCS is hardware and software based digital process control and data acquisition based system. The DCS is based on a data highway and has a modular, distributed, but integrated architecture. Each module performs a specific dedicated task such as the operator interface/analog or loop control/digital control. There is normally an interface unit situated on the data highway allowing easy connection to other devices such as PLCs and supervisory computer devices.

1.9   Programmable logic controllers (PLCs)

PLCs were developed in the late sixties to replace collections of electromagnetic relays, particularly in the automobile manufacturing industry. They were primarily used for sequence control and interlocking with racks of on/off inputs and outputs, called digital I/O. They are controlled by a central processor using easily written 'ladderlogic' type programs. Modern PLCs now include analog and digital I/O modules as well as sophisticated programming capabilities similar to a DCS e.g. PID loop programming. High speed inter-PLC links are also available, such as 10 and 100 Mbps Ethernet. A diagram of a typical PLC system is given in Figure 1.4.

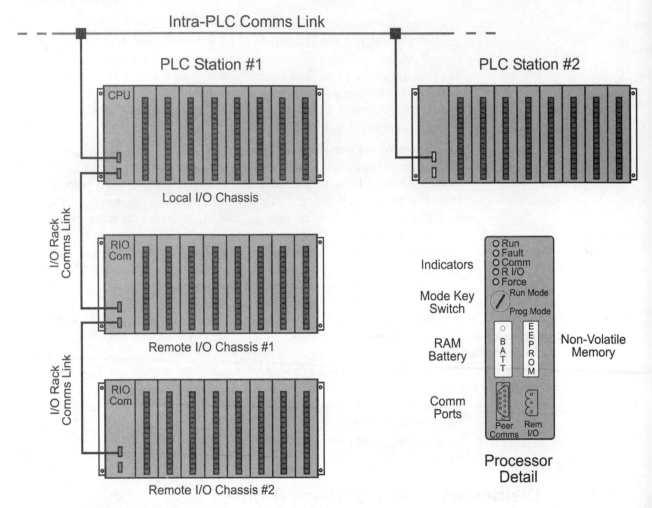

Figure 1.4
A typical PLC system

1.10   Impact of the microprocessor

The microprocessor has had an enormous impact on instrumentation and control systems. Historically, an instrument had a single dedicated function. Controllers were localized and, although commonly computerized, they were designed for a specific purpose.

It has become apparent that a microprocessor, as a general-purpose device, can replace localized and highly site-specific controllers. Centralized microprocessors, which can analyze and display data as well as calculate and transmit control signals, are capable of greater efficiency, productivity, and quality gains.

Currently, a microprocessor connected directly to sensors and a controller, requires an interface card. This implements the hardware layer of the protocol stack and in conjunction with appropriate software, allows the microprocessor to communicate with other devices in the system. There are many instrumentation and control software and hardware packages; some are designed for particular proprietary systems and others are more general-purpose. Interface hardware and software now available for microprocessors cover virtually all the communications requirements for instrumentation and control.

As a microprocessor is relatively cheap, it can be upgraded as newer and faster models become available, thus improving the performance of the instrumentation and control system.

1.11 Smart instrumentation systems

In the 1960s, the 4–20 mA analog interface was established as the *de facto* standard for instrumentation technology. As a result, the manufacturers of instrumentation equipment had a standard communication interface on which to base their products. Users had a choice of instruments and sensors, from a wide range of suppliers, which could be integrated into their control systems.

With the advent of microprocessors and the development of digital technology, the situation has changed. Most users appreciate the many advantages of digital instruments. These include more information being displayed on a single instrument, local and remote display, reliability, economy, self tuning, and diagnostic capability. There is a gradual shift from analog to digital technology.

There are a number of intelligent digital sensors, with digital communications, capability for most traditional applications. These include sensors for measuring temperature, pressure, levels, flow, mass (weight), density, and power system parameters. These new intelligent digital sensors are known as 'smart' instrumentation.

The main features that define a 'smart' instrument are:

- Intelligent, digital sensors
- Digital data communications capability
- Ability to be multidropped with other devices

There is also an emerging range of intelligent, communicating, digital devices that could be called 'smart' actuators. Examples of these are devices such as variable speed drives, soft starters, protection relays, and switchgear control with digital communication facilities.

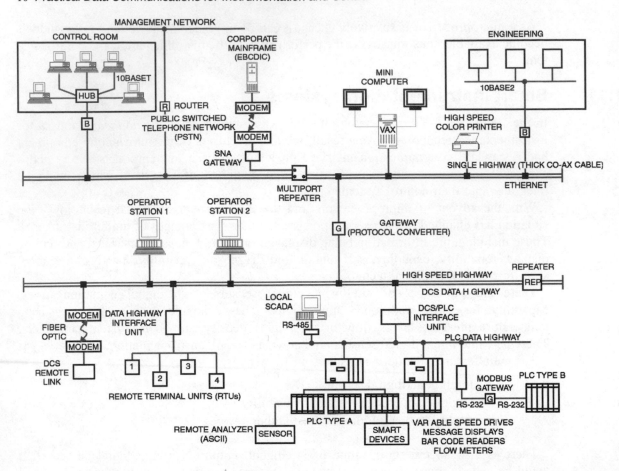

Figure 1.5
Graphical representation of data communications

2

Basic principles

The aim of this chapter is to lay the groundwork for the more detailed information presented in the following chapters.

Objectives

When you have completed study of this chapter you will be able to:

- Explain the basics of the binary numbering system – bits, bytes and characters
- Describe the factors that affect transmission speed:
    - Bandwidth
    - Signal-to-noise ratio
    - Data throughput
    - Error rate
- Explain the basic components of a communication system
- Describe the three communication modes
- Describe the message format and error detection in asynchronous communication systems
- List and explain the most common data codes:
    - Baudot
    - ASCII
    - EBCDIC
    - 4-bit binary code
    - Gray code
    - Binary coded decimal (BCD)
- Describe the message format and error detection in synchronous communication systems
- Describe the universal asynchronous transmitter/receiver

2.1   Bits, bytes and characters

A computer uses the binary numbering system, which has only two digits, 0 and 1. Any number can be represented by a string of these digits, known as bits (from binary digit). For example, the decimal number 5 is equal to the binary number 101.

Bit	1 or a 0
Dibit	two bits (10)
Nibble	four bits (1001 or one Hex character)
Byte	eight bits or two nibbles (11000001, C1 Hex)
Word	the width of the bus of the computer

Table 2.1
Different sets of bits

As a bit can have only two values, it can be represented by a voltage that is either on (1) or off (0). This is also known as logical 1 and logical 0. Typical values used in a computer are 0 V for logical 0 and +5 V for logical 1, although it could also be the other way around i.e. 0 V for 1 and +5 V for 0.

A string of eight bits is called a 'byte' (or octet), and can have values ranging from 0 (0000 0000) to 255_{10} (1111 1111_2). Computers generally manipulate data in bytes or multiples of bytes.

Decimal	Hexadecimal	Binary
0	0	0000
1	1	0001
2	2	0010
3	3	0011
4	4	0100
5	5	0101
6	6	0110
7	7	0111
8	8	1000
9	9	1001
10	A	1010
11	B	1011
12	C	1100
13	D	1101
14	E	1110
15	F	1111

Table 2.2
The hexadecimal table

Programmers use 'hexadecimal' notation because it is a more convenient way of defining and dealing with bytes. In the hexadecimal numbering system, there are 16 digits (0–9 and A–F) each of which is represented by four bits. A byte is therefore represented by two hexadecimal digits.

A 'character' is a symbol that can be printed. The alphabet, both upper and lower case, numerals, punctuation marks and symbols such as '*' and '&' are all characters. A computer needs to express these characters in such a way that they can be understood by other computers and devices. The most common code for achieving this is the American Standard Code for Information Interchange (ASCII) described in section 2.8.

2.2   Communication principles

Every data communications system requires:

- A source of data (a transmitter or line driver), which converts the information into a form suitable for transmission over a link
- A receiver that accepts the signal and converts it back into the original data
- A communications link that transports the signals. This can be copper wire, optical fiber, and radio or satellite link

In addition, the transmitter and receiver must be able to understand each other. This requires agreement on a number of factors. The most important are:

- The type of signaling used
- Defining a logical '1' and a logical '0'
- The codes that represent the symbols
- Maintaining synchronization between transmitter and receiver
- How the flow of data is controlled, so that the receiver is not swamped
- How to detect and correct transmission errors

The physical factors are referred to as the 'interface standard'; the other factors comprise the 'protocols'.

The physical method of transferring data across a communication link varies according to the medium used. The binary values 0 and 1, for example, can be signaled by the presence or absence of a voltage on a copper wire, by a pair of audio tones generated and decoded by a modem in the case of the telephone system, or by the use of modulated light in the case of optical fiber.

2.3   Communication modes

In any communications link connecting two devices, data can be sent in one of three communication modes. These are:

- Simplex
- Half duplex
- Full duplex

A simplex system is one that is designed for sending messages in one direction only. This is illustrated in Figure 2.1.

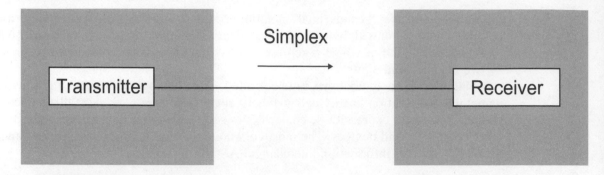

Figure 2.1
Simplex communications

A duplex system is designed for sending messages in both directions.

Half duplex occurs when data can flow in both directions, but in only one direction at a time (Figure 2.2).

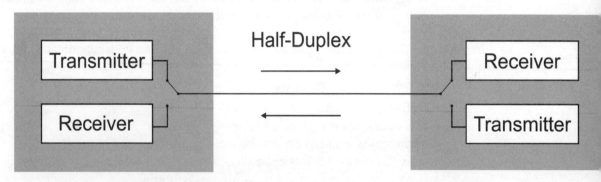

Figure 2.2
Half-duplex communications

In a full-duplex system, the data can flow in both directions simultaneously (Figure 2.3).

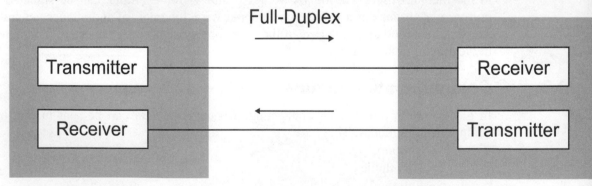

Figure 2.3
Full duplex communications

2.4   Asynchronous systems

An asynchronous system is one in which each character or byte is sent within a frame. The receiver does not start detection until it receives the first bit, known as the 'start bit'.

The start bit is in the opposite voltage state to the idle voltage and allows the receiver to synchronize to the transmitter for the following data in the frame.

The receiver reads in the individual bits of the frame as they arrive, seeing either the logic 0 voltage or the logic 1 voltage at the appropriate time. The 'clock' rate at each end must be the same so that the receiver looks for each bit at the time the transmitter sends it. However, as the clocks are synchronized at the start of each frame, some variation can be tolerated at lower transmission speeds. The allowable variation decreases as data transmission rates increase, and asynchronous communication can have problems at high speeds (above 100 kbps).

Message format

An asynchronous frame may have the following format:

Start bit: Signals the start of the frame
Data: Usually 7 or 8 bits of data, but can be 5 or 6 bits
Parity bit: Optional error detection bit
Stop bit(s):   Usually 1, 1.5 or 2 bits. A value of 1.5 means that the level is held for 1.5 times as long as for a single bit

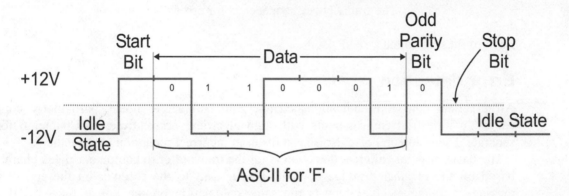

Figure 2.4
Asynchronous frame format

An asynchronous frame format is shown in Figure 2.4. The transmitter and receiver must be set to exactly the same configuration so that the data can be correctly extracted from the frame. As each character has its own frame, the actual data transmission speed is less than the bit rate. For example, with a start bit, seven data bits, one parity bit and one stop bit, there are ten bits needed to send seven bits of data. Thus the transmission of useful data is 70% of the overall bit rate.

2.5   Synchronous systems

In synchronous systems, the receiver initially synchronizes to the transmitter's clock pulses, which are incorporated in the transmitted data stream. This enables the receiver to maintain its synchronization throughout large messages, which could typically be up to 4500 bytes (36 000 bits). This allows large frames to be transmitted efficiently at high data rates. The synchronous system packs many characters together and sends them as a continuous stream, called a packet or a frame.

Message format

A typical synchronous system frame format is shown below in Figure 2.5.

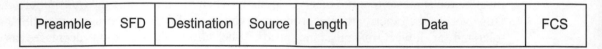

Preamble	SFD	Destination	Source	Length	Data	FCS

Figure 2.5
Typical synchronous system frame format

Preamble: This comprises one or more bytes that allow the receiving unit to synchronize with the frame.
SFD: The start of frame delimiter signals the beginning of the frame.
Destination: The address to which the frame is sent.
Source: The address from which the frame originated.
Length: The number of bytes in the data field.
Data: The actual message.
FCS: The frame check sequence is for error detection.

Each of these is called a field.

2.6   Error detection

All practical data communications channels are subject to noise, particularly copper cables in industrial environments with high electrical noise. Refer to Chapter 6 for a separate discussion on noise. Noise can result in incorrect reception of the data.

The basic principle of error detection is for the transmitter to compute a check character based on the original message content. This is sent to the receiver on the end of the message and the receiver repeats the same calculation on the bits it receives. If the computed check character does not match the one sent, we assume an error has occurred. The various methods of error detection are covered in Chapter 4.

The simplest form of error checking in asynchronous systems is to incorporate a parity bit, which may be even or odd.

Even parity requires the total number of data bits at logic 1 plus the parity bit to equal an even number. The communications hardware at the transmission end calculates the parity required and sets the parity bit to give an even number of logic 1 bits.

Odd parity works in the same way as even parity, except that the parity bit is adjusted so that the total number of logic 1 bits, including the parity bit, equals an odd number.

The hardware at the receiving end determines the total number of logic 1 bits and reports an error if it is not an appropriate even or odd number. The receiver hardware also detects receiver overruns and frame errors.

Statistically, use of a parity bit has only about a 50% chance of detecting an error on a high speed system. This method can detect an odd number of bits in error and will not detect an even number of bits in error. The parity bit is normally omitted if there are more sophisticated error checking schemes in place.

2.7 Transmission characteristics

Signaling rate (or baud rate)

The signaling rate of a communications link is a measure of how many times the physical signal changes per second and is expressed as the baud rate. An oscilloscope trace of the data transfer would show pulses at the baud rate. For a 1000 baud rate, pulses would be seen at multiples of 1 ms.

With asynchronous systems, we set the baud rate at both ends of the link so that each physical pulse has the same duration.

Data rate

The data rate or bit rate is expressed in bits per second (bps), or multiples such as kbps, Mbps and Gbps (kilo, mega and gigabits per second). This represents the actual number of data bits transferred per second. An example is a 1000 baud RS-232 link transferring a frame of 10 bits, being 7 data bits plus a start, stop and parity bit. Here the baud rate is 1000 baud, but the data rate is 700 bps.

Although there is a tendency to confuse baud rate and bit rate, they are not the same. Whereas baud rate indicates the number of signal changes per second, the bit rate indicates the number of bits represented by each signal change. In simple baseband systems such as RS-232, the baud rate equals the bit rate. For synchronous systems, the bit rate invariably exceeds the baud rate. For ALL systems, the data rate is less than the bit rate due to overheads such as stop, stand, and parity bits (synchronous systems) or fields such as address and error detection fields in synchronous system frames.

There are sophisticated modulation techniques, used particularly in modems that allow more than one bit to be encoded within a signal change. The ITU V.22bis full duplex standard, for example, defines a technique called quadrature amplitude modulation, which effectively increases a baud rate of 600 to a data rate of 2400 bps. Irrespective of the methods used, the maximum data rate is always limited by the bandwidth of the link. These modulation techniques used with modems are discussed in Chapter 7.

Bandwidth

The single most important factor that limits communication speeds is the bandwidth of the link. Bandwidth is generally expressed in hertz (Hz), meaning cycles per second. This represents the maximum frequency at which signal changes can be handled before attenuation degrades the message. Bandwidth is closely related to the transmission medium, ranging from around 5000 Hz for the public telephone system to the GHz range for optical fiber cable.

As a signal tends to attenuate over distance, communications links may require repeaters placed at intervals along the link, to boost the signal level.

Calculation of the theoretical maximum data transfer rate uses the Nyquist formula and involves the bandwidth and the number of levels encoded in each signaling element, as described in Chapter 4.

Signal to noise ratio

The signal to noise (S/N) ratio of a communications link is another important limiting factor. Sources of noise may be external or internal, as discussed in Chapter 6.

The maximum practical data transfer rate for a link is mathematically related to the bandwidth, S/N ratio and the number of levels encoded in each signaling element. As the S/N decreases, so does the bit rate. See Chapter 4 for a definition of the Shannon-Hartley Law that gives the relationships.

Data throughput

As data is always carried within a protocol envelope, ranging from a character frame to sophisticated message schemes, the data transfer rate will be less than the bit rate. As explained in Chapter 9, the amount of redundant data around a message packet increases as it passes down the protocol stack in a network. This means that the ratio of non-message data to 'real' information may be a significant factor in determining the effective transmission rate, sometimes referred to as the throughput.

Error rate

Error rate is related to factors such as S/N ratio, noise, and interference. There is generally a compromise between transmission speed and the allowable error rate, depending on the type of application. Ordinarily, an industrial control system cannot allow errors and is designed for maximum reliability of data transmission. This means that an industrial system will be comparatively slow in data transmission terms. As data transmission rates increase, there is a point at which the number of errors becomes excessive. Protocols handle this by requesting a retransmission of packets. Obviously, the number of retransmissions will eventually reach the point at which a high apparent data rate actually gives a lower real message rate, because much of the time is being used for retransmission.

2.8 Data coding

An agreed standard code allows a receiver to understand the messages sent by a transmitter. The number of bits in the code determines the maximum number of unique characters or symbols that can be represented. The most common codes are described on the following pages.

Baudot code

Although not in use much today, the Baudot code is of historical importance. It was invented in 1874 by Maurice Emile Baudot and is considered to be the first uniform-length code. Having five bits, it can represent 32 (2^5) characters and is suitable for use in a system requiring only letters and a few punctuation and control codes. The main use of this code was in early teleprinter machines.

A modified version of the Baudot code was adopted by the ITU as the standard for telegraph communications. This uses two 'shift' characters for letters and numbers and was the forerunner for the modern ASCII and EBCDIC codes.

ASCII code

The most common character set in the western world is the American Standard Code for Information Interchange, or ASCII (see Table 2.3).

This code uses a 7-bit string giving 128 (2^7) characters, consisting of:

- Upper and lower case letters

- Numerals 0 to 9
- Punctuation marks and symbols
- A set of control codes, consisting of the first 32 characters, which are used by the
- Communications link itself and are not printable

For example: D = ASCII code in binary 1000100.

A communications link setup for 7-bit data strings can only handle hexadecimal values from 00 to 7F. For full hexadecimal data transfer, an 8-bit link is needed, with each packet of data consisting of a byte (two hexadecimal digits) in the range 00 to FF. An 8-bit link is often referred to as 'transparent' because it can transmit any value. In such a link, a character can still be interpreted as an ASCII value if required, in which case the eighth bit is ignored.

The full hexadecimal range can be transmitted over a 7-bit link by representing each hexadecimal digit as its ASCII equivalent. Thus the hexadecimal number 8E would be represented as the two ASCII values 38 45 (hexadecimal) ('8' 'E'). The disadvantage of this technique is that the amount of data to be transferred is almost doubled, and extra processing is required at each end.

ASCII control codes can be accessed directly from a PC keyboard by pressing the Control key [Ctrl] together with another key. For example, Control-A (^A) generates the ASCII code start of header (SOH).

The ASCII Code is the most common code used for encoding characters for data communications. It is a 7-bit code and, consequently, there are only $2^7 = 128$ possible combinations of the seven binary digits (bits), ranging from binary 0000000 to 1111111 or hexadecimal 00 to 7F.

Each of these 128 codes is assigned to specific control codes or characters as specified by the following standards:

- ANSI-X3.4
- ISO-646
- ITU alphabet #5

The ASCII Table is the **reference** table used to record the bit value of every character defined by the code. There are many different forms of the table, but all contain the same basic information according to the standards. Two types are shown here.

Table 2.3 shows the condensed form of the ASCII Table, where all the characters and control codes are presented on one page. This table shows the code for each character in hexadecimal (HEX) and binary digits (BIN) values. Sometimes the decimal (DEC) values are also given in small numbers in each box.

This table works like a matrix, where the MSB (most significant bits – the digits on the left-hand side of the written HEX or BIN codes) are along the top of the table and the LSB (least significant bits – the digits on the right-hand side of the written HEX or BIN codes) are down the left-hand side of the table. Some examples of the HEX and BIN values are given below:

Table 2.4 and Table 2.5 show the form commonly used in printer manuals, sometimes also called the ASCII Code Conversion Table, where each ASCII character or control code is cross referenced to:

- BIN :   A 7-bit binary ASCII code
- DEC :   An equivalent 3 digit decimal value (0 to 127)
- HEX :   An equivalent 2 digit hexadecimal value (00 to 7F)

HEX	HEX	0	1	2	3	4	5	6	MSB 7
HEX	BIN	000	001	010	011	100	101	110	111
0	0000	(NUL)	(DLE)	Space	0	@	P	`	p
1	0001	(SOH)	(DC1)	!	1	A	Q	a	q
2	0010	(STX)	(DC2)	"	2	B	R	b	r
3	0011	(ETX)	(DC3)	#	3	C	S	c	s
4	0100	(EOT)	(DC4)	$	4	D	T	d	t
5	0101	(ENQ)	(NAK)	%	5	E	U	e	u
6	0110	(ACK)	(SYN)	&	6	F	V	f	v
7	0111	(BEL)	(ETB)	'	7	G	W	g	w
8	1000	(BS)	(CAN)	(	8	H	X	h	x
9	1001	(HT)	(EM)	)	9	I	Y	i	y
A	1010	(LF)	(SUB)	*	:	J	Z	j	z
B	1011	(VT)	(ESC)	+	;	K	[	k	{
C	1100	(FF)	(FS)	,	<	L	\	l	|
D	1101	(CR)	(GS)	-	=	M	]	m	}
E	1110	(SO)	(RS)	.	>	N	^	n	~
F	1111	(SI)	(US)	/	?	O	_	o	DEL
LSB									

Table 2.3
The ASCII table

DEC	HEX	BIN	ASCII	DEC	HEX	BIN	ASCII
0	00	000 0000	(NUL)	32	20	010 0000	Space
1	01	000 0001	(SOH)	33	21	010 0001	!
2	02	000 0010	(STX)	34	22	010 0010	"
3	03	000 0011	(ETX)	35	23	010 0011	#
4	04	000 0100	(EOT)	36	24	010 0100	$
5	05	000 0101	(ENQ)	37	25	010 0101	%
6	06	000 0110	(ACK)	38	26	010 0110	&
7	07	000 0111	(BEL)	39	27	010 0111	'
8	08	000 1000	(BS)	40	28	010 1000	(
9	09	000 1001	(HT)	41	29	010 1001	)
10	0A	000 1010	(LF)	42	2A	010 1010	*
11	0B	000 1011	(VT)	43	2B	010 1011	+
12	0C	000 1100	(FF)	44	2C	010 1100	,
13	0D	000 1101	(CR)	45	2D	010 1101	-
14	0E	000 1110	(SO)	46	2E	010 1110	.
15	0F	000 1111	(SI)	47	2F	010 1111	/
16	10	001 0000	(DLE)	48	30	011 0000	0
17	11	001 0001	(DC1)	49	31	011 0001	1
18	12	001 0010	(DC2)	50	32	011 0010	2
19	13	001 0011	(DC3)	51	33	011 0011	3
20	14	001 0100	(DC4)	52	34	011 0100	4
21	15	001 0101	(NAK)	53	35	011 0101	5
22	16	001 0110	(SYN)	54	36	011 0110	6
23	17	001 0111	(ETB)	55	37	011 0111	7
24	18	001 1000	(CAN)	56	38	011 1000	8
25	19	001 1001	(EM)	57	39	011 1001	9
26	1A	001 1010	(SUB)	58	3A	011 1010	:
27	1B	001 1011	(ESC)	59	3B	011 1011	;
28	1C	001 1100	(FS)	60	3C	011 1100	<
29	1D	001 1101	(GS)	61	3D	011 1101	=
30	1E	001 1110	(RS)	62	3E	011 1110	>
31	1F	001 1111	(US)	63	3F	011 1111	?

Table 2.4
ASCII code conversion table

DEC	HEX	BIN	ASCII	DEC	HEX	BIN	ASCII
64	40	100 0000	@	96	60	110 0000	`
65	41	100 0001	A	97	61	110 0001	a
66	42	100 0010	B	98	62	110 0010	b
67	43	100 0011	C	99	63	110 0011	c
68	44	100 0100	D	100	64	110 0100	d
69	45	100 0101	E	101	65	110 0101	e
70	46	100 0110	F	102	66	110 0110	f
71	47	100 0111	G	103	67	110 0111	g
72	48	100 1000	H	104	68	110 1000	h
73	49	100 1001	I	105	69	110 1001	i
74	4A	100 1010	J	106	6A	110 1010	j
75	4B	100 1011	K	107	6B	110 1011	k
76	4C	100 1100	L	108	6C	110 1100	l
77	4D	100 1101	M	109	6D	110 1101	m
78	4E	100 1110	N	110	6E	110 1110	n
79	4F	100 1111	O	111	6F	110 1111	o
80	50	101 0000	P	112	70	111 0000	p
81	51	101 0001	Q	113	71	111 0001	q
82	52	101 0010	R	114	72	111 0010	r
83	53	101 0011	S	115	73	111 0011	s
84	54	101 0100	T	116	74	111 0100	t
85	55	101 0101	U	117	75	111 0101	u
86	56	101 0110	V	118	76	111 0110	v
87	57	101 0111	W	119	77	111 0111	w
88	58	101 1000	X	120	78	111 1000	x
89	59	101 1001	Y	121	79	111 1001	y
90	5A	101 1010	Z	122	7A	111 1010	z
91	5B	101 1011	[	123	7B	111 1011	{
92	5C	101 1100	\	124	7C	111 1100		
93	5D	101 1101	]	125	7D	111 1101	}
94	5E	101 1110	^	126	7E	111 1110	~
95	5F	101 1111	_	127	7F	111 1111	(DEL)

Table 2.5
ASCII code conversion table (cont.)

Character		Control	7-Bit Binary Code	Hex	Decimal
NUL	Null	^@	000 0000	00	0
SOH	Start of Header	^A	000 0001	01	1
STX	Start of Text	^B	000 0010	02	2
ETX	End of Text	^C	000 0011	03	3
EOT	End of Transmission	^D	000 0100	04	4
ENQ	Enquiry	^E	000 0101	05	5
ACK	Acknowledge	^F	000 0110	06	6
BEL	Bell	^G	000 0111	07	7
BS	Backspace	^H	000 1000	08	8
HT	Horizontal Tabulation	^I	000 1001	09	9
LF	Line feed	^J	000 1010	0A	10
VT	Vertical Tabulation	^K	000 1011	0B	11
FF	Form Feed	^L	000 1100	0C	12
CR	Carriage return	^M	000 1101	0D	13
SO	Shift Out	^N	000 1110	0E	14
SI	Shift In	^O	000 1111	0F	15
DLE	Data Link Escape	^P	001 0000	10	16
DC1	Device Control 1	^Q	001 0001	11	17
DC2	Device Control 2	^R	001 0010	12	18
DC3	Device Control 3	^S	001 0011	13	19
DC4	Device Control 4	^T	001 0100	14	20
NAK	Negative Acknowledgement	^U	001 0101	15	21
SYN	Synchronous Idle	^V	001 0110	16	22
ETB	End of Trans Block	^W	001 0111	17	23
CAN	Cancel	^X	001 1000	18	234
EM	End of Medium	^Y	001 1001	19	25
SUB	Substitute	^Z	001 1010	1A	26
ESC	Escape	^[	001 1011	1B	27
FS	File Separator	^\	001 1100	1C	28
GS	Group Separator	^]	001 1101	1D	29
RS	Record Separator	^|	001 1110	1E	30
US	Unit Separator	^_	001 1111	1F	31
DEL	Delete, Rubout		111 1111	7F	127

Table 2.6
Table of control codes for the ASCII

Least significant bits

Bit positions				4	0	0	0	0	0	0	0	0	1	1	1	1	1	1	1	1
				3	0	0	0	0	1	1	1	1	0	0	0	0	1	1	1	1
				2	0	0	1	1	0	0	1	1	0	0	1	1	0	0	1	1
				1	0	1	0	1	0	1	0	1	0	1	0	1	0	1	0	1
8	7	6	5																		
0	0	0	0		NUL	SOH	STX	ETX	PF	HT	LC	DEL			SMM	VT	FF	CR	SO	SI
0	0	0	1		DLE	DC$_1$	DC$_2$	DC$_3$	RES	NL	BS	IL	CAN	EM	CC		IFS	IGS	IRS	IUS
0	0	1	0		DS	SOS	FS		BYP	LF	EOB	PRE			SM			ENQ	ACK	BEL
0	0	1	1				SYN		PN	RS	UC	EOT					DC4	NAK		SUB
0	1	0	0		SP										¢	.	<	(	+	|
0	1	0	1		&										!	$	*	)	;	_
0	1	1	0		-	/									`	%		-	>	?
0	1	1	1												:	#	@	'	=	"
1	0	0	0			a	b	c	d	e	f	g	h	i						
1	0	0	1			j	k	l	m	n	o	p	q	r						
1	0	1	0				s	t	u	v	w	x	y	z						
1	0	1	1																	
1	1	0	0			A	B	C	D	E	F	G	H	I						
1	1	0	1			J	K	L	M	N	O	P	Q	R						
1	1	1	0				S	T	U	V	W	X	Y	Z						
1	1	1	1		0	1	2	3	4	5	6	7	8	9						

Most significant bits

Table 2.7
EBCDIC code table

Control codes are often difficult to detect when troubleshooting a data system, unlike printable codes, which show up as a symbol on the printer or terminal. Digital line analyzers can be used to detect and display the unique code for each of these control codes to assist in the analysis of the system.

To represent the word DATA in binary form using the 7-bit ASCII code, each letter is coded as follows:

		Binary	*Hex*
D	:	100 0100	44
A	:	100 0001	41
T	:	101 0100	54
A	:	100 0001	41

Referring to the ASCII table, the binary digits on the right-hand side of the binary column change by one digit for each step down the table. Consequently, the bit on the far right has become known as **least significant bit (LSB)** because it changes the overall value so little. The bit on the far left has become known as **most significant bit (MSB)** because it changes the overall value so much.

According to the reading conventions in the western world, words and sentences are read from left to right. When looking at the ASCII code for a character, we would read the MSB (most significant bit) first, which is on the left-hand side. However, in data communications, the convention is to **transmit the LSB of each character FIRST,** which is on the right-hand side and the **MSB last.** However, the **characters** are still usually sent in the conventional reading sequence in which they are generated. For example, if the word D-A-T-A is to be transmitted, the characters are transferred in that sequence, but the 7 bit ASCII code for each character is 'reversed'.

Consequently, the bit pattern that is observed on the communication link will be as follows, reading each bit in order from right to left.

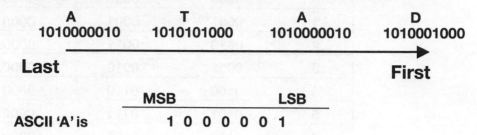

Adding the stop bit (1) and parity bit (1 or 0) and the start bit (0) to the ASCII character, the pattern indicated above is developed with even parity. For example, an ASCII 'A' character is sent as:

STOP	P	MSB						LSB	START
1	0	1	0	0	0	0	0	1	0

EBCDIC

Extended binary coded data interchange code (EBCDIC), originally developed by IBM, uses 8 bits to represent each character. EBCDIC is similar in concept to the ASCII code, but specific bit patterns are different and it is incompatible with ASCII. When IBM introduced its personal computer range, they decided to adopt the ASCII Code, so EBCDIC does not have much relevance to data communications in the industrial environment. Refer to the EBCDIC Table 2.7.

4-bit binary code

For purely numerical data a 4-bit binary code, giving 16 characters (2^4), is sometimes used. The numbers 0–9 are represented by the binary codes 0000 to 1001 and the remaining codes are used for decimal points. This increases transmission speed or reduces the number of connections in simple systems. The 4-bit binary code is shown in Table 2.8.

Decimal	Binary Code	Gray Code	BCD Code
0	0000	0000	0000
1	0001	0001	0000
2	0010	0011	0000
3	0011	0010	0000
4	0100	0110	0000
5	0101	0111	0000
6	0110	0101	0000
7	0111	0100	0000
8	1000	1100	0000
9	1001	1101	0000
10	1010	1111	0001 0000
11	1011	1110	0001 0001
12	1100	1010	0001 0010
13	1101	1011	0001 0011
14	1110	1001	0001 0100
15	1111	1000	0001 0101

Table 2.8
4-bit binary code

Gray code

Binary code is not ideal for some types of devices because multiple digits have to change every alternate count as the code increments. For incremental devices, such as shaft position encoders, which give a code output of shaft positions, the Gray code can be used. The advantage of this code over binary is that only one bit changes every time the value is incremented. This reduces the ambiguity in measuring consecutive angular positions. The Gray code is shown in Table 2.9.

Decimal	Gray Code
0	0000
1	0001
2	0011
3	0010
4	0110
5	0111
6	0101
7	0100
8	1100
9	1101
10	1111
11	1110
12	1010
13	1011
14	1001
15	1000

Table 2.9
Gray code

Binary coded decimal

Binary coded decimal (BCD) is an extension of the 4-bit binary code. BCD encoding converts each separate digit of a decimal number into a 4-bit binary code. Consequently, the BCD uses 4 bits to represent one decimal digit. Although 4 bits in the binary code can represent 16 numbers (from 0 to 15) only the first 10 of these, from 0 to 9, are valid for BCD.

Decimal	Binary Code	Gray Code	BCD Code
0	0000	0000	0000
1	0001	0001	0001
2	0010	0011	0010
3	0011	0010	0011
4	0100	0110	0100
5	0101	0111	0101
6	0110	0101	0110
7	0111	0100	0111
8	1000	1100	1000
9	1001	1101	1001
10	1010	1111	0001 0000
11	1011	1110	0001 0001
12	1100	1010	0001 0010
13	1101	1011	0001 0011
14	1110	1001	0001 0100
15	1111	1000	0001 0101

Table 2.10
Comparison of Binary, Gray and BCD codes

BCD is commonly used on relatively simple systems such as small instruments, thumbwheels, and digital panel meters. Special interface cards and integrated circuits (ICs) are available for connecting BCD components to other intelligent devices. They can be connected directly to the inputs and outputs of PLCs.

A typical application for BCD is the setting of a parameter on a control panel from a group of thumbwheels. Each thumbwheel represents a decimal digit (from left to right; thousands, hundreds, tens and units digits). The interface connection of each digit to a PLC requires 4 wires plus a common, which would mean a total of 20 wires for a 4-digit set of thumbwheels. The number of wires, and their connections to a PLC, can be reduced to 8 by using a time division multiplexing system as shown in Figure 2.6. Each PLC output is energized in turn, and the binary code is measured by the PLC at four inputs. A similar arrangement is used in reverse for the digital display on a panel meter, using a group of four 7-segment LCD or LED displays.

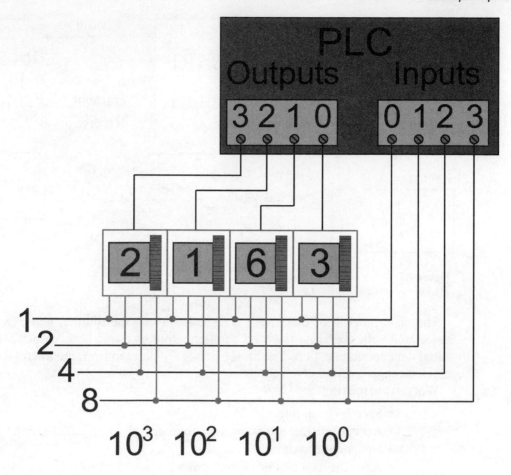

Figure 2.6
BCD Thumbwheel switches and connections to PLC

2.9   The universal asynchronous receiver/transmitter (UART)

The start, stop and parity bits used in asynchronous transmission systems are usually physically generated by a standard integrated circuit (IC) chip that is part of the interface circuitry between the microprocessor bus and the line driver (or receiver) of the communications link. This type of IC is called a **UART (universal asynchronous receiver/transmitter)** or sometimes an ACE (asynchronous communications element).

Various forms of UART are also used in synchronous data communications, called **USRT**. Collectively, these are all called **USART**s. The outputs of a UART are not designed to interface directly with the communications link. Additional devices, called line drivers and line receivers, are necessary to give out and receive the voltages appropriate to the communications link.

8250, 16450, 16550 are examples of UARTs, and 8251 is an example of a USART.

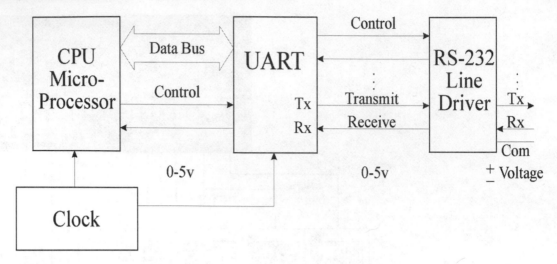

Figure 2.7
Typical connection details of the UART

The main purpose of the UART is to look after all the routine 'housekeeping' matters associated with preparing the 8 bit parallel output of a microprocessor for asynchronous serial data communication. The timing pulses are derived from the microprocessor master clock through external connections.

When **transmitting**, the UART:

- Sets the baud rate
- Accepts character bits from microprocessor as a parallel group
- Generates a start bit
- Adds the data bits in a serial group
- Determines the parity and adds a parity bit (if required)
- Ends transmission with a stop bit (sometimes 2 stop bits)
- Then signals the microprocessor that it is ready for the next character
- Coordinates handshaking when required

The UART has a separate signal line for transmit (TX) and one for receive (RX) so that it can operate in the full-duplex or a half-duplex mode. Other connections on the UART provide hardware signals for handshaking, the method of providing some form of 'interlocking' between two devices at the ends of a data communications link. Handshaking is discussed in more detail in Chapter 3.

When **receiving**, the UART:

- Sets the baud rate at the receiver
- Recognizes the start bit
- Reads the data bits in a serial group
- Reads the parity bit and checks the parity
- Recognizes the stop bit(s)
- Transfers the character as a parallel group to the microprocessor for further processing
- Coordinates handshaking when required
- Checks for data errors and flags the error bit in the status register

This removes the burden of programming the above routines in the microprocessor and, instead, they are handled transparently by the UART. All the program does with serial data is to simply write/read bytes to/from the UART.

The UART transmitter

A byte received from the microprocessor for transmission is written to the I/O address of the UART's transmission sector. The bits to be transmitted are loaded into a shift register, then shifted out on the negative transition of the transmit data clock. This pulse rate sets the baud rate. When all the bits have been shifted out of the transmitter's shift register, the next packet is loaded and the process is repeated. The word 'packet' is used to indicate start, data, parity and stop bits all packaged together. Some authors refer to the packet as a serial data unit (SDU).

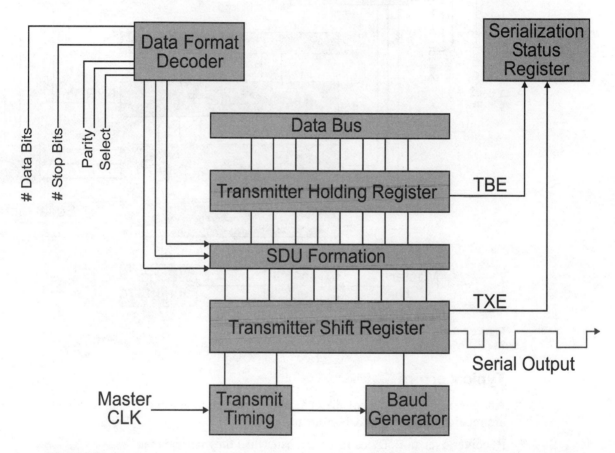

Figure 2.8
The UART transmitter

Between the transmitter holding register and the shift register is a section called the SDU (serial data unit) formation. This section constructs the actual packet to be loaded into the shift register.

In full duplex communications, the software needs to only test the value of the transmitter buffer empty (TBE) flag to decide whether to write a byte to the UART. In half-duplex communications, the modem must swap between transmitter and receiver states. Hence, the software must check both the transmitter buffer and the transmitter's shift register, as there may still be some data there.

The UART receiver

The UART receiver continuously monitors the incoming serial line waiting for a start bit. When the start bit is received, the receiver line is monitored at the selected baud rate and the successive bits are placed in the receiver's shift register. This takes place according to the format described in the user programmable data format register. After assembly of the byte, it is moved into a FIFO (first in first out) buffer. At this stage, the RxRDY (receiver ready) flag is set **true** and remains true until all the contents of the FIFO buffer are empty.

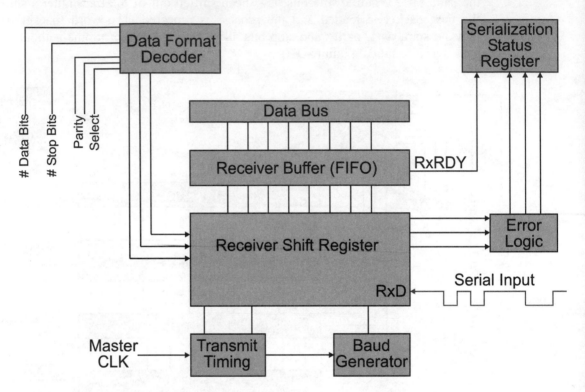

Figure 2.9
The UART receiver

Typical errors

Another major function of the UART is to detect errors in the data received. Most errors are receiver errors. Typical errors are:

Receiver overrun: Bytes received faster than they can be read
Parity error: Parity bit disagreement
Framing error: This occurs if the detected bits do not fit into the frame selected
Break error: This occurs if a start bit is detected for more than a frame time.

Break detect

To gain the attention of a receiver, a transmitter may hold the data line in a space condition (+voltage) for a period of time longer than that required for a complete character. This is called a break, and receivers can be equipped with a break detect to detect this condition. It is useful for interrupting the receiver, even in the middle of a stream of characters being sent. The break detect time is a function of the baud rate.

Serialization errors are reported in the serialization status register as shown in Figure 2.8 and Figure 2.9.

Receiver timing

It is necessary to have separate clock signals for the UART's internal operations and to control the shifting operations in the transmitter and receiver sections. The frequency of the master signal is designed to be many times higher than that of the baud rate. This ratio of master serial clock to baud rate is called the clocking factor (typically 16). Instead of sampling the input line at the baud rate frequency, the improved start bit detector samples the incoming line at the rate of the master clock. This minimizes the possibility of an error due to slippage of sampling a stream of serial bits and sampling the wrong bit.

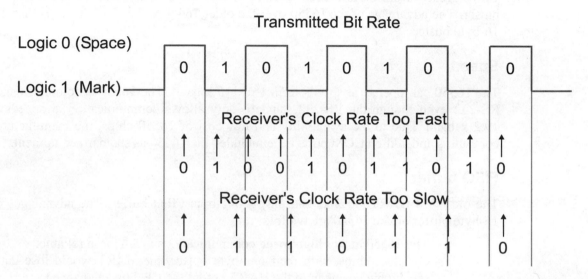

Figure 2.10
Example of incorrect timing between source and receiver

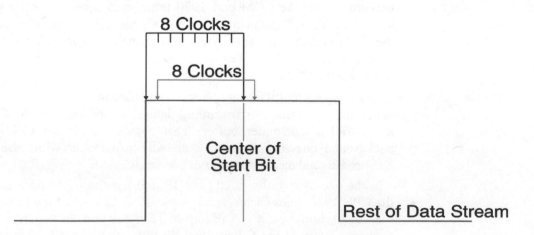

Figure 2.11
Minimization of error with a clocking factor of 16

The earliest serial ports used 8250 or 8251 chips, which interrupted the main processor for every character to be transmitted or received. This worked well for the speeds of that time. This has since been replaced by the 16450 chip which works in a similar fashion but supported faster PC bus speeds, and later by 16550 which has a 16-byte buffer thereby reducing the number of CPU interruptions by a factor of 16. A more recent development is to use an enhanced serial port, which provides a buffer of about 1000 bytes and has its own processor to reduce the interruptions to the main CPU by a factor of 1000.

2.10 The high speed UART (16550)

The 16550 is a high speed serial universal asynchronous receiver transmitter (UART). It is the default UART used on all IBM compatible computers and COM ports sold today. It varies greatly from the old 8250 UARTs in two ways: speed and the size of the FIFO buffer. The advantage of the 16550 over the older 16450 and 8250 UARTs is that it has a 16-byte buffer.

Speed

The 16550 can operate at speeds from 1 to 115 k baud. The 16550 is commonly used on RS-232 even though the RS-232 standard only allows communication at speeds up to 19.2 k baud. Due to the availability and low cost of 16550 chips, the manufacturers of computers and add-on COM ports have included the 16550 as standard equipment.

FIFO buffer

The old 8250 UART (19.2 k) had only a one-byte FIFO buffer. The advantages of the 16-byte buffer on the 16550 are twofold:

- The 16550 makes high speed communications much more reliable.
  On older chips, with their one-byte buffer, the UART would lose data if a second byte came in to the UART before the CPU had a chance to retrieve the first byte. The 16550, with its 16-byte buffer, gives the CPU up to 16 chances to retrieve the data before a character is lost. To realize what this means, if a UART is running at 19 200 bps (with a 10-bit character frame) the CPU will need to service the COM port 1920 times each second or once every .0005 seconds. If the CPU happens to take .0006 seconds to get around to servicing the COM port then the first byte is lost in the one-byte buffer UART. On the 16550 chip, with 16 bytes of buffer space, you can have up to 0.008 seconds to service the COM port.

- It helps make a multitasking system more efficient.
  When the COM port is transmitting data, it has to interrupt the CPU and fill the UART's transmitter buffer. That means that if the CPU is doing a background directory scan the scan will take longer while the COM port attempts to send data out to the outside world.

  In the one-byte buffer UARTs at 19 200 bps the COM port must interrupt the CPU 1920 times each second just to send data out of the COM port. With the 16550, however, it can put up to 16 bytes into the buffer at a time and therefore interrupt the CPU only 120 times each second. This increases the performance of the CPU to COM port system.

3

Serial communication standards

This chapter discusses the main physical interface standards associated with data communications for instrumentation and control systems. It includes information on balanced and unbalanced transmission lines, current loops, and serial interface converters.

Objectives

When you have completed studying this chapter you will be able to:

- List and explain the function of the important standards organizations
- Describe and compare the serial data communications interface standards:
    - RS-232
    - RS-449
    - RS-423
    - RS-422
    - RS-485
    - RS/TIA-530A
    - RS/TIA-562

- Explain troubleshooting in serial data communication circuits
- Describe commonly used serial interface techniques:
    - 20 mA current loop
    - Serial interface converters
    - Interface to serial printers
- Describe the most important parallel data communication interface standards:
    - General purpose interface bus
    - Centronics

3.1　Standards organizations

There are seven major organizations worldwide involved in drawing up standards or recommendations, which affect data communications. These are:

- ISO:　　International Standards Organization
- ITU-T:　International Telecommunications Union (ITU formerly CCITT)
- IEEE:　Institute of Electrical and Electronic Engineers
- IEC:　　International Electrotechnical Commission
- RS:　　Electronic Industries Association
- ANSI:　American National Standards Institute
- TIA:　　Telecommunication Industries Association

ANSI is the principal standards body in the USA and is that country's member body to the ISO. ANSI is a non-profit, non-governmental body supported by over 1000 trade organizations, professional societies, and companies.

The International Telecommunications Union (ITU) is a specialist agency of the United Nations Organization (UNO). It consists of representatives from the Postal, Telephony, and Telegraphy organizations (PTTs), common carriers and manufacturers of telecommunications equipment. In Europe, administrations tend to follow the ITU defined recommendations closely. Although the US manufacturers did not recognize them in the past, they are increasingly conforming to ITU recommendations.

The ITU defines a complete range of standards for interconnecting telecommunications equipment. The standards for data communications equipment are generally defined by the ITU-T 'V' series recommendations.

The two ITU-T physical interface standards are:

- V.24:　equivalent to RS-232 for low speed asynchronous serial circuits
- V.35:　equivalent to RS-449 for wide bandwidth circuits

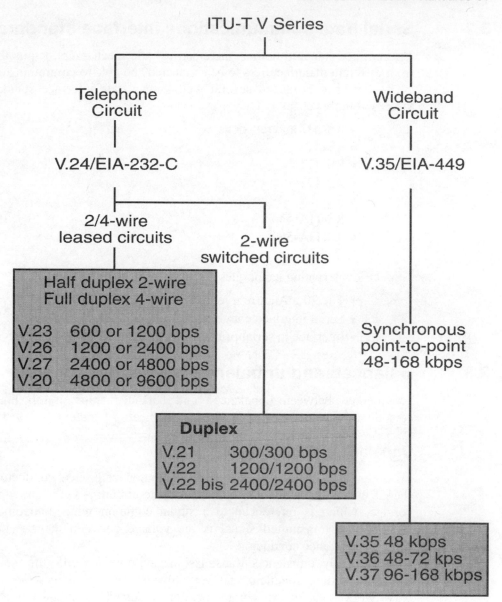

Figure 3.1
ITU-T V series

The RS is a voluntary standards organization in the USA, specializing in the electrical and functional characteristics of interface equipment. It mainly represents the manufacturers of electronic equipment. Since the RS and the TIA merger in 1988, the TIA represents the telecommunications sector of the RS and its initials appear on certain RS standard documents.

The IEC is an international standards body, affiliated to ISO. It concentrates on electrical standards. The IEC developed in Europe and is used by most Western countries, except the USA or those countries closely affiliated with the USA.

The IEEE is a professional society for electrical engineers in the USA and issues its own standards and codes of practice. The IEEE is a member of ANSI and ISO.

The ISO draws members from all countries of the world and concentrates on coordination of standards internationally.

3.2   Serial data communications interface standards

An interface standard defines the electrical and mechanical details that allow equipment from different manufacturers to be connected and able to communicate.

The RS have produced several well known data interface standards, which will be discussed in this chapter. They are:

- RS-232 and revisions
- RS-449
- RS-423
- RS-422
- RS-485
- RS/TIA-530A
- RS/TIA-562

Specific interfacing techniques discussed here also include:

- The 20 mA current loop
- Serial interface converters
- Interface to serial printers

3.3   Balanced and unbalanced transmission lines

The choice between unbalanced and balanced transmission lines is an important consideration when selecting a data communications system.

Unbalanced transmission

In an unbalanced system, the signal common reference conductor is simultaneously shared by many signals and other electronic circuitry. Only one wire carries the signal voltage, which is referenced to a signal common wire, sometimes called the signal ground. The transmitted signal is the voltage between the signal conductor and the common reference conductor.

Theoretically, unbalanced transmission should work well if the signal currents are small and the common conductor has very low impedance. In practice, unbalanced systems only work over short communication links. The signal common conductor has characteristics similar to other conductors (resistance, inductance and capacitance) and is not a perfect reference point. For long communication distances, the common conductor does not have the same zero voltage at all points along its length or at its ends. The common conductor can also pick up noise and have other voltages superimposed on it. Sometimes the shield conductor is used as the common reference wire. This practice can introduce excessively high noise-levels and should be avoided. Unbalanced transmission is used in the RS-232 and RS-423 interfaces.

The fact that the common reference conductor may carry superimposed interference voltages means that the voltages V1, V2, and V3 measured at the receiver will be affected (Figure 3.2).

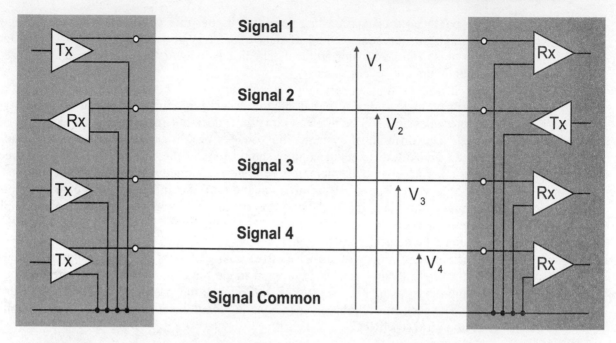

Figure 3.2
Data communication with unbalanced interfaces

Balanced transmission

Balanced communication interfaces require two conductors to transmit each signal. The voltage at the receiving end is measured as the voltage difference between these two wires. This is known as a balanced or differential system. This eliminates many of the interference problems associated with the common reference wire.

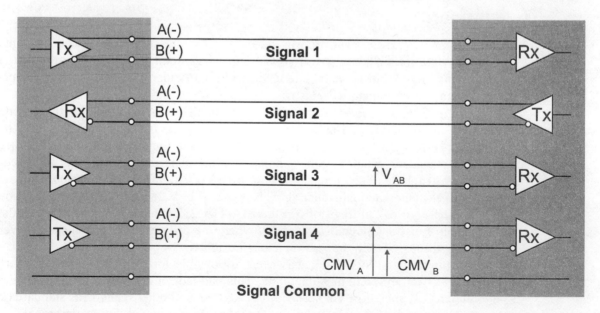

Figure 3.3
Data communications with balanced interfaces

The balanced transmission line permits a higher rate of data transfer over longer distances. The differential method of data transfer is preferable in industrial applications where noise can be a major problem. The disadvantage is that a balanced system requires two conductors for every signal.

The successful transfer of voltage signals across two conductors in the presence of, say noise or voltage drops is based on the assumption that the conductors have similar characteristics and will be affected equally. It does not mean that noise does not exist in the balanced differential system. The voltages on both conductors should rise and fall together, and the differential voltage should remain the same. The voltage between the signal conductor and the common reference conductor is called the common mode voltage (CMV). The CMV is an indication of the induced voltage or noise on the communication link. Ideally, the CMV on the two wires will cancel out completely. However, the greater the CMV, the higher the likelihood of output voltage distortion and damage to the device.

The receiver circuitry of a 2-wire differential system is designed to ignore or reject the CMV, using a technique called common mode rejection (CMR). The effect of noise on the signal is measured as the ratio of the voltage after passing through the receiver to the CMV. The success of the receiver in rejecting the noise is measured as the common mode rejection ratio (CMRR).

$$\text{CMRR(Db)} = 20 \log\left[\frac{dV_{\text{OUT}}}{dV_{\text{CM}}}\right]$$

Balanced transmission is used in most of the fast interfaces such as RS-422 and RS-485.

3.4   RS/TIA-232 interface standard (CCITT V.24 interface standard)

The RS-232 interface standard was developed for the single purpose of the interface between data terminal equipment (DTE) and data circuit terminating equipment (DCE) employing serial binary data interchange. In particular, RS-232 was developed for interfacing data terminals to modems.

The RS-232 interface standard was issued in the USA in 1969 by the engineering department of the RS. Almost immediately minor revisions were made and today's standard – RS-232-C – was issued. RS-232 was originally named RS-232, (recommended standard) which is still in popular usage. The prefix 'RS' was superseded by 'EIA/TIA' in 1988. The current revision is EIA/TIA-232E (1991), which brings it into line with the international standards ITU V.24, ITU V.28, and ISO-2110. The common convention is to call all revisions of the EIA/TIA 232 standards as EIA-232, as they are effectively functionally equivalent. Only where the differences between specific versions are being discussed, will the version letters be added.

Poor interpretation of RS-232 has been responsible for many problems in interfacing equipment from different manufacturers. This had led some users to dispute whether it is a 'standard'. It should be emphasized that RS-232 and other related RS standards define the electrical and mechanical details of the interface and do not define a protocol.

The RS-232 interface standard specifies the method of connection of two devices – the DTE and DCE.

DTE: Data terminal equipment, for example, a computer or a printer. A DTE device communicates with a DCE device. A DTE device transmits data on pin 2 and receives data on pin 3.

DCE: Data communications equipment, for example a modem, now also called data circuit-terminating equipment in RS-232E. A DCE device receives data from the DTE and retransmits via another data communications link, such as the telephone system. A DCE device transmits data on pin 3 and receives data on pin 2.

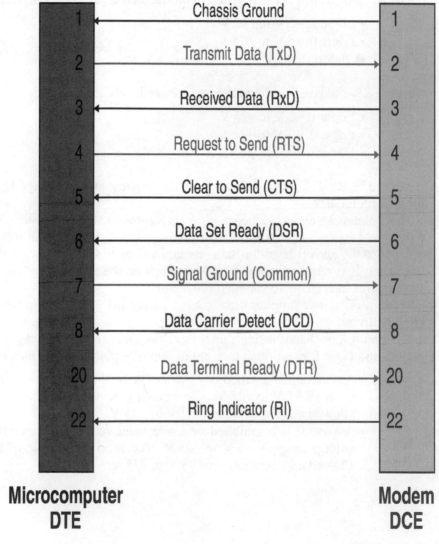

Figure 3.4
Connections between the DTE and the DCE

The major elements of RS-232

The RS-232 standard consists of three major parts, which define:

- Electrical signal characteristics
- Mechanical characteristics of the interface
- Functional description of the interchange circuits

Electrical signal characteristics

RS-232 defines electrical signal characteristics such as the voltage levels and grounding characteristics of the interchange signals and associated circuitry for an unbalanced system.

The RS-232 transmitter is required to produce voltages in the range:

- Logic 1: −5 V to −25 V
- Logic 0: +5 V to +25 V
- Undefined logic level: +5 V to −5 V

At the RS-232 receiver the following voltage levels are defined:

- Logic 1: −3 V to −25 V
- Logic 0: +3 V to +25 V
- Undefined logic level: −3 V to +3 V

Note: The RS-232 transmitter requires the slightly higher voltage to overcome voltage drop along the line.

The voltage levels associated with a microprocessor are 0 V to +5 V for transistor–transistor Logic (TTL). A line driver is required at the transmitting end to adjust the voltage to the correct level for the communications link. Similarly, at the receiving end a line receiver is required to translate the voltage on the communications link to the correct voltages for interfacing to the microprocessor.

Modern PC power supplies usually have a standard +12 V output that could be used for the line driver.

The control, or 'handshaking', lines have the same range of voltages as transmission of logic 0 and logic 1, except that they are of opposite polarity. This means that:

- A control line asserted or made active by the transmitting device has a voltage range of +5 V to +25 V. The receiving device connected to this control line is allowed a voltage range of +3 V to +25 V.
- A control line inhibited or made inactive by the transmitting device has a voltage range of −5 V to −25 V. The receiving device of this control line is allowed a voltage range of −3 V to −25 V.

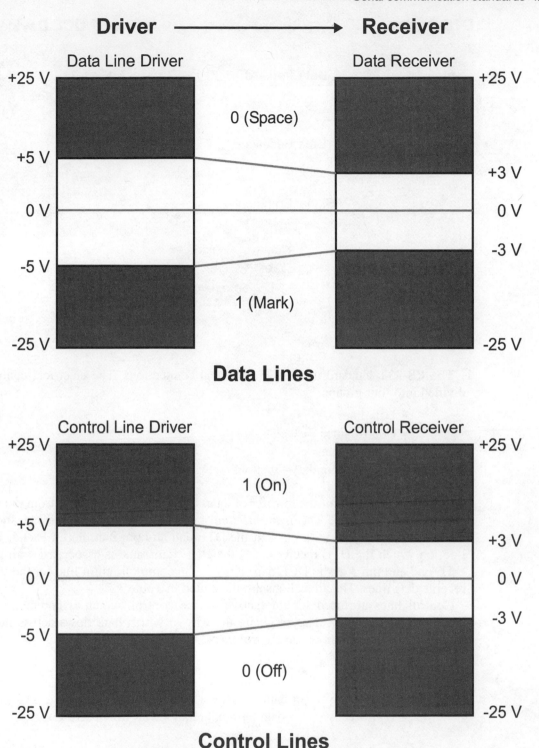

Figure 3.5
Voltage levels for RS-232

At the receiving end, a line receiver is necessary in each data and control line to convert the line voltage levels back to the 0 V and +5 V logic levels required by the internal electronics.

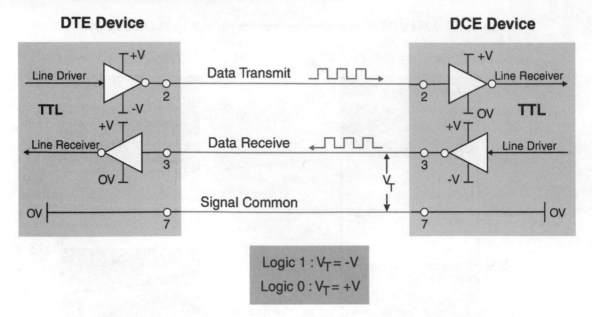

Figure 3.6
RS-232 transmitters and receivers

The RS-232 standard defines 25 electrical connections. The electrical connections are divided into four groups:

- Data lines
- Control lines
- Timing lines
- Special secondary functions

Data lines are used for the transfer of data. Data flow is designated from the perspective of the DTE interface. The *transmit line,* on which the DTE transmits and the DCE receives, is associated with pin 2 at the DTE end and pin 2 at the DCE end. The receive line, on which the DTE receives, and the DCE transmits, is associated with pin 3 at the DTE end and pin 3 at the DCE end. Pin 7 is the common return line for the transmit and receive data lines. The allocations are illustrated in Table 3.2.

Control lines are used for interactive device control, which is commonly known as hardware *handshaking.* They regulate the way in which data flows across the interface. The four most commonly used control lines are:

- RTS: request to send
- CTS: clear to send
- DCE Ready: or data set ready (DSR in RS-232-C)
- DTE Ready   or data terminal ready (DTR in RS-232-C)

It is important to remember that with the handshaking lines, the enabled state means a positive voltage and the disabled state means a negative voltage.

Hardware handshaking is the cause of most interfacing problems. Manufacturers sometimes omit control lines from their RS-232 equipment or assign unusual applications to them. Consequently, many applications do not use hardware handshaking but instead use only the three data lines (transmit, receive and signal common ground) with some form of software handshaking. The control of data flow is then part of the application program. Most of the systems encountered in data communications for instrumentation and control use some sort of software-based protocol in preference to hardware handshaking. These protocols are discussed in Chapter 8.

There is a relationship between the allowable speed of data transmission and the length of the cable connecting the two devices on the RS-232 interface. As the speed of data transmission increases, the quality of the signal transition from one voltage level to another, for example from –25 V to +25 V, becomes increasingly dependent on the capacitance and inductance of the cable.

The rate at which voltage can 'slew' from one logic level to another depends mainly on the cable capacitance, and the capacitance increases with cable length. The length of the cable is limited by the number of data errors acceptable during transmission. The RS-232 D&E standard specifies the limit of total cable capacitance to be 2500 pF. With typical cable capacitance having improved from around 160 pF/m to only 50 pF/m, the maximum cable length has extended from around 15 meters (50 feet) to about 50 meters (166 feet).

The common data transmission rates used with RS-232 are 110, 300, 600, 1200, 2400, 4800, 9600 and 19 200 bps. Based on field tests, Table 3.1 shows the practical relationship between selected Baud rates and maximum allowable cable length, indicating that much longer cable lengths are possible at lower Baud rates. Note that the achievable speed depends on the transmitter voltages, cable capacitance (as discussed above) as well as the noise environment.

Baud Rate	Cable Length (metres)
110	850
300	800
600	700
1200	500
2400	200
4800	100
9600	70
19 200	50
115 K	20

Table 3.1
Demonstrated maximum cable lengths with RS-232 interface

Mechanical characteristics of the interface

RS-232 defines the mechanical characteristics of the interface between the DTE and the DCE. This section dictates that the interface must consist of a plug and socket and that the socket will normally be on the DCE. The familiar DB-25 connector is specified together with a smaller 26 pin alternative connector.

Although not specified by RS-232C, the DB-25 connector (25 pin, D-type) is closely associated with RS-232 and became the *de facto* standard with revision D. Revision E formally specifies a new connector in the 26-pin alternative connector (known as the ALT A connector). This connector supports all 25 signals associated with RS-232. ALT A is physically smaller than the DB-25 and satisfies the demand for a smaller connector suitable for modern computers. Pin 26 is not currently used. On most RS-232 compatible equipment, where little or no handshaking is required, the DB-9 connector (9 pin, D-type) is common. This practice originated when IBM decided to make a combined serial/parallel adapter for the AT personal computer. A small connector format was needed to allow both interfaces to fit onto the back of a standard ISA interface card. Subsequently, the DB-9 connector has also became an industry standard to reduce the wastage of pins. The pin allocations commonly used with the DB-9 and DB-25 connectors for the RS-232 interface are shown in Table 3.2. The pin allocation for the DB-9 connector is not the same as the DB-25 and often traps the unwary.

The data pins of a DB-9 IBM connector are usually allocated as follows:

- Data transmit pin 3
- Data receive pin 2
- Signal common pin 5

Pin No. DTE	DB-9 Connector IBM Assignment	DB-25 Connector EIA-232 Pin Assignment	DB-25 Connector EIA-530 Pin Assignment
1	Received Line Signal	Shield	Shield
2	Received Data	Transmitted Data	Transmitted Data (A)
3	Transmitted Data	Received Data	Received Data (A)
4	DTE Ready	Request to Send	Request to Send (A)
5	Signal/Common Ground	Clear to Send	Clear to Send (A)
6	DCE Ready	DCE Ready	DCE Ready (A)
7	Request to Send	Signal/Common Ground	Signal/Common Ground
8	Clear to Send	Received Line Signal	Received Line Signal (A)
9	Ring Indicator	+Voltage (testing)	Receiver Signal DCE Element Timing (B)
10		-Voltage (testing)	Received Line (B)
11		Unassigned	Transmitter Signal DTE Element Timing (B)
12		Sec Received Line Signal Detector/Data Signal	Transmitter Signal DCE Element Timing
13		Sec Clear to Send	Clear to Send (B)
14		Sec Transmitted Data	Transmitted Data (B)
15		Transmitter Signal DCE Element Timing	Transmitter Signal DCE Element Timing (A)
16		Sec Received Data	Received Data (B)
17		Receiver Signal DCE Element Timing	Receiver Signal DCE Element Timing (A)
18		Local Loopback	Local Loopback
19		Sec Request to Send	Request to Send (B)
20		DTE Ready	DTE Ready (A)
21		Remote Loopback/Signal Quality Detector	Remote Loopback
22		Ring Indicator	DCE Ready (B)
23		Data Signal Rate	DTE Ready (B)
24		Transmit Signal DTE Element Timing	Transmitter Signal DTE Element Timing (A)
25		Test Mode	Test Mode

Table 3.2
Common DB-9 and DB-25 pin assignments for RS-232 and EIA/TIA-530 (often used for RS-422 and RS-485)

Functional description of the interchange circuits

RS-232 defines the function of the data, timing, and control signals used at the interface of the DTE and DCE. However, very few of the definitions in this section are relevant to applications for data communications for instrumentation and control.

The circuit functions are defined with reference to the DTE as follows:

- Protective ground (Shield)

  The protective ground ensures that the DTE and DCE chassis are at equal potentials. (Remember that this protective ground could cause problems with circulating earth currents.)

- Transmitted data (TxD)

  This line carries serial data from the DTE to the corresponding pin on the DCE. The line is held at a negative voltage during periods of line idle.

- Received data (RXD)

  This line carries serial data from the DCE to the corresponding pin on the DTE.

- Request to send (RTS)

  (RTS) is the request to send hardware control line. This line is placed active (+V) when the DTE requests permission to send data. The DCE then activates (+V) the CTS (clear to send) for hardware data flow control.

- Clear to send (CTS)

  When a half duplex modem is receiving, the DTE keeps RTS inhibited. When it is the DTE's turn to transmit, it advises the modem by asserting the RTS pin. When the modem asserts the CTS, it informs the DTE that it is now safe to send data.

- DCE ready

  Formerly called data set ready (DSR) – the DTE Ready line is an indication from the DCE to the DTE that the modem is ready.

- Signal ground (Common)

  This is the common return line for all the data transmit and receive signals and all other circuits in the interface. The connection between the two ends is always made.

- Data carrier detect (DCD)

  This is also called the received line signal detector. It is asserted by the modem when it receives a remote carrier and remains asserted for the duration of the link.

- DTE ready (data terminal ready)

  Formerly called data terminal ready (DTR) – DTE ready enables, but does not cause, the modem to switch onto the line. In originate mode, DTE ready must be asserted in order to auto dial. In answer mode, DTE ready must be asserted to auto answer.

- Ring indicator

  This pin is asserted during a ring voltage on the line.

- Data signal rate selector (DSRS)

  When two data rates are possible, the higher is selected by asserting DSRS, however, this line is not used much these days.

Pin No.	CCITT No.	Circuit	Description	Circuit Direction
1	-	-	Shield	To DCE
2	103	BA	Transmitted Data	To DCE
3	104	BB	Received Data	From DCE
4	105/133	CA/CJ	Request to Send/Ready for Receiving	To DCE
5	106	CB	Clear to Send	From DCE
6	107	CC	DCE Ready	From DCE
7	102	AB	Signal Common	-
8	109	CF	Received Line Signal Detector	From DCE
9	-	-	Reserved for Testing	-
10	-	-	Reserved for Testing	-
11	126	See Note	Unassigned	-
12	122/112	SCF/CI	Secondary Received Line Signal Detector/ Data Signal Rate Selector	From DCE
13	121	SCB	Secondary Clear to Send	From DCE
14	118	SBA	Secondary Transmitted Data	To DCE
15	114	DB	Transmitter Signal Element Timing (DTE Source)	From DCE
16	119	SBB	Secondary Received Data	From DCE
18	141	LL	Local Loopback	To DCE
19	120	SCA	Secondary Request to Send	To DCE
20	108/112	CD	DTE Ready	To DCE
21	140/110	RL/CG	Remote Loopback/Signal Quality Detector	T/F DCE
22	125	CE	Ring Indicator	From DCE
23	111/112	CH/CI	Data Signal Rate Selector (DTE/DCE Source)	T/F DCE
24	113	DA	Transmit Signal Element Timing (DTE Source)	To DCE
25	142	TM	Test Mode	From DCE
26		None	(Alt A Connector) No Connection at this Time	

Table 3.3
ITU-T V24 pin assignment (ISO 2110)

The half duplex operation of the RS-232 interface

The following description of one particular operation of the RS-232 interface is based on a half duplex data interchange. The description encompasses the more generally used full duplex operation.

Figure 3.7 shows the operation with the initiating user terminal, DTE, and its associated modem, DCE, on the left of the diagram, the remote computer, and its modem on the right.

The following sequence of steps occurs when a user sends information over a telephone link to a remote modem and computer.

- The initiating user manually dials the number of the remote computer.
- The receiving modem asserts the ring indicator line (RI) in a pulsed ON/OFF fashion reflecting the ringing tone. The remote computer already has its data terminal ready (DTR) line asserted to indicate that it is ready to receive calls. Alternatively, the remote computer may assert the DTR line after a few rings. The remote computer then sets its request to send (RTS) line to ON.
- The receiving modem answers the phone and transmits a carrier signal to the initiating end. It asserts the DCE Ready line after a few seconds.
- The initiating modem asserts the data carrier detect (DCD) line. The initiating terminal asserts its DTR, if it is not already high. The modem responds by asserting its DTE ready line.
- The receiving modem asserts its clear to send (CTS) line, which permits the transfer of data from the remote computer to the initiating side.
- Data is transferred from the receiving DTE (transmitted data) to the receiving modem. The receiving remote computer then transmits a short message to indicate to the originating terminal that it can proceed with the data transfer. The originating modem transmits the data to the originating terminal.
- The receiving terminal sets its request to send (RTS) line to OFF. The receiving modem then sets its clear to send (CTS) line to OFF.
- The receiving modem switches its carrier signal OFF.
- The originating terminal detects that the data carrier detect (DCD) signal has been switched OFF on the originating modem and switches its RTS line to the ON state. The originating modem indicates that transmission can proceed by setting its CTS line to ON.
- Transmission of data proceeds from the originating terminal to the remote computer.
- When the interchange is complete, both carriers are switched OFF and in many cases the DTR is set to OFF. This means that the CTS, RTS, and DCE Ready lines are set to OFF.

Full duplex operation requires that transmission and reception occur simultaneously. In this case, there is no RTS/CTS interaction at either end. The RTS line and CTS line are left ON with a carrier to the remote computer.

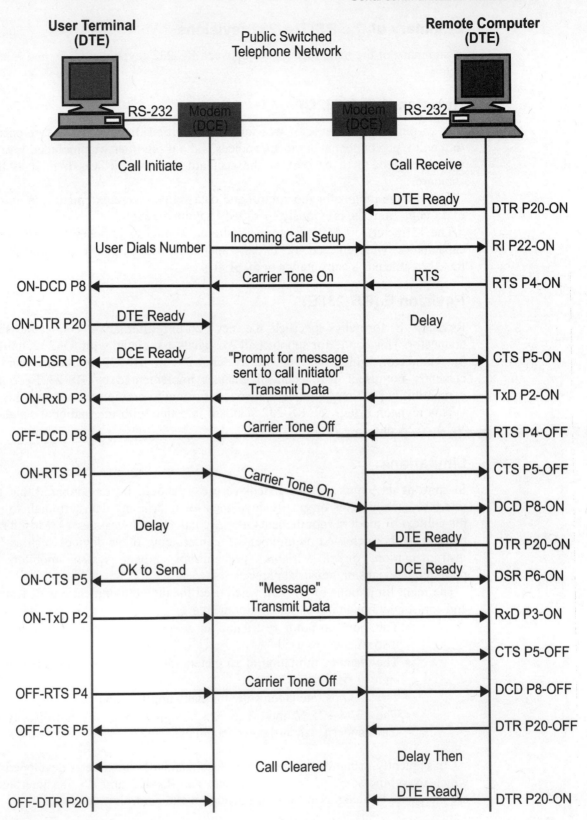

Figure 3.7
Half duplex operational sequence of RS-232

Summary of the RS/TIA-232 revisions

A summary of the main differences between RS-232 revisions, C, D, and E are discussed below.

Revision D (RS-232D)

The 25-pin D type connector was formally specified. In revision C, reference was made to the D type connector in the appendices and a disclaimer was included revealing that it was not intended to be part of the standard, however it was treated as the *de-facto* standard.

The voltage ranges for the control and data signals were extended to a maximum limit of 25 volts from the previously specified 15 volts in revision C.

The 15 meter (50 foot) distance constraint, implicitly imposed to comply with circuit capacitance, was replaced by 'circuit capacitance shall not exceed 2500 pF'. (Standard RS-232 cable has a capacitance of 50 pF/ft.)

Revision E (RS-232E)

Revision E formally specifies the new 26-pin alternative connector, the ALT A connector. This connector supports all 25 signals associated with RS-232, unlike the 9-pin connector, which has become associated with RS-232 in recent years. Pin 26 is currently not used. The technical changes implemented by RS-232E do not present compatibility problems with equipment confirming to previous versions of RS-232.

This revision brings the RS-232 standard into line with international standards CCITT V.24, V.28, and ISO 2110.

Limitations

In spite of its popularity and extensive use, it should be remembered that the RS-232 interface standard was originally developed for interfacing data terminals to modems. In the context of modern requirements, RS-232 has several weaknesses. Most have arisen as a result of the increased requirements for interfacing other devices such as PCs, digital instrumentation, digital variable speed drives, power system monitors and other peripheral devices in industrial plants.

The main limitations of RS-232 when used for the communications of instrumentation and control data in an industrial environment are:

- The point-to-point restriction is a severe limitation when several 'smart' instruments are used
- The distance limitation of 15 meters (50 feet) end to end is too short for most control systems
- The 20 kbps rate is too slow for many applications
- The –3 to –25 V and +3 to +25 V signal levels are not directly compatible with modern standard power supplies.

Consequently, a number of other interface standards have been developed by the RS, which overcome some of these limitations. The RS-422 and RS-485 interface standards are increasingly being used for instrumentation and control systems.

3.5 Troubleshooting serial data communication circuits

When troubleshooting a serial data communications interface, you need to adopt a logical approach in order to avoid frustration and wasting many hours. A procedure similar to that outlined below is recommended:

- Check the basic parameters. Are the baud rate, stop/start bits, and parity set identically for both devices? These are sometimes set on DIP switches in the device. However, the trend is towards using software, configured from a terminal, to set these basic parameters.
- Identify which is DTE or DCE. Examine the documentation to establish what actually happens at pins 2 and 3 of each device. On the 25-pin DTE device, pin 2 is used for transmission of data and should have a negative voltage (mark) in idle state, whilst pin 3 is used for the receipt of data (passive) and should be at approximately 0 volts. Conversely, at the DCE device, pin 3 should have a negative voltage, whilst pin 2 should be around 0 Volts. If no voltage can be detected on either pin 2 or 3, then the device is probably not RS-232 compatible and could be connected according to another interface standard, such as RS-422, RS-485, etc.

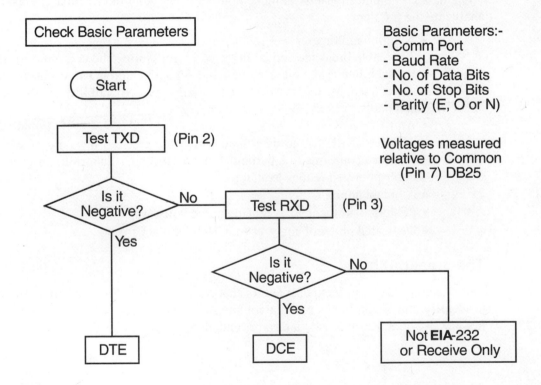

Figure 3.8
Flowchart to identify an RS-232 device as either a DTE or DCE

- Clarify the needs of the hardware handshaking when used. Hardware handshaking can cause the greatest difficulties and the documentation should be carefully studied to yield some clues about the handshaking sequence. Ensure all the required wires are correctly terminated in the cables.

- Check the actual protocol used. This is seldom a problem but, when the above three points do not yield an answer; it is possible that there are irregularities in the protocol structure between the DCE and DTE devices.
- Alternatively, if software handshaking is utilized, make sure both have compatible application software. In particular, check that the same ASCII character is used for XON and XOFF.

3.6   Test equipment

From a testing point of view, section 2.1.2 in the RS-232-E interface standard states that:

'...The generator on the interchange circuit shall be designed to withstand an open circuit, a short circuit between the conductor carrying that interchange circuit in the interconnecting cable and any other conductor in that cable... including signal ground, without sustaining damage to itself or its associated equipment...'

In other words, any pin may be connected to any other pin, or even earth, without damage and, theoretically, one cannot blow anything up! This does not mean that the RS-232 interface cannot be damaged. The incorrect connection of incompatible external voltages can damage the interface, as can static charges.

If a data communication link is inoperable, the following devices may be useful when analyzing the problem:

- A digital multimeter
  Any cable breakage can be detected by measuring the continuity of the cable for each line. The voltages at the pins in active and inactive states can also be ascertained by the multimeter to verify its compatibility to the respective standards.
- An LED
  The use of LED is to determine which the asserted lines are or whether the interface conforms to a particular standard. This is laborious and accurate pin descriptions should be available.
- A breakout box
- PC-based protocol analyzer (including software)
- Dedicated protocol analyzer (e.g. Hewlett Packard)

The breakout box

The breakout box is an inexpensive tool that provides most of the information necessary to identify and fix problems on data communications circuits, such as the serial RS-232, RS-422, RS-423, RS-485, etc., interfaces and also on parallel interfaces.

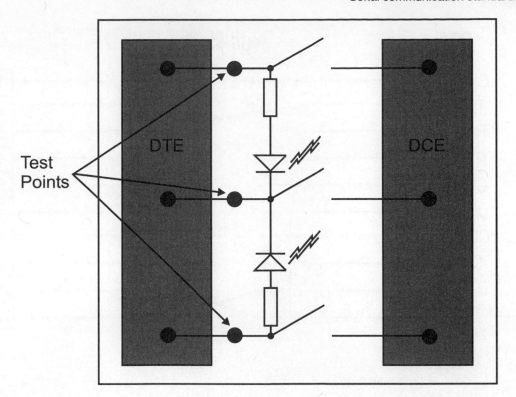

Figure 3.9
Breakout box showing test points

A breakout box is connected into the data cable, to bring out all 25 (or 9, 37, 50, etc.) conductors in the cable to accessible test points. Many versions of this equipment are available on the market, from the 'homemade' using a back-to-back pair of male and female DB-25 sockets, to fairly sophisticated test units with built in LEDs, switches and test points.

Breakout boxes usually have a male and a female socket and by using 2 standard serial cables the box can be connected in series with the communication link. The 25 test points can be monitored by LEDs, a simple digital multimeter, an oscilloscope, or a protocol analyzer. In addition, a switch in each line can be opened or closed while trying to identify where the problem is.

The major weakness of the breakout box is that, while one can interrupt any of the data lines, it does not help much with the interpretation of the flow of bits on the data communication lines. A protocol analyzer is required for this purpose.

Null modem

Null modems look like DB-25 'through' connectors and are used when interfacing two devices of the same gender (e.g. DTE–DTE, DCE–DCE) or devices from different manufacturers with different handshaking requirements. A null modem has appropriate internal connections between handshaking pins that 'trick' the terminal into believing conditions are correct for passing data. A similar result can be achieved by soldering extra loops inside the DB-25 Plug. Null modems generally cause more problems than they cure and should be used with extreme caution and preferably avoided.

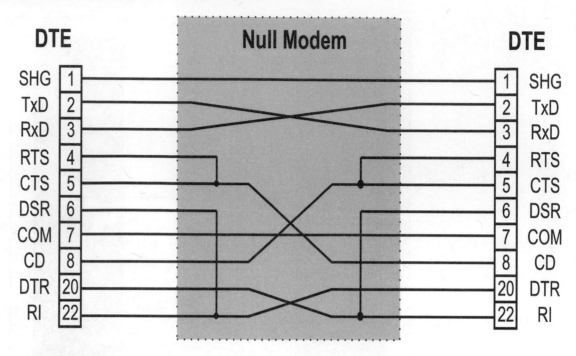

Figure 3.10
Null modem connections

Note that the null modem may inadvertently connect pins 1 together, as in Fig 3.10. This is an undesirable practice and should be avoided.

Loop back plug

This is a hardware plug, which loops back the transmit data pin to receive data pin and similarly for the hardware handshaking lines. This is another quick way of verifying the operation of the serial interface without connecting to another system.

Protocol analyzer

A protocol analyzer is used to display the actual bits on the data line, as well as the special control codes, such as STX, DLE, LF, CR, etc. The protocol analyzer can be used to monitor the data bits as they are sent down the line and compared with what should be on the line. This helps to confirm that the transmitting terminal is sending the correct data and that the receiving device is receiving it. The protocol analyzer is useful in identifying incorrect setting of baud rate, parity, stop bit, noise or incorrect wiring and connection. It also makes it possible to analyze the format of the message and look for protocol errors.

When the problem has been shown not to be due to the connections, baud rate, bits, or parity, then the content of the message will have to be analyzed for errors or inconsistencies. Protocol analyzers can quickly identify these problems.

Purpose built protocol analyzers are expensive devices and it is often difficult to justify the cost when it is unlikely that the unit will be used very often. Fortunately, software has been developed that enables a normal PC to be used as a protocol analyzer. The use of a PC as a test device for many applications is a growing field, and one way of connecting a PC as a protocol analyzer is shown in Figure 3.11.

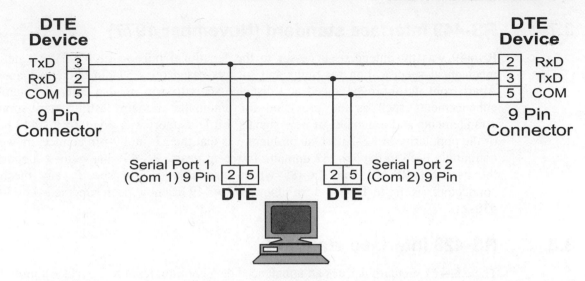

Figure 3.11
Protocol analyzer connection

Typical RS-232 problems

Below is a list of typical RS-232 problems, which can arise because of inadequate interfacing. These problems could equally apply to two PCs connected to each other or to a PC connected to a printer.

Problem	Probable Cause of Problem
Garbled or lost data	Baud rates of both connecting ports may be different
	Connecting cables could be defective
	Data Formats may be inconsistent (Stop Bit/Parity/No of data bits)
	Flow control may be inadequate
	High error rate due to electrical interference
	Buffer size of receiver is inadequate
First characters garbled	The receiving port may not be able to respond quickly enough. Precede the first few characters with the ASCII (DEL) code to ensure frame synchronization.
No data communications	Power for both devices may not be ON
	Transmit and receive lines of cabling may be incorrect
	Handshaking lines of cabling may be incorrectly connected
	Baud rates for both ports may not match
	Data format may be inconsistent
	Earth loop may have formed for EIA-232 line
	Extremely high error rate due to electrical interference for transmitter and receiver
	Protocols may be inconsistent Intermittent communications
	Intermittent interference on cable
ASCII data has incorrect spacing	There is a mismatch between 'LF' and 'CR' characters generated by transmitting device and for receiving device.

Table 3.4
A list of typical RS-232 problems

3.7　RS-449 interface standard (November 1977)

RS-449 was the intended successor to the functional portion of RS-232. It defined a mechanical specification for plugs and sockets based on a 37-pin and 9-pin assembly. Apart from its improved speed and distance specification, it also offered a number of enhancements such as the provision for automatic modem testing, new grounding arrangements and a number of new signals. Little support was given to RS-449 because of the popularity of RS-232. One problem was that the 37- and 9-pin connectors were not commonly used in the RS-232 domain. In recent years, RS-449 has gained support from the users of RS-422 and RS-485 whose standards do not specify any mechanical connectors. (Refer to Table 3.5 for pinouts). RS-449 has now been superseded by RS/TIA 530-A.

3.8　RS-423 interface standard

The RS-423 standard defines an unbalanced data communications interface similar to RS-232, but with some improvements. It allows an increase in cable length between devices, improved data transmission rates and multiple receivers on a line. RS-423 permits reliable communication for:

- Distances of up to 1200 meters (4000 feet)
- Data rates of up to 100 kbps
- Only one line driver on a line, but the driver current rating has been increased to permit multiple receivers
- Up to 10 line receivers, with lower current requirements, to be driven by the line driver

The improvements in performance have mainly been achieved by reducing the voltages to

- Logic 1:　　　-3.6 V to -6 V
- Logic 0:　　　$+3.6$ V to $+6$ V

Compared to RS-232, the total voltage slew is reduced by a factor of 4, with 12 V swings compared to 50 V swings. The effect of the line capacitance is reduced which allows faster data rates. Like the RS-232 interface, the data link is unbalanced and requires 3 wires for a full duplex signal path. Figure 3.12 illustrates the transmit connections.

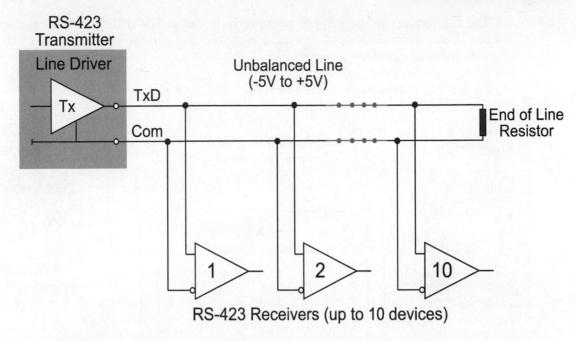

Figure 3.12
The RS-423 unbalanced line driver connection

The ability to support up to 10 receivers is achieved by increasing the current capacity of the line drivers, reducing the current drain of the line receivers, and connecting the receivers in a differential mode.

The RS-423 standard does not specify the mechanical connections or assign pin numbers. It is common to use the DB-25 connector with pin allocations as specified by RS-232.

3.9   The RS-422 interface standard

The RS-422 standard introduced in the early '70s defines a balanced, or differential, data communications interface using two separate wires for each signal. This permits very high data rates and minimizes problems with varying ground potentials because the ground is not used as a voltage reference, as in RS-232 and RS-423. RS-422 is an improvement on RS-423 and allows:

- Data to be communicated at distances of up to 1200 m (4000 feet, similar to RS-423)
- Data rates of up to 10 Mbps (increase of 100 times)
- Only one line driver on a line
- Up to 10 line receivers to be driven by one line driver

The differential voltages between the A and B lines are specified as:

- –2 V to –6V with respect to the B line for a binary 1 (MARK or OFF) state
- +2 V to +6V with respect to the B line for a binary 0 (SPACE or ON) state

The specification refers to the lines as A and B, but there are also called A(–) and B(+), or TX+ and TX–.

The line driver for the RS-422 interface produces a 5 V differential voltage on two wires. These voltage levels allows the transmitters and drivers to be supplied by the 5 V supply, common in today's computers.

Figure 3.13 illustrates the connection of RS-422 devices.

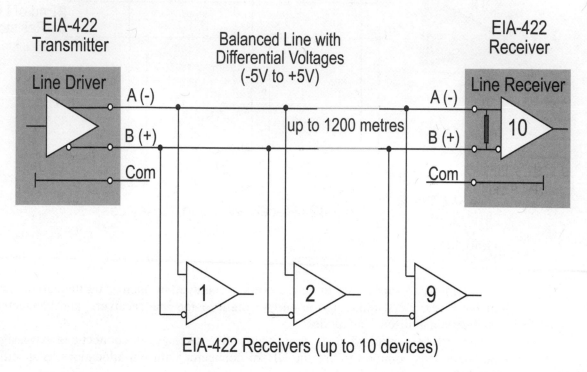

Figure 3.13
The RS-422 balanced line driver connections

As the differential receiver is only sensitive to the difference between two signals on its inputs, common noise signals picked up in both wires will have little effect on the operation of the receiver. Differential receivers are therefore said to have good common mode rejection (CMR) characteristics.

The major feature of the RS-422 standard is the differential voltage signal, which allows an increase in speed and provides higher noise immunity. Each signal is transferred on one pair of wires and is the voltage difference between them. The penalty is that two wires are required for each signal, compared to one wire for RS-232 and RS-423. A common ground wire is preferred to aid noise rejection. Consequently, 3 wires are required for a half duplex, and 5 wires for a full duplex system.

The balanced line driver can also have an input signal called an *enable* signal. The purpose of this signal is to connect the driver to its output terminals, A & B. If the enable signal is off, one can consider the driver as disconnected from the transmission line or in a *high impedance* state. (This *tri-state* approach is discussed under RS-485.)

The differential lines of the RS-422 are normally terminated with a resistor equal to the characteristic impedance (Z_0) of the line. This will prevent signal distortion due to reflections from the end of line. A typical value of Z_0 would be in the order of 120 Ω, for a twisted pair line.

The RS-422 standard does not specify mechanical connections or assign pin numbers. It is common practice to use the pin assignment of RS-449 for a DB-37 connector (see Table 3.5).

Pin No	DB-9 Connector Common for EIA-422 & EIA-485		DB-37 Connector EIA-449 Pin Assignment	
1	Shield		Shield	
2	Send Data	B+	Signaling Rate Indic	
3	Receive Data	B+		
4	Request to Send	B+	Send Data	A-
5	Clear to Send	B+	Send Timing	A-
6	Send Data	A-	Receive Data	A-
7	Receive Data	A-	Request to Send	A-
8	Request to Send	A-	Receive Timing	A-
9	Clear to Send	A-	Clear to Send	A-
10			Local Loopback	
11			Data Mode	A-
12			Terminal Ready	A-
13			Receiver Ready	A-
14			Remote Loopback	
15			Incoming Call	
16			Select Frequency	
17			Terminal Timing	A-
18			Test Mode	
19			Signal Ground	
20			Receive Common	
21				
22			Send Data	B+
23			Send Timing	B+
24			Receive Data	B+
25			Request to Send	B+
26			Receive Timing	B+
27			Clear to Send	B+
28			Terminal in Service	
29			Data Mode	B+
30			Terminal Ready	B+
31			Receiver Standby	B+
32			Select Standby	
33			Signal Quality	
34			New Signal	
35			Terminal Timing	B+
36			Standby/Indicator	
37			Send Common	

Table 3.5
Common DB-9 pin assignments for RS-422 and RS-485 and DB-37 pin assignments specified according to RS-449

3.10   The RS-485 interface standard

The RS-485 standard is the most versatile of the four RS interface standards discussed in this chapter. It is an extension of RS-422 and allows the same distance and data speed but increases the number of transmitters and receivers permitted on the line.

RS-485 permits a 'multidrop' network connection on 2 wires and allows reliable serial data communication for:

- Distances of up to 1200 m (4000 feet, same as RS-422)
- Data rates of up to 10 Mbps (same as RS-422)
- Up to 32 line drivers on the same line
- Up to 32 line receivers on the same line

Note: You can have 32 transceivers on a RS-485 network. If you require more than 32 devices, you would have to use repeaters (which is not defined in the RS-485 standards).

The differential voltages between the A and B lines are specified as:

- −1.5 V to −6V with respect to the B line for a binary 1 (MARK or OFF) state
- +1.5 V to +6V with respect to the B line for a binary 0 (SPACE or ON) state

The specification refers to the lines as A and B, but there are also called A(−) and B(+), or TX+ and TX−.

As with RS-422, the line driver for the RS-485 interface produces a ±5V differential voltage on two wires.

The major enhancement of RS-485 is that a line driver can operate in three states called tri-state operation:

- Logic 1
- Logic 0
- High-impedance

In the high impedance state, the line driver draws virtually no current and appears not to be present on the line. This is known as the 'disabled' state and can be initiated by a signal on a control pin on the line driver integrated circuit. Tri-state operation allows a multidrop network connection and up to 32 transmitters can be connected on the same line, although only one can be active at any one time. Each terminal in a multidrop system must be allocated a unique address to avoid conflicting with other devices on the system. RS-485 includes current limiting in cases where contention occurs.

The RS-485 interface standard is very useful for systems where several instruments or controllers may be connected on the same line. Special care must be taken with the software to coordinate which devices on the network can become active. In most cases, a master terminal, such as a PC or computer, controls which transmitter/receiver will be active at any one time.

The 2-wire data transmission line does not normally require special termination unless required by the manufacturer. On long lines, the leading and trailing edges of data pulses will be much sharper if terminating resistors approximately equal to the characteristic impedance (Z_0) of the line are fitted at the extreme ends. This is indicated in Figure 3.14. For twisted pair systems, the resistor used is typically 120 Ω.

Figure 3.14 shows a typical two wire multidrop network. Note that the transmission line is terminated on both ends of the line but not at drop points in the middle of the line. The signal ground line is also recommended in an RS-485 system to keep the common mode voltage that the receiver must accept within the −7 to +12 volt range.

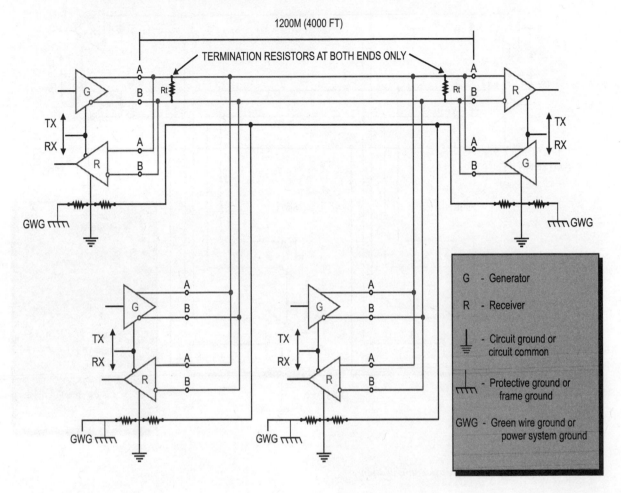

Figure 3.14
Typical two wire multidrop network

An RS-485 network can also be connected as a four wire configuration as shown in Figure 3.15. In this type of connection, it is necessary that one node be a master node and all others be slaves. The master node communicates to all slaves, but a slave node can communicate only to the master. Since the slave nodes never listen to another slave's response to the master, a slave node cannot reply incorrectly to another slave node. This is an advantage in a mixed protocol environment.

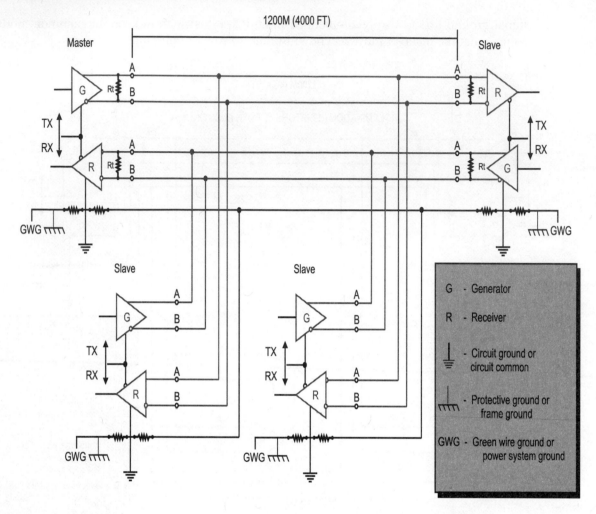

Figure 3.15
Four wire network configuration

During normal operation, there are periods when all RS-485 drivers are off, and the communications lines are in the idle, high impedance state. In this condition the lines are susceptible to noise pick up, which can be interpreted as random characters on the communications line. If a specific RS-485 system has this problem, it should incorporate 10 kΩ bias resistors as indicated in Figure 3.16. These resistors will maintain the data lines in a mark condition (idle) when the system is in the high impedance state.

The ground resistors shown in figure 3.15 are recommended in the specification and should be 100 ohms ½ watt. Their purpose is to reduce any loop currents if the earth potentials are significantly different.

The bias resistors are chosen in such a way that the B line will be kept at least 200 mV HIGHER than the A line with no input signal (i.e. all transmitters in the high impedance state). For the purpose of the calculation, remember that the two 120 Ω terminating resistors appear in parallel for this purpose. This particular example uses bias resistors on only one node.

Some systems employ bias resistors on all nodes, in which case the values of the bias resistors will be significantly higher since they appear in parallel.

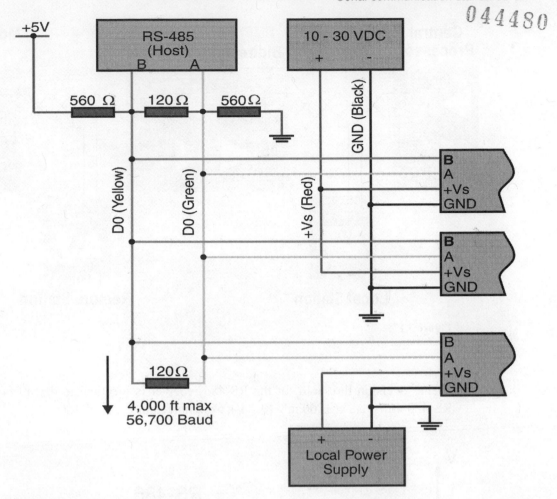

Figure 3.16
Suggested installation of resistors to minimize noise

RS-485 line drivers are designed to handle 32 nodes. This limitation can be overcome by employing an RS-485 repeater connected to the network. When data occurs in either side of the repeater, it is transmitted to the other side. The RS-485 repeater transmits at full voltage levels, consequently another 31 nodes can be connected to the network. A diagram for the use of RS-485 with a bi-directional repeater is given in Figure 3.17.

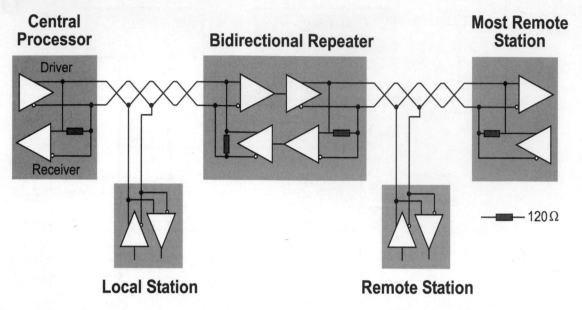

Figure 3.17
RS-485 used with repeaters

The 'decision threshold' of the RS-485 receiver is identical to that of both RS-422 & RS-423 receivers at ±200 mV (0.2 V), as indicated in Figure 3.18.

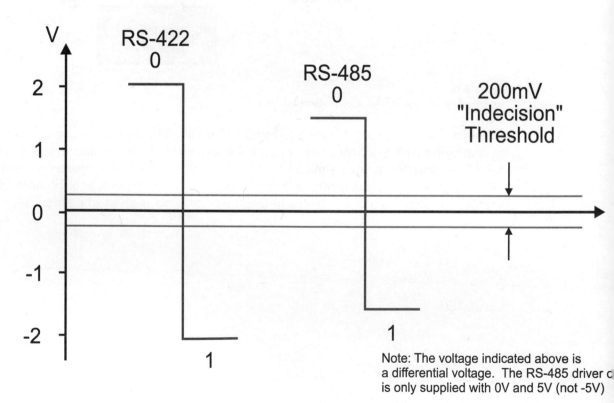

Note: The voltage indicated above is a differential voltage. The RS-485 driver c is only supplied with 0V and 5V (not -5V)

Figure 3.18
RS-485/422 & 423 receiver sensitivities

3.11   Troubleshooting and testing with RS-485

A few suggestions and testing procedures for RS-485 systems are outlined below. Both hardware and software testing will be discussed. There are also some examples of real problems and their solutions.

Hardware test procedure

- Check cabling connections for proper and complete connections. This should be done off-line (no power to the circuit). It maybe helpful to do a continuity check at this point, while the power is off.
- Check the polarity of the connections between each device on the RS-485 network. They should be A to A and B to B. Usually the black wire is B and the white wire is A, but this can be different for the various systems on the market. In most systems, the red wire of the voltmeter is to go on A and the black lead goes on the B. Check for your system by following the manufacturer's instructions.
- Check the voltage of the RS-485 system when the system is in an idle state. The idle state on most RS-485 systems is when the transmitters are not sending any data. Most RS-485 systems bias their transmitters to a minus voltage output (a mark or '1') using the voltage divider network indicated in an earlier discussion. However, this is not always the case. Some systems bias the lines with a positive voltage (a space or a '0') via the software while others turn the transmitters off (and have no voltage divider network connected externally) with a resultant zero voltage. Remember the true output of the RS-485 driver chip is usually an open collector output. Biasing keeps the lines from floating.
- Check that the shield (if there is one) is connected to ground at only one point. Ensure that there are no other connections to earth or ground. This is of course more important if the RS-485 devices are far away from each other. It would be wise if the devices were far away (100 meters/300 feet or more), to visually check the shield at each of the locations. The shields of different pairs must be connected.
- Check that the single shield ground is properly connected to an independent, clean earth ground. A clean ground is an earth ground that is as free as possible from noise.
- Check for common mode voltage problems. This is a difficult problem to detect, as it can be intermittent and sporadic. Usual symptoms may include intermittent failures and/or unusual events such as every other unit failing to communicate. The common mode voltage can be measured from each line to ground using a hand-held voltmeter. These voltages should not approach the maximum voltage of the driver chip.
- Ensure that the lines are terminated with the correct resistance terminators (if needed or required). The manufacturer's requirements must be followed.

Solution

The manufacturer's recommended termination resistors were added, one at each end. The reflections disappeared and the system worked correctly.

Note 1: Termination resistors on an RS-485 system are *not necessarily* required for operation but are preferable to reduce reflections (especially above cable lengths longer than 100 m and baud rates exceeding 40 kbaud).

Note 2: Reducing the baud rate (if possible) is another possible solution.

Note 3: Adding the terminating resistors effectively loads down the RS-485 line drivers. This can cause problems of its own. Do not put a resistor in any location except at the ends of the line. *Never* install a lower resistance than the manufacturer's specifications.

Example 2

A two wire master/slave RS-485 system had errors in the responses. The situation was:

- Master dispatched request to slave
- The slave responded with the information. Unfortunately, the master had given up and sent out another request.
- The problem was the delay in the slave responding.
- The master then recorded this as a communication error

Solution

The timeout for the master, before sending out another request to the slave for information, was extended.

Alternative solution

An increase in the baud rate may have improved the situation slightly (minor fix).

Example 3

A two wire master/slave RS-485 system had errors in the responses. The situation was:

- Master dispatched request to slave.
- Slave responded very quickly.
- Master transmitter lines had not gone into the high impedance state.
- Hence, the response from the slave was 'flattened' by being transmitted into low impedance.
- The master then recorded this as a communication error.

Solution

The master was put into the high impedance state quicker by disabling the RTS line. Obviously, ensure that this does not worsen the situation by disabling the transmitter before it has completed transmitting the signal.

Alternative solution

Slow down the response from the slave so that the master has a chance to go into the high impedance state.

3.12 EIA/TIA-530A interface standard (May 1992)

This standard supersedes RS-449. It is intended that the standard be used for applications where a balanced system is required such as RS-422 and RS-485.

3.13 EIA/TIA-562 interface standard (June 1992)

EIA/TIA-562 supports the new 3.3 V technology, which enables systems to have higher clock speeds, faster data communication rates, lower energy consumption, and to be smaller and lighter. The EIA/TIA-562 standard allows 64 kbps operation compared with the RS-232 maximum limit of 20 kbps.

Typical features of the EIA-562 standard are:

- 64 kbps operation (maximum)
- Stringent wave form specifications (ripple no larger than 5% of voltage swing)
- Maximum slew rate of 30 V/microsecond
- Ability to interface to RS-232

One of the disadvantages of the EIA-562 standard, compared to RS-232 is the reduction of the noise margin from 2 V to 0.7 V.

3.14   Comparison of the EIA interface standards

The main features of the four most common EIA interface standards are compared below:

Transmitter		EIA-232	EIA-423	EIA-422	EIA-485
Mode of operation		Unbalanced	Unbalanced	Differential	Differential
Max No. of Drivers & Receivers on line		1 Driver 1 Receiver	1 Driver 10 Receivers	1 Driver 10 Receivers	32 Drivers 32 Receivers
Recommended cable length		75 m	1,200 m	1,200 m	1,200 m
Maximum Data Rate		20 kbps	100 kbps	10 Mbps	10 Mbps
Maximum Common Mode Voltage		±25 V	±6 V	±6 V to −0.25 V	+12 V to −7 V
Driver Output Signal		±5.0 V min ±25 V max	±3.6 V min ±6.0 V max	±2.0 V min ±6.0 V max	±1.5 V min ±6.0 V max
Driver Load		> 3 ohm	> 450 ohm	100 ohm	60 ohm
Driver Output Resistance	Power On	n/a	n/a	n/a	100 µA −7 V ≤ Vcm ≤12V
(high-Z state)	Power Off	300 ohm	100 µA @ ±6 V	100 µA −0.25 V ≤Vcm ≤6 V	100 µA −7V ≤ Vcm ≤ 12 V
Receiver input resistance		3 kohm to 7 kohm	> 4 kohm	> 4 kohm	> 12 kohm
Receiver sensitivity		±3.0 V	±200 mV	±200 mV −7 Vcm 7 V	±200 mV −12 V ≤ Vcm ≤12 V

Table 3.6
Comparison of main features of RS-232, RS-423, RS-422, and RS-485

The data signaling rate versus cable length for balanced interface using 24 AWG twisted pair cable is shown in Figure 3.19.

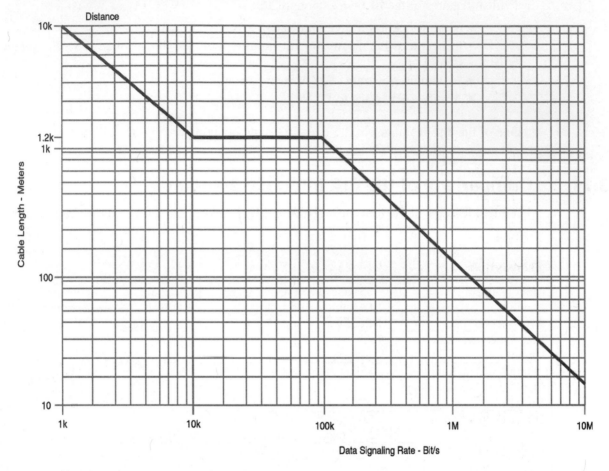

Figure 3.19
Data signaling rate vs. cable length

From a practical point of view, many RS-422/485 systems run up to 5000 meters (16 000 feet) at 1200 bps without any problems.

3.15   The 20 mA current loop

Another commonly used interface technique is the *current loop*. This uses a current signal rather than a voltage signal, employing a separate pair of wires for the transmitter current loop and receiver current loop.

A current level of 20 mA, or up to 60 mA, is used to indicate logic 1 and 0 mA logic 0. The use of a constant current signal enables a far greater separation distan e to be achieved than with a standard RS-232 voltage connection. This is due to the higher noise immunity of the 20 mA current loop which can drive long lines of up to 1 km, but at reasonably slow bit rates. Current loops are mainly used between printers and terminals in the industrial environment. Figure 3.20 illustrates the current loop interface.

Device #1   **Device #2**

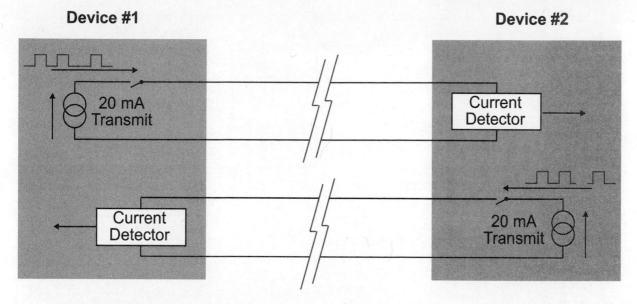

Figure 3.20
The 20 mA current loop interface

3.16   Serial interface converters

Serial interface converters are becoming increasingly important with the move away from RS-232 to industrial standards such as RS-422 and RS-485. Since many industrial devices still use RS-232 ports, it is necessary to use converters to interface a device to other physical interface standards. Interface converters can also be used to increase the effective distance between two RS-232 devices.

The most common converters are:

- RS-232/422
- RS-232/485
- RS-232/current loop

Figure 3.21 is a block diagram of an RS-232/RS-485 converter. Figure 3.22 shows a circuit wiring diagram.

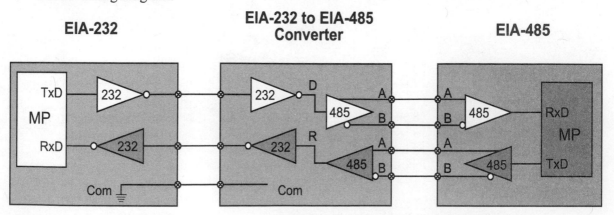

Figure 3.21
Block structure of RS-232/RS-485 converter

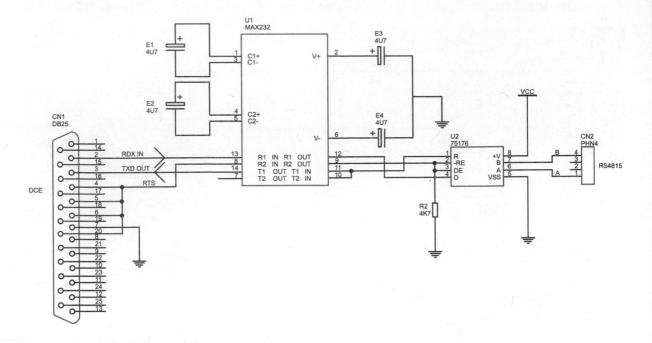

Figure 3.22
RS-232/485 converter

The RS-232/422 and RS-232/485 interface converters are very similar and provide bi-directional full-duplex conversion for synchronous or asynchronous transfer between RS-232 and RS-485 ports. These converters may be powered from an external AC source, such as 240 V, or smaller units can be powered at 12 V DC from pins 9 and 10 of the RS-232 port. For industrial applications, externally powered units are recommended. The RS-232 standard was designed for communications, not as a power supply unit!

The connections for a typical, externally powered RS-232/485 converter (Black Box Corporation) are shown below. Black Box does not recommend operating both ports on the converter at both DCE and DTE. LEDs are provided to show the status of incoming signals from both EIA-232 and EIA-485.

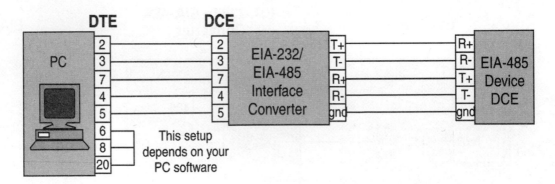

Figure 3.23
Wiring diagram for EIA-232/485 converter

When operating over long distances, a useful feature of interface converters is optical isolation. This is especially valuable in areas prone to lightning. Even if the equipment is not directly hit by lightning, the earth potential rise (EPR) in the surrounding area may be sufficient to damage the communications equipment at both ends of the link. Some specifications quote over 10 kV isolation, but these figures are seldom backed up with engineering data and should be treated with some caution.

Typical specifications for the RS-232/422 or RS-232/485 converters are:

- Data transfer rate of up to 1 Mbps
- DCE/DTE switch selectable
- Converts all data and control signals
- LEDs for status of data and control signals
- Powered from AC source
- Optically isolated (optional)
- DB-25 connector (male or female)
- DB-37 connector (male or female)

Typical specifications for the RS-232/current loop converters are:

- 20 mA or 60 mA operation
- DCE/DTE, full/half-duplex selectable
- Active or passive loops supported
- Optically isolated (optional)
- Powered from AC source
- Data rates of up to 19 200 kbps over 3 km (10 000 feet)
- DB-25 connector (male or female)
- Current loop connector – 5 screw

3.17   Interface to serial printers

It is important that the serial interface for serial printers is set up correctly. The following points should provide a guide:

- Select the correct terminator for the end of each block of characters (or bytes) sent down from the PC to the printer: This can either be a CR or LF (although CR is more common).
- Check what sort of flow control is being supported between the PC and the printers. If XON/XOFF is selected on the printer (as opposed to DTR), the printer cable must have the transmit and receive data pins connected and the PC software must support the XON/XOFF flow control. The PC's hardware and firmware itself does not directly support XON/XOFF. Ensure that the ASCII characters used to represent XON and XOFF are the same for both the printer and the PC.
- Select the appropriate settings for the buffer full (XOFF) and buffer empty (XON) codes on the printer. Most PCs have a maximum line buffer length of 256 bytes. The printer should send an XOFF when the buffer has 256 free bytes of storage remaining and XON when the buffer has emptied to less than 256 bytes. Typical printers have at least 2 K bytes of buffer memory.
- Check that the option of using pin 20 or pin 11 for the indication of DTR or printer ready line has been set correctly.
- Check that the correct selection has been made for either parallel or serial connections to the PC if this is optional on the printer.

- Ensure that the baud rate is not set too high, which do not employ any form of handshaking or flow control at all, as there could be loss of data.

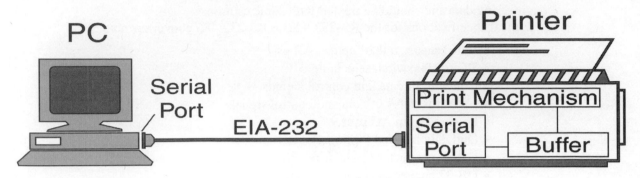

Figure 3.24
PC to printer serial connection

3.18   Parallel data communications interface standards

The two most important parallel data communication interface standards are:

- The general purpose interface bus (GPIB) or IEEE-488, used mainly for scientific purposes and automatic testing/measurement
- The Centronics standard – used mainly for cabling between PCs and printers.

3.19   General purpose interface bus (GPIB) or IEEE-488 or IEC-625

The GPIB was originally developed for automatic testing and for use with scientific equipment in laboratories, manufacturing, and other industrial and power system applications. In the early days of automatic testing, it became clear to manufacturers of digital test and scientific equipment that a universally accepted data communications interface was required between the computer controlling the testing sequence, the test equipment and recording devices, such as plotters and printers.

The standard used today was originally defined by Hewlett-Packard in 1965 as a digital data interface standard for the interconnection of engineering test instruments and was initially called the Hewlett-Packard Interface Bus (HPIB). This standard was adopted by other manufacturers and was published in 1975 as the IEEE-488 standard. IEEE-488 was updated in 1978 and issued internationally as IEC-625. There have been further revisions to the standard since 1978. IEC-625 is the common designation for GPIB in Europe.

The GPIB is an interface design that allows the simultaneous connection of up to 15 devices or instruments on a common parallel data communications bus. This allows instruments to be controlled or data to be transferred to a controller, printer, or plotter. It defines methods for the orderly transfer of data, addressing of individual units, standard bus management commands and defines the physical details of the interface.

Physical connection configurations

The devices on the GPIB can be connected in one of two ways:

- A star configuration
- A chain (linear) configuration

A star configuration is one in which each instrument is connected directly to the controller by means of a separate GPIB cable. The connectors are all connected to the same port as the controller, as shown in Figure 3.25. A drawback to this simple arrangement is that all devices on the same bus must be relatively close to the controller because of the length limitation of each cable.

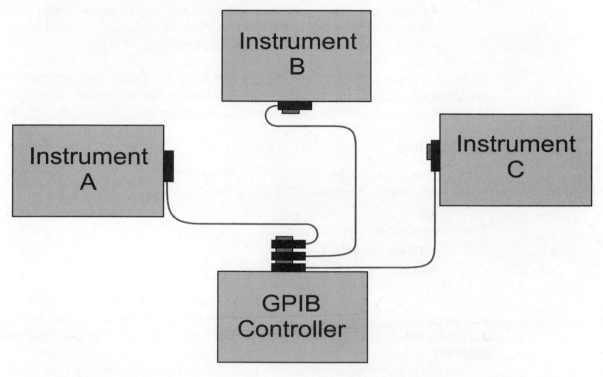

Figure 3.25
The GPIB star connection

The chain configuration, each device, including the controller, is connected to the next in a chain. The controller does not have to be the first or last device in the chain, but can be linked in anywhere as shown in Figure 3.26. A controller is a controller in the sense that it coordinates the events on the bus. Physically and electrically, it is similar to any other device connected to the GPIB. This configuration is usually the most convenient way to connect equipment.

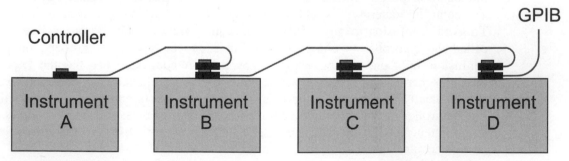

Figure 3.26
The GPIB chain (linear) configuration

Although the star and chain configurations are suggested for GPIB, connections can be made in any other way, if the following rules are observed:

- All devices are connected to the bus
- No more than 15 devices, including the controller, may be on the bus with no less than two thirds powered on
- Cable length between any two devices may not exceed 4 meters (13.33 feet) with an average separation of 2 meters (6.66 feet) over the entire bus
- Total cable length may not exceed 20 meters (66.66 feet)

A single device on the GPIB can transfer data to up to 14 other devices. The GPIB uses asynchronous handshaking, so the actual data transfer rate is determined by the devices themselves. Generally, the hardware limits the maximum data rate to about 250 kbps.

Device types

There are many thousands of different types of GPIB-compatible devices available for various applications. There are 4 different groups of devices:

- Talkers
- Listeners
- Talkers/listeners
- Controllers

A talker is a one-way communicating device that can only send data to other devices. It does not receive data. The talker waits for a signal from the controller and then places its data on the bus. Common examples are simple digital voltmeters (DVMs) and some A/D converters.

A listener is a one-way communicating device that can only receive data from another device. It does not send data. It receives data when the controller signals it to read the bus. Common examples are printers, plotters, and recorders.

A talker/listener has the combined characteristics of both talkers and listeners, with the limitation that it is never a talker and a listener at the same time. A common example is a programmable one, which is a listener while its range is being set by the controller, and a talker while it sends the results back to the controller. Most modern digital instruments are talker/listeners as this is the most flexible configuration.

A controller manages and controls everything that happens on the GPIB. It is usually an intelligent or programmable device, such as a PC or a microprocessor controlled device. The controller determines which devices will send and which will receive data and when. To avoid confusion in any GPIB application, there can only be one active controller, called the controller in charge (CIC). There can be several controllers, but to avoid confusion, only one can be active at any time. A controller also has the features of a talker/listener. In some cases, when several PCs are simultaneously connected on a GPIB, one is usually configured as the controller and the others configure as talkers/listeners. The controller needs to be involved in every transfer of data. It needs to address a talker and a listener before the talker can send its message to a listener. After the message is sent, the controller unaddresses both units.

Some GPIB configurations do not require a controller, for example, when only one talker is connected to one or more listener. A controller is necessary when the active or addressed talker or listener must be changed.

Electrical or mechanical characteristics

The GPIB bus is carried inside a shielded 24-wire cable with standard connectors at each end. The standard connector used is the 'Amphenol' as shown in Figure 3.27. Adding a new device to the bus is done by connecting a new cable in a star or chain configuration. Screws hold one connector securely to the next. Since the 24-pin connectors are usually stackable, it is easy to connect or disconnect devices to the bus.

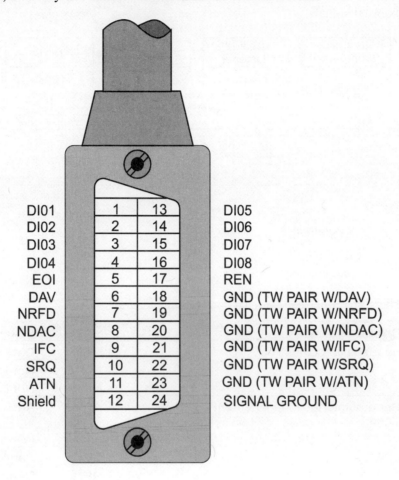

DIO1	1	13	DIO5
DIO2	2	14	DIO6
DIO3	3	15	DIO7
DIO4	4	16	DIO8
EOI	5	17	REN
DAV	6	18	GND (TW PAIR W/DAV)
NRFD	7	19	GND (TW PAIR W/NRFD)
NDAC	8	20	GND (TW PAIR W/NDAC)
IFC	9	21	GND (TW PAIR W/IFC)
SRQ	10	22	GND (TW PAIR W/SRQ)
ATN	11	23	GND (TW PAIR W/ATN)
Shield	12	24	SIGNAL GROUND

Figure 3.27
GPIB connector and pin assignment

The 24 lines in each cable consist of 8 data lines and 8 pairs (16) of control and bus management lines. The data lines are used exclusively to carry data, in a parallel configuration (byte by byte), along the bus. The control and bus management lines are used for various bus management tasks that synchronize the flow of data. When data or commands are sent down the bus, the bus management lines distinguish between the two. Detailed knowledge of how these management lines interact is useful but not necessary to effectively use the GPIB.

In the RS-232, the UART is used to coordinate the 'housekeeping' activities associated with the serial interface. The full capacity of the microprocessor can then be directed to other duties. In a similar way, the coordination of the GPIB parallel interface is controlled by a GPIB IC. The most common GPIB ICs used for this purpose are the Texas Instruments TMS9914A and the NEC-7210.

GPIB bus structure

The GPIB consists of 8 data lines (DI01–DI08) and 8-pairs of control lines. Three of the eight control line pairs are the handshaking lines, which coordinate the transfer of data (DAV, NRFD and NDAC). The other five pairs are for bus control and management (ATN, REN, IFC, SRQ and EOI). The 8 'ground' wires provide electronic shielding and prevent bus control signals from interfering with one another or from being influenced by external signals.

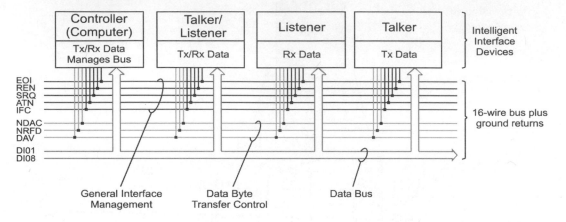

Figure 3.28
The GPIB bus structure

The signal lines can be separated into three groups:

Data Bus Lines	D101	D108 Data Bus Lines	
Handshaking Lines		DAV	Data Available
		NRFD	Not Ready for Data
		NDAC	Not Data Accepted
General Interface Management Lines		ATN	Attention
		IFC	Interface Clear
		SRQ	Service Request
		REN	Remote Enable
		EOI	End or Identify

Table 3.7
Three groups of signal lines

The eight data lines DI01 to DI08 carry both data and command messages. All commands and most data use the 7-bit ASCII code, in which case the eighth bit, DI08, is either unused or used for parity.

The GPIB uses binary signals to represent the information that is carried on the lines of the bus. It uses the symbols 'true' and 'false' to represent the two states of voltages on the lines. The GPIB uses the logic convention called 'low-true' or negative logic, where the lower voltage state is 'true', and the higher voltage states are 'false'. Standard TTL voltage levels are used. For example, when DAV is 'true', the TTL voltage level is low (±0.8 V), when DAV is false, the TTL level is high (±2.0 V). Despite low being 'true', no line can be high (i.e. 'false'), unless all devices on that line allow it to go high. This is convenient when there are several listeners accepting data. The 'not data accepted' (NDAC) line cannot go to the 'false' state, indicating data accepted, until the last listener has accepted the data. Consequently, the handshaking process waits for the slowest listener on the bus.

Each device connected to the GPIB has a unique device address and must be designed with enough intelligence to identify whether the data or command sent down the data line is meant for it or for another device. Device addresses are arbitrary and are set by the user, usually on a DPI switch on the back of the device, or by programming the device software. Each connected device is identified in the software of the controllers program. The only limitation in choosing a device address is that it must be an integer number in the range 1 to 30.

GPIB handshaking

Data is transmitted asynchronously on the GPIB parallel interface one byte at a time. The transfer of data is coordinated by the handshake voltage signals on the 3 bus control lines (DAV, NDAC and NRFD), called a three-wire interlocked handshake. Handshaking ensures that a talker will put a data byte on the bus only when all listeners are ready and will keep the data on the bus until it has been read by all listeners. It also ensures that listeners will accept data only when a valid byte is available on the bus.

The talker must wait for the NRFD line to go high (false) before any data can be put onto the bus. The NRFD line is controlled by the listeners. Only when NRFD voltage is high (false) are all listeners ready to receive data. The talker then asserts DAV 'true' (voltage low) and when the listeners detect the low level on DAV, they read the byte on the data lines. As each listener accepts the data, it releases NDAC. After the last listener has accepted the data, the NDAC line voltage goes high (false) and this signals the talker that the data has been accepted. Only when the data byte has been accepted by all the listeners can the talker allow DAV voltage to go high (false) and remove its data from the bus. Figure 3.29 illustrates this handshaking sequence.

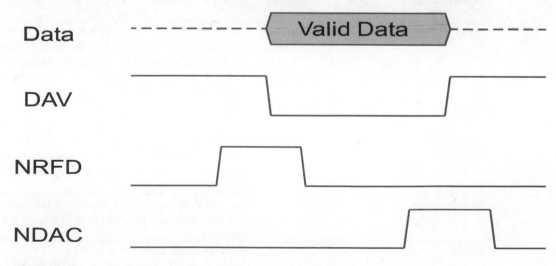

Figure 3.29
The GPIB handshaking timing diagram

Interface management lines

The other five lines manage the flow of information across the interface:

- ATN
- REN
- IFC
- SRQ
- EOI

Data transfers

Device addresses are any integer in the range of 1 to 30. There are 2 kinds of addresses for a device:

- Talk address (maximum of 15)
- Listen address (maximum of 15)

If a device sees its talk address on the bus, it knows it will act as a talker and will be required to send data. Conversely, when it sees its listen address on the bus, it will be required to act as a listener and receive data.

Both command/response with polling and interrupt driven protocols are allowed under the IEEE-488 specification.

3.20   The Centronics interface standard

The parallel printer or Centronics interface standard is used primarily to interface printers to computers or other intelligent devices and includes a 36-pin connector. This interface has a limited distance capability because of its low level +5 V signals. Full signal definitions are given in Table 3.8.

Signal Name	Signal Pin	Return Pin	Signal Definition
*DSTB	1	19	Low level pulse of 0.5 microseconds or more, used to strobe the DATA signals into the printer. The printer reads the data at the Low Level of this signal. Ensure an Acknowledge has been returned before using the next Data Strobe. Data Strobe is ignored if the BUSY is high.
*DATA 1-8	2-9	20-27	8 data lines from the host. High level represents binary 1, Low level represents binary 0. DATA 8 is the most significant bit. Signal must be High at least 0.5 microseconds before the falling edge of the Data Strobe signal and held at least 0.5 microseconds after the rising edge.
Acknowledge	10	28	Low level pulse of 2 to 6 microseconds indicates input of a character into the print data buffer, or the end of an operation.
BUSY	11	29	High level indicates the printer cannot receive data. Typical conditions that cause a High BUSY level are buffer full or ERROR condition.
PE(Paper Empty)	12		High Level indicates that the printer is out of paper.
SLCT (Select)	13		High Level indicates that the printer is ON LINE.

*AUTO FEED XT	14		Low Level indicates LF (Line Feed) occurs after each CT (carriage return) code.
No Connection	15		Reserved Signal Line.
Signal Ground		16	Logic/Signal Ground Level (0 V)
Frame Ground		17	Printer Cabinet/Frame Ground line
No Connection			Reserved Signal Line
Signal Ground		19-30	Twisted Pair cable return lines.
*INIT Initialize	31		Low Level pulse of 50 microseconds or more, resets the buffer and initializes the printer.
ERROR	32		Low Level indicates the printer is OFF LINE, has a PAPER OUT or has sensed an ERROR condition.
Signal Ground	33		Logic/Signal Ground Level (0 V)
No Connection	34		Not used.
+5 V Regulate	35		Connected to the +5 V source through a $3.n$ k Ohm Resistor.
*SLCT IN (-Select In)	36		Low level indicates the printer is placed ON LINE (Selected) when the power is turned ON.

Table 3.8
Centronics pin assignment

* Indicates that the signal is generated by the host system, for example, the PC.771.

3.21   The universal serial bus (USB)

In September 1998, Microsoft, Intel, Compaq, and NEC developed revision 1.1 of the universal serial bus. The objective was to standardize the input/output connections on the IBM PC for devices like printers, mice, keyboards, and speakers. Data acquisition (DAQ) devices were not envisioned to be connected to the USB system. However, that does not mean that the USB cannot be used for DAQ. In many ways, the USB is well suited for DAQ systems in the laboratory or other small-scale systems.

Small-scale DAQ systems have traditionally suffered from the need for an easy to use and standardized bus system for connecting smart DAQ devices. The nearest thing was the IEEE 488 GPIB system. The GPIB system can be expensive and is not supported on every PC without purchasing additional hardware. There is a need for an easy to operate, inexpensive, and standardized bus system to connect small-scale DAQ devices. The USB can fill those needs. With its plug-and-play ability, it is extremely easy to implement and use. In addition, it is now standard on all IBM compatible PCs. Although it is not in any way as cheap as say an RS-232 connection, it is affordable.

The USB is limited by its very nature for its application to DAQ systems. The biggest problem is the maximum cable distance. The low-speed version is limited to 3 meters (10 feet) and the high-speed version is limited to 5 meters (16 feet) in total cable length. This requirement reduces the ability of USB to be used in the large factory or plant environment. Typically, the DAQ systems in these industries need to cover distances of up to 1 km (0.6 miles). Due to the timing requirements of the USB, the length of the cable **cannot** be increased with repeaters. This limits the use of the USB to the laboratory or bench top systems.

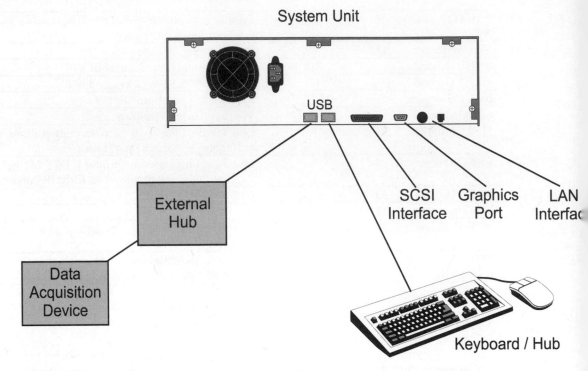

Figure 3.30
Connecting the USB

USB – overall structure

The USB is a master/slave, half-duplex, timed communication bus system designed to connect close peripherals and hubs to an IBM compatible PC. It runs at either 1.5 Mbps (low-speed) or 12 Mbps (high-speed). The PC's software program (using device drivers) creates packets of information that are going to be sent to devices connected on the USB bus. The USB drivers in the computer allocate a certain time within a frame for the information. The packet is then placed in this 1 ms frame that can contain many packets. One frame might contain information for many devices or it may contain information for only one device. The frame is then sent to the physical layer via the USB drivers, and then on to the bus.

The device receives its part of the packet and if necessary formulates a response. It places this response on the bus. The USB drivers in the PC detect the response on the bus and verify that the frame is correct using a CRC checking method. If the CRC indicates that the frame is correct, the software in the PC accepts the response.

The devices connected to the USB bus can also be powered off the bus cable. Devices can use no more than 500 mA. This works well for small scale DAQ devices, larger DAQ devices usually use external power supplies. Both power and communications are on the same cable and connector.

There are many parts in the USB system that make the communication possible. These include:

- Host hubs
- External hubs
- Type A connector
- Type B connector
- Low-speed cables

- High-speed cables
- USB devices
- Host hub controller hardware and driver
- USB software driver
- Device drivers

Topology

The USB uses a pyramid-shaped topology with everything starting at the host hub. The host hub usually consists of two USB ports on the back of the PC. These ports are basically in parallel with each other.

Each port is a four-pin socket with two pins reserved for power and two for communications. The cables from external hubs or USB devices are plugged into the host hub ports. One or both of the ports can be used. It does not matter which one is used if only one connection is being made. If the external device or hub has a removable cable then a 'type A or type B' cable is used to make the connection. The 'A' plug goes into the back of the PC (host hub) and the 'B' plug goes into the device or external hub. If the external hub or device has an in-built cable then the 'A' plug is plugged into the host hub port. The socket on the host hub is keyed so the plug will only go in one way. 'B' plugs will not go into 'A' sockets and vice-versa.

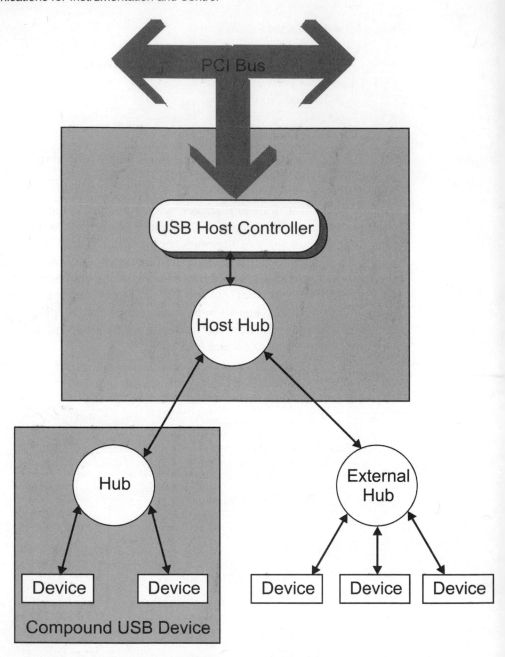

Figure 3.31
USB topology

The limitations on cable lengths are very important for the USB system. All cables, even if they come out of repeater hubs, must be counted in the total length of the cables.

Host hubs

The controller chips for the host hub usually reside on the motherboard inside the PC, although the hub could be a PCB in a PCI slot. The host controller does the parallel to serial and serial to parallel conversion from the PCI bus to the USB connectors. Sometimes a pre-processor is used to improve efficiency of the USB system. This host controller and connector combination is called the root hub or host hub. The host hub's function is to pass the information to and from the PCI bus to the data lines (+D and –D) on the USB socket. The host controller can control the speed at which the USB operates. It also connects power lines (+5 V and ground) to a USB device via the USB cable. The external USB device may be another USB Hub or a USB type device like a printer

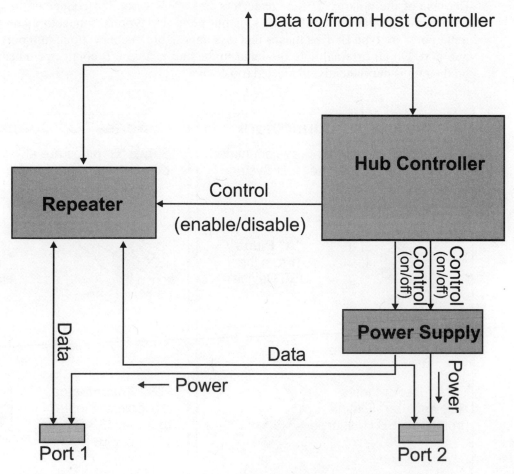

Figure 3.32
Host hub block diagram

The host hub has complete control over the USB ports. This control would include:

- Initialization and configuration
- Enabling and disabling the ports
- Recognizing the speed of devices
- Recognizing that a device has been connected
- Getting information from the application software
- Creating a packet and then frame
- Sending the information on to the bus
- Waiting and recognizing a response
- Error correction
- Recognizing that a device has been disconnected
- Using the port as a repeater

The connectors (Type A and B)

There are two types of connectors, type A and type B. The reason there are two types is that some devices have built in cables while others have removable cables. If the cables were the same, it would be possible to connect a host hub port to another host hub port. Because of the polarity of the connectors, the +5 V would be connected to ground. To keep this from happening the hub's output ports use type A connectors and the device input ports are type B. This means that it is impossible to connect one hub port to another hub port. On an external hub, the input to the hub is a type B connector unless the cable on the hub is permanently connected (no connector).

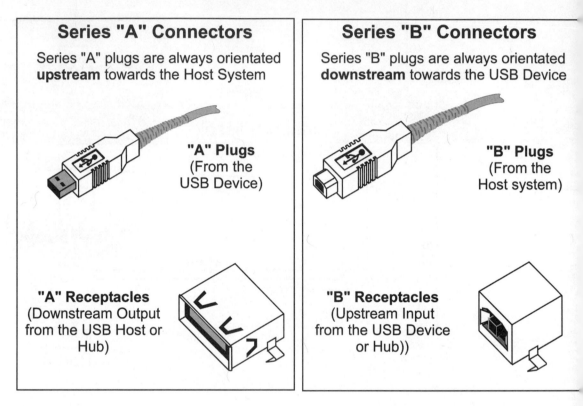

Figure 3.33
USB connectors

Low-speed cables and high-speed cables

The USB standard states that the USB will run at either 1.5 Mbps (slow-speed) or 12 Mbps (fast speed). The USB must have low-speed cables and high-speed cables. This is due to the impedance difference caused by the different frequencies of data transfer. The low-speed cables use untwisted unshielded cable. The data pair is 28 AWG and the power pair is 20–28 AWG. The low-speed cable is used on devices like keyboards and mice. The maximum distance for low-speed cabling is 3 meters (10 feet). The high-speed cables use twisted shielded cable pairs. The data pair is 28 AWG and the power pair is 20–28 AWG. The maximum propagation delay must be less than 30 ns. The maximum distance for high-speed USB is 5 meters (16 feet).

External hubs

The external hubs are used to increase the amount of devices connected to the system. Usually they have four USB output ports and either one type B input connector or a dedicated cable. This cable has a type A plug. It is usually connected to a host hub, but could be connected to the output socket (type A) of another external hub. Even though the external hub is a repeater, it cannot extend the overall length of the system. This is because of the timing requirements of the USB standard.

The external hub is an intelligent device that can control the communication lines and power lines on its USB ports. It is a bi-directional repeater for information coming from the host hub and from USB devices. It talks to and even acts like an external USB device to the host hub. It plays an integral part in the configuration of devices at start up. There is no physical limit to the number of hubs.

USB devices

The USB system supports every existing peripheral that can be connected to a PC. It also can and has been adapted to devices that are not usually considered peripherals. This would include data acquisition devices such as digital I/O modules and analogue input/output modules. All USB devices must be intelligent devices. Smart devices obviously cost more than 'dumb' RS-232 and RS-485 connected devices. With this increased cost, the user gets more functions, ease of use and the ability to connect more devices to the PC. With the old non-USB system, the computer was limited to a few devices. The USB system allows 127 devices to be connected to the PC at the same time.

There are two types of USB device:

- Low-speed, and
- High-speed

The low-speed devices are not only limited in their speed but also in features. These devices include keyboards, mice, and digital joysticks. Since these devices put out small amounts of information, they are polled less frequently and are slower than other devices. When high-speed devices access the USB bus, the low-speed device communication is disabled. Turning off the low-speed device ports at the root or external hubs disables the low-speed devices. The hubs re-enable the low-speed ports after receiving a special preamble packet.

High-speed devices like printers, CD–ROMs, and speakers need the speed of the 12 Mbps bus to transfer the large amount of data required for these devices. All high-speed devices see all traffic on the bus. They are never disabled like the low-speed devices. When a device like a microphone is 'connected' to the speakers most of the traffic and therefore packets will be used by the audio system. Other traffic like keyboard and mouse functions will have to wait. The host hub controller driver decides who has to wait and how long.

Host hub controller hardware and driver

The host hub controller hardware and software driver controls all transactions. The host hub controller hardware does the physical connections from the PCI bus to the USB connectors. It enables and initializes the host ports one at a time. It determines the speed and direction of data transfer on both host ports. The host controller in conjunction with the host hub software driver determines the frame contents, prioritization of the devices and how many frames are needed for a particular transfer.

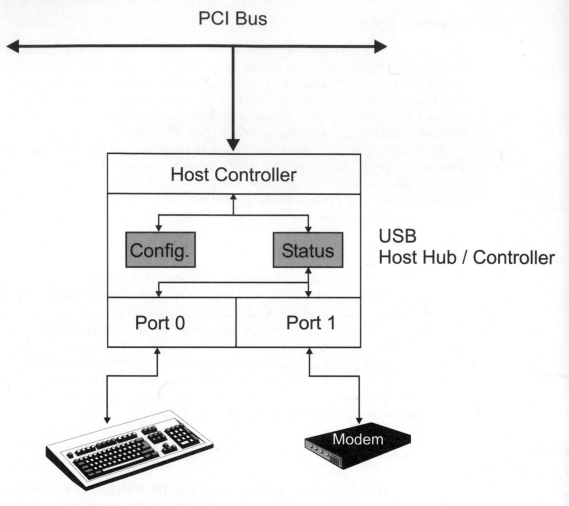

Figure 3.34
Host hub controller diagram

USB software driver

The USB software driver handles the interface between the USB devices, the device drivers, and the host hub driver. When it receives a request from a device driver in the PC to access a certain device, it coordinates the request with other device requests from the application software in the PC. It works with the host hub controller driver to prioritize packets before they are loaded into a frame. The USB software driver gets information from the USB devices during device configuration. It uses this information to tell the host hub controller how to communicate to the device.

Device drivers

For each USB device, a device driver must be loaded into the PC. This device driver is a software interface between the external USB device and the application software, the USB software driver and the host hub controller driver. It has information for the other drivers about that particular device's needs. This information is used to determine things like the type, speed (although that information can be determined physically by the hub ports), priority, and function of the device, as well as the size of packet needed for the transfer of data.

Communication flow

As mentioned before, the USB system is a master/slave, half duplex, timed communication bus system designed to connect peripherals and external hubs. This means that the peripherals cannot initiate a communication on the USB bus. The master (or host) hub has complete control over the transaction. It initiates all communications with hubs and devices. The USB is timed because all frames are sent within a 1 ms time slot. More than one device can place a packet of information inside that 1 ms frame. The host hub driver, in conjunction with the USB software driver determines the size of the packet and how much time each device is allocated in one frame.

If the applications software wants to send or receive some information from a device, it initiates a transfer via the device driver. Either the manufacturer of the device supplies this device driver or it comes with the operating system. The USB driver software then takes the request and places it in a memory location with other requests from other device drivers. Working together the USB driver, the host hub driver and the host hub controller place the request, data, and packets from the device drivers into a 1 ms wide frame. The host controller then transfers the data serially to the host hub ports. Since all the devices are in parallel on the USB bus, all devices 'hear' the information (except low-speed devices, unless it is a low-speed transfer. Low-speed devices are turned off when they are not being polled.) If necessary, the host waits for a response. The remote USB device then responds with an appropriate packet of information. If a device does not see any bus activity for 3 ms, it will go into the suspend mode.

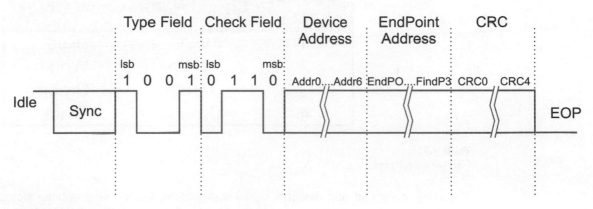

Figure 3.35
Example of an IN packet

There are four types of IN packets (reading information from a device) and three types of OUT packets (sending information out to a device).

Certain devices like mice and keyboards need to be polled (IN packets), but not too often. The USB software driver knows about these devices and schedules a regular poll for them. Included in the response are three levels of error correction. This type of transfer is very reliable. The peripherals are usually low-speed devices and therefore need a distinct low-speed packet to enable them. This packet is called a preamble packet. The preamble packet is sent out before the poll. The low-speed devices are disabled until they receive this preamble packet. Once they are enabled, they hear the poll and respond. Only one device can be polled at a time and therefore only one device will respond. USB has no provisions for multiple responses from devices.

On the other hand, there are devices that need constant attention but polling is not possible. These would be devices like microphones (IN packets), speakers (OUT packets), and CD-ROMs (both types of packets IN and OUT). The transfer rate is very important to these devices. Obviously, they would use the high-speed transfer rate and they would use a large portion of the frame (up to 90%). The receiving device does **not** respond to the data transfer. This transfer is a one-way data transfer (simplex). This means that error correction is effectively turned off for these types of transfers.

The physical layer

The physical layer of the Universal Serial Bus is based on a differential +/– 3 V dc communication system. It is in some ways very similar to the RS-485 voltage standard. Unfortunately, it does not have the range of RS-485. This is not because of the type of wire used or because of the USB voltage standard itself, but because of the timing requirements of the USB protocol. In order to fit in everything the peripherals do on a USB bus, it was necessary to put very strict time requirements on the USB.

The USB physical standard has many benefits to the user. It is fast – 12 MHz – is very resistant to noise and is very reliable as long as the cabling rules are followed. With standardized cables and connectors, it is very hard for the user to get things wrong when cabling the USB system.

Contact Number	Signal Name	Cable Color
1	VCC	Red
2	-Data	White
3	+Data	Green
4	Ground	Black

Table 3.9
USB connector pins

The story goes that one day Bill Gates was watching his new computer being installed. When he saw the number of wires coming out of the back of the computer, he called the CEO at Intel and said, 'We have to get rid of this mess of cables and connectors'. And as they say, the rest is history.

Connectors

The plugs and sockets on the USB have two wires for data communication and two wires for power. Using bus-powered devices is optional. The pins on the plug are not the same lengths. The power pins are 7.41 mm long and the communication pins are 6.41 mm long. This means that if a cable is plugged in 'hot' the power will be applied to the device before the communications lines. More importantly, it also means that when a cable is unplugged the data communications lines will be disconnected before the power. This reduces the possibility of back EMF voltage damaging the equipment. There are two types of connectors for the USB, type A and type B.

Type A is a flat semi-rectangular keyed connector that is used on the host ports, external hubs, and devices. The type B keyed connector is half-round and smaller than the type A connector is. Note that both type A and B plugs have the USB symbol on the **top** of the connector. This is for orientation purposes.

The hubs and devices all have female sockets, while the cables have a type A male plug on one end and a type B on the other end. This is because if there were a type A on both ends it would be possible to connect two host hub sockets or external hub sockets. Cables that are not removable from the device or external hub only have a type A plug on one end.

Cables

The cables for the USB are specified as either low- or high-speed cables. Both the low- and high-speed cables can use type A connectors, but only a high-speed device can use type B connectors. Detachable cables are therefore always high-speed cables.

Due to that fact that the impedance of a cable is determined in part by the frequency of the signal, the two speeds need two different cables. External hubs are always high-speed units, but they accept low- and high-speed cables. Low-speed devices like keyboards only connect to other low-speed devices using low-speed cables. The ports on the hub can detect the speed of the device on the other end. If the D+ line is pulled high (+3.0 V dc to +3.6 V dc) then the device is considered high-speed. If the D– line is pulled high then the device is considered low-speed.

The low-speed (1.5 Mbps) cable is an unshielded, untwisted data cable. The communication pair is 28 AWG gauge but due to the lack of shielding and twisting, the overall diameter of the cable is smaller than a high-speed cable. The maximum distance for the low-speed cable is 3 meters (10 feet). This includes all host hub ports to external hub as well as the external hub to device cables. Usually on data communication systems, slower data speeds mean longer distances. In this case, the cable is unprotected against noise and because of the FCC restrictions on 1 to 16 Mbps communication the distance is severely limited.

The high-speed (12 Mbps) cable uses shielded twisted pair 28 AWG gauge wire. The maximum distance for high-speed cables is 5 meters (16 feet). Again, this includes all hub-to-hub and hub-to-device connections. The shield is internally connected to chassis ground at both ends. Usually on data communication systems, the ground is connected at only one end, but because the distances are short, this is not a problem.

Note: It is recommended to measure the chassis to chassis ground difference between both devices before making the connection.

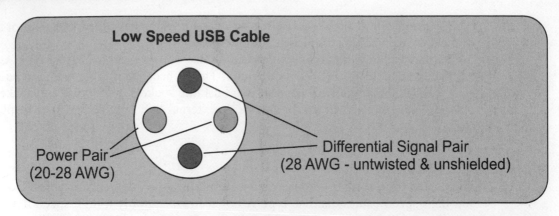

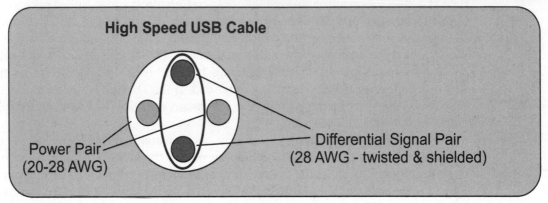

Figure 3.36
Low-speed and high-speed cables

The power pair on both low- and high-speed cables is 20 to 28 AWG gauge. The power pair supplies between 500 and 100 mA to external devices at +5 V dc. Every port on a hub provides this power to the devices if enabled by the hub. All hubs can decide if a port has power applied to the connector. If an external hub is itself powered by the bus then it divides the 500 mA up into 100 mA or so per port.

Signaling

When a device is plugged in to a hub, the port on the hub immediately determines the speed of the device. The port looks at the voltage on the D+ and D– lines. If the D+ line goes positive, the port knows that the device is a high-speed device. If the D– line goes positive, the port knows that the device is a low-speed device.

If both D+ and D– voltages fall below 0.8 V dc for more than 2.5 ms, the hub sees this as the device having been disconnected. If the voltage on either line is raised above 2 V dc for more than 2.5 ms, the port sees this as the device is plugged in.

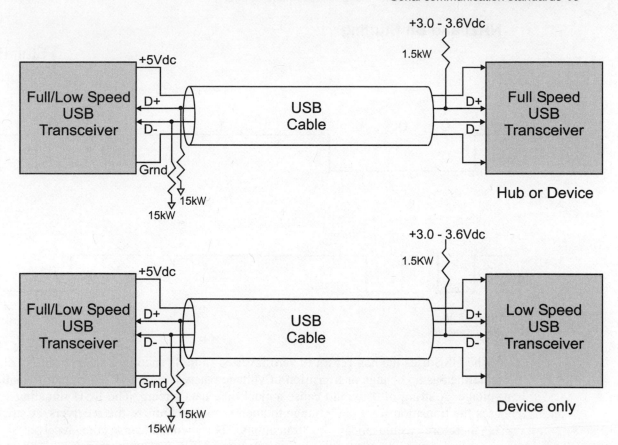

Figure 3.37
USB wiring diagram

The idle states for low and high-speed devices are opposite each other. For the low-speed device the idle state is the D+ line is a 0 V and the D– is a positive voltage. The idle state for the high-speed devices is such that the D+ is a positive voltage and the D– is 0 V at idle. In most data communications, a positive voltage indicates a zero (0) condition and a one (1) is minus voltage. In the USB system, it is not possible to say this because it uses an encoding system called NRZI.

The voltages used for the differential balanced signaling are:

- Maximum voltage transmitted +3.6 V dc
- Minimum voltage transmitted +2.8 V dc
- Minimum voltage needed to sense a transition +/– 0.2 V dc
- Typical line voltage as seen from the receiver +/– 3 V dc

NRZI and Bit Stuffing

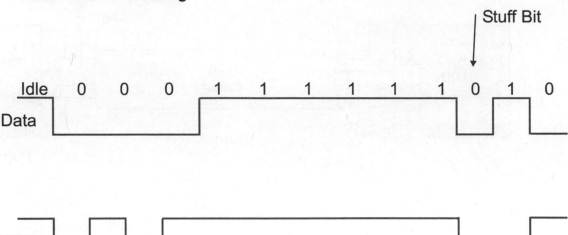

Figure 3.38
NRZI example

The USB uses the non-return to zero inverted (NRZI) encoding scheme. In NRZI a '1' is defined as no change or transition of voltage whereas a "0" is a change or transition of voltage. A string of 0s would cause a clock-like data stream. The USB signaling system uses the transition from one voltage to another to synchronize the receivers. A stream of 1s therefore would mean no transitions. This would cause the receiver to lose synchronization. To overcome this problem the USB system uses a 6 of 7 bit stuffing technique. If six or more 1s are to be transmitted in a row, the transmitter stuffs in a 0 (a transition). If the receiver sees six 1s in a row, it knows that the next transition (zero) is to be ignored.

Power distribution

Devices like keyboards and mice need power to operate. This power is supplied by the USB system through the cables and hubs. External hubs can be either self-powered or powered off the bus. The voltage supplied by a USB hub is +5 V dc. The hubs must be able to supply minimum of 100 mA and maximum of 500 mA through each port. If an external hub with four ports is powered off the bus it divides the 500 mA supplied off the bus between the ports. Four times 100 mA equals 400 mA. This leaves 100 mA to run the hub. It is not possible to connect two bus-powered hubs together unless the devices connected to the last hub are self-powered. If the external hub is self-powered (i.e. mains-powered), it should be able to supply 500 mA to each of the ports.

Data link layer

The data link layer within the USB specification defines the USB as a master/slave, half duplex, timed communication bus system designed to connect close peripherals and external hubs. The hardware and software devices such as the host hub controller hardware and driver, USB software driver and device drivers all contribute to the data link layer of the USB.

With all these devices working together, the data link layer accomplishes the following:

- Collects data off the PCI bus via the device drivers
- Processes the information or data
- Verifies, determines and processes the different transfer types
- Calculates and checks for errors in the packets and frames
- Puts the different packets into 1 ms frames
- Checks for start of frame delimiters
- Sends the packets to the physical layer
- Receives packets from the physical layer

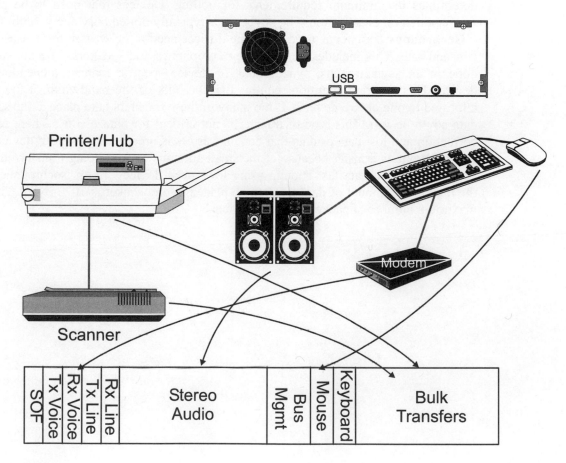

Figure 3.39
USB data link layer block diagram

Transfer types

A good place to start when looking at the data link layer of the USB is with the four different transfer types. The wide range of devices that the USB has to deal with requires that there be multiple transfer types. These are:

- Interrupt transfer
- Isochronous transfer
- Control transfers
- Bulk transfers

As stated before, two speeds can be used in the USB system. For the most part the data link layer is the same, but there are some differences. The low-speed devices do not support bulk and isochronous transfers. The reason for this will become apparent in the following transfer descriptions.

The **interrupt transfer** is used for devices that traditionally used IRQ lines. Devices like keyboards, mice, and DAQ cards use the IRQ lines to tell the computer that they needed service. The USB does not support devices that initiate requests to the computer. To overcome this problem the USB driver initiates a poll of those devices that it knows need periodical attention. This poll must be frequent enough so that data does not get lost, but not too frequent, as not to use up much needed bandwidth. When installed, the device determines its minimum requirements for polling. Devices that need to be polled are rarely polled on every frame. The keyboard is typically polled only every 100th frame.

Isochronous transfer is used when the devices need to be written to or read from at a constant rate. This includes devices like microphones and speakers. The transfer can be done in an asynchronous, synchronous or device specific manner, depending on the device. This constant attention requires that the bulk of the bandwidth of the frame be allocated to one or two devices. If too many of these transfers take place at the same time, data could be lost. This type of transfer is not critical for data quality. There is no error correction and lost data or data that contains errors is ignored. Low-speed devices cannot use isochronous transfer because of the small amounts of data being transferred. It is not possible to move data fast enough using low-speed devices. In an isochronous transfer, the maximum amount of data that can be placed in one packet is 1023 bytes. There is no maximum number of packets that can be sent.

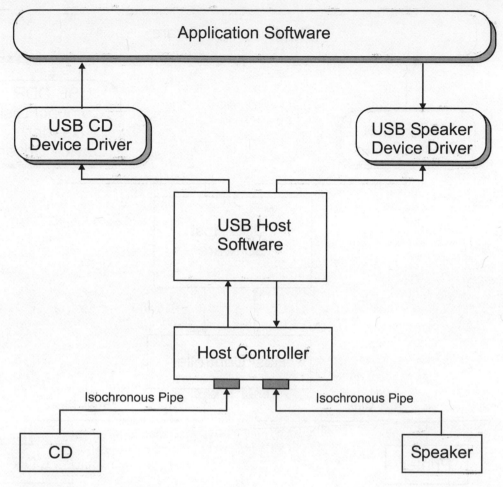

Figure 3.40
Isochronous transfer example

Control transfers are used to transfer specific requests and information to specific devices. This method is used mostly during the configuration and initialization cycles. These transfers are very data critical and require a response or acknowledgment from the device. Full error correction is in force for this type of transfer. All devices use this type of transfer at one time or another. These transfers use very little bandwidth but because the device must respond back to the host hub, the frames are dedicated to this one transfer.

Bulk transfers are used to transfer large blocks of data to devices that are not time dependent but where data quality is important. A typical device that would use the bulk transfer method would be a writeable CD or printer. These devices need large amounts of data but there is no time constraint like there is for a speaker. Whether the data get there in this 10 ms block or the next is not a problem. However, they do need correct data, so this type of transfer includes handshaking and full error correction.

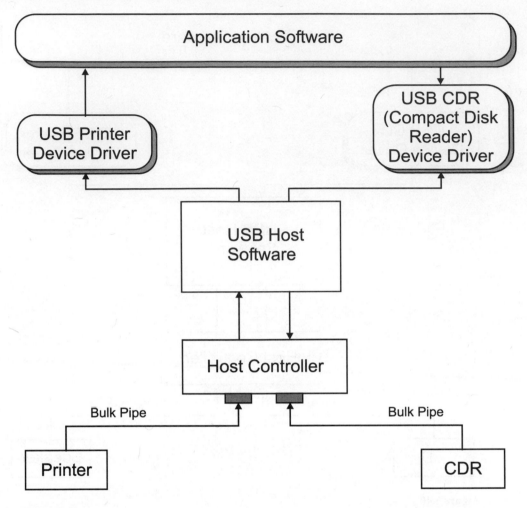

Figure 3.41
Bulk transfer example

Packets and frames

The USB protocol can and often does use a multi-packeted frame format. The USB frame is made up of up to three parts. One frame equals one transaction.

The three parts of the USB frame are:

- The token packet
- The data packet
- Handshaking

Every frame starts with a token packet. The token packet includes the other smaller packets. These include the synchronization pattern, packet type ID and token packet type.

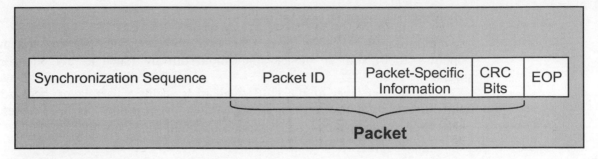

Synchronization Sequence	Packet ID	Packet-Specific Information	CRC Bits	EOP

Packet

Figure 3.42
Packet format

There are four types of token packets:

- Start of frame packets
- In packets
- Out packets
- Setup packets

The start of frame token packet indicates the start of the packet. This tells the receiver that this is the beginning of the 1 ms frame. The 'in' packets transfer data in from the devices to the PC. The 'out' packets transfer data out from the PC to the device. The 'set-up' packet is used to ask the devices or hubs for startup information. They have information for the devices or hubs.

A special packet is only used on low-speed transfers. It is called the preamble packet. It is a shorter packet than the high-speed frame, only holds up to 64 bytes of data and always uses handshaking. It only has three variations, in packet, out packet and setup packet.

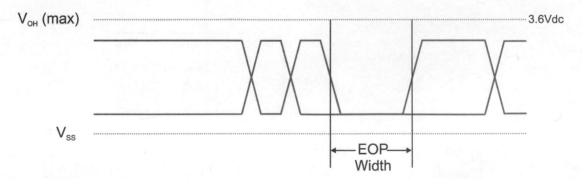

Figure 3.43
End of packet waveform

At the end of all packets, except for isochronous frames, there is an error correction packet. On high-speed frames, this is a 16 bit CRC, low-speed devices use a 5 bit CRC because of their smaller packets. If a device or host hub sees an end of frame message, it checks the CRC. If the CRC is correct, it assumes that this is the end of the message. If the CRC is not correct and the timeout limit has not been reached, the receiver waits. If the CRC is not correct and the timeout has been reached, the receiver assumes that the frame is not correct.

Application layer (user layer)

The application layer can be divided into two sub-layers, the operating system, (such as Windows 2000) and the device application software (such as a modem application program).

The application layer of the USB standard is really a user layer, because the USB standard does not define a true application layer. What it does define is a user layer that can be used (by an application programmer) to build an application layer.

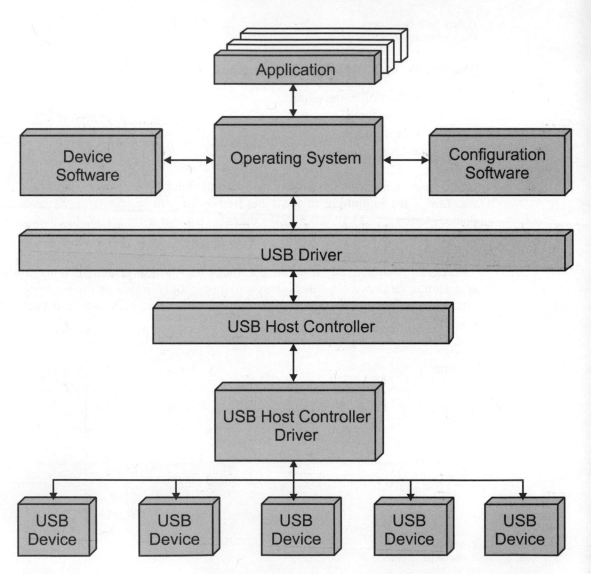

Figure 3.44
Application software diagram

The operating system user layer includes:

- Commands
- Software drivers
- Hub configuration
- Bandwidth allocation

Device applications would use:

- Commands
- Device drivers
- Device configuration

Specific user layer information can be found in the universal serial bus specifications at the USB Implementers Forum web page at http://www.usb.org.

Conclusion

Designed as a peripheral connection system for the PC, the USB can be adapted to be used on data acquisition systems. Now that the DAQ industry is developing increasingly intelligent data acquisition and control systems, the USB is easily adaptable to modern DAQ. The devices can be either low- or high-speed devices and very quickly and easily connected to a PC. There are many devices on the market now and it is bound to grow in the future. With the plug and play, system incorporated in USB the user does not have to spend hours or even days configuring the DAQ system. These time savings often offset the extra cost of the devices.

The target speed of USB 2.0 is 480 Mbps, as announced by the USB 2.0 promoter group, consisting of Compaq, Hewlett-Packard, Intel, Lucent, Microsoft, NEC, and Philips. The target speed announcement coincides with the release of the USB 2.0 specification draft to industry developers.

Acknowledgments

Information from the following sources has been included in this section:

- Universal Serial Bus Specification, USB Implementers Forum web page at *http://www.usb.org*
- Universal Serial Bus System Architecture, MindShare Inc., by Don Anderson
- Intel USB Product Specifications Intel 8x930 and 8x931 USB Peripheral Controllers at:
  http://www.intel.com/design/usb/prodbref/29776501.htm
- Other Web sites of interest:
  http://www.lucent.com/micro/suite/usb.html
  http://www-us.semiconductors.philips.com/usb/

4

Error detection

Errors in data communications occur when the value of a bit is altered from 1 to 0 or vice versa. This chapter looks at how errors are produced and the types of error detection, control, and correction available.

Objectives

When you have completed studying this chapter you will be able to:

- Describe the origin of errors
- List and explain the factors affecting the propagation of signals
- List and explain the methods of feedback error control
- Explain forward error correction

4.1 Origin of errors

Errors are produced by one or more of the following three phenomena:

- Static events
- Thermal noise
- Transient events

A static event is caused by a predictable process such as high frequency alternations, bias distortion, or radio frequency interference. They can generally be minimized by good design and engineering.

Thermal noise is caused by natural fluctuations in the physical transmission medium.

Transient events are difficult to predict because they are caused by natural phenomena such as electrical interference (e.g. lightning), dropouts, and crosstalk. It is not always possible to eliminate errors resulting from transient events.

4.2   Factors affecting signal propagation

A signal transmitted across any form of transmission medium can be practically affected by:

- Attenuation
- Limited bandwidth
- Delay distortion
- Noise

Attenuation

Signal attenuation is the decrease in signal amplitude, which occurs as a signal is propagated through a transmission medium.

A limit needs to be set on the maximum length of cable allowable before one or more amplifiers, or repeaters, must be inserted to restore the signal to its original level. The attenuation of a signal increases for higher frequency components. Devices such as equalizers can be employed to equalize the amount of attenuation across a defined band of frequencies.

Limited bandwidth

Essentially, the larger the bandwidth of the medium the closer the received signal will be to the transmitted one.

The Hartley Law is used to determine the maximum data transfer rate of a transmission line, in the absence of noise:

Max. transfer rate (bps) = $2 B \log_2 M$

where:

B is the bandwidth in Hertz
M is the number of levels per signaling element

For example:
A modem using Phase QAM, four levels per signaling element and a bandwidth on the public telephone network of 3000 Hz, has a maximum data transfer rate calculated by:

$$\text{Maximum data transfer rate} \quad = 2 \times 3000 \log_2 4$$
$$= 12\,000 \text{ bits per second}$$

Delay distortion

When transmitting a digital signal, the different frequency components arrive at the receiver with varying delays between them. The received signal is affected by delay distortion. Intersymbol interference occurs when delays become sufficiently large that frequency components from different discrete bits interfere with each other. As the bit rate increases, delay distortion can lead to an increasingly incorrect interpretation of the received signal.

Noise

An important parameter associated with the transmission medium is the concept of signal to noise ratio (S/N ratio). The signal and noise levels will often differ by many orders of magnitude, so it is common to express the S/N ratio in decibels where:

S/N ratio = $10 \log_{10} S/N$ dB

where:

S is the signal power in watts

N is the noise power in watts

For example:

an S/N ratio of 1 000 000, is referred to as 60 dB

To calculate the maximum theoretical data rate of a transmission medium we use the Shannon-Hartley Law, which states:

Max data Rate = $B \log_2 (1 + S/N)$ bps

where:

B is the bandwidth in Hz

For example:

with an S/N ratio of 100, and a bandwidth of 3000 Hz, the maximum theoretical data rate that can be obtained is given by:

Maximum information rate   = $3000 \log_2 (1 + 100)$

                                          = 19 963 bits per second

4.3   Types of error detection, control, and correction

There are two approaches for dealing with errors in a message:

- Feedback error control,
- Forward error correction

Feedback error control

Feedback error control is where the receiver is able to detect the presence of errors in the message sent by the transmitter. The detected error cannot be corrected but its presence is indicated. This allows the receiver to request a retransmission of the message as defined by a specific protocol. The majority of industrial systems use this approach.

The three most important mechanisms for error detection within feedback error control are:

- Character redundancy: parity check
- Block redundancy: longitudinal parity check, arithmetic checksum
- Cyclic redundancy check (CRC)

Character redundancy checks (parity)

Before transmission of a character, the transmitter uses the agreed mechanism of even or odd parity to calculate the necessary parity bit to append to a character.

For example:

If odd parity has been chosen, then ASCII 0100001 becomes 10100001 to ensure that there are an odd number of 1s in the byte.

For even parity, the above character would be represented as 00100001. At the receiving end, parity for the 7 bit data byte is calculated and compared to the parity bit received. If the two do not agree, an error has occurred.

However: If two of the bits in the character 0100001 had changed the character to 00111001, the parity error reporting scheme would not have indicated an error, when in fact there had been a substantial error.

Parity checking provides only minimal error detection, catching only around 60% of errors on high speed systems.

Parity has been popular because:

- It is low in cost and simple to implement electronically
- It allows a quick check on data accuracy
- It is easy to mentally calculate by the engineer verifying the performance of a system

Although parity has significant weaknesses it is still used where the application is not critical, such as transmitting data to a printer, or communicating between adjacent components in a common electrical system where the noise level is low. Parity is appropriate where the noise burst length is expected to not exceed one bit, i.e. only single bit errors can be expected. This means it is only effective for slow systems. Parity error detection is not used much today for communication between different computer and control systems. Sophisticated algorithms, such as block redundancy, longitudinal parity check, and cyclic redundancy check (CRC), are preferred where the application is more critical.

Block redundancy checks

The parity check on individual characters can be supplemented by a parity check on a block of characters. There are two block check methods:

Longitudinal redundancy check (vertical parity and column parity)

In the vertical redundancy check (VRC), block check strategy, message characters are treated as a two dimensional array. A parity bit is appended to each character. After a defined number of characters, a block check character (BCC), representing a parity check of the columns, is transmitted. Although the VRC, which is also referred to as column parity, is better than character parity error checking, it still cannot detect an even number of errors in the rows. It is acceptable for messages up to 15 characters in length.

Transmitted		Received	
Message MSB LSB PB		Double-bit Error in 1 Row	Double-bit Error in 2 Rows
A	1000001 **0**	1000001 **0**	1000001 **0**
B	1000010 **0**	1000**100** 0	1000**100** 0
z	1111010 **1**	1111010 **1**	1111**100** 1
checksum	1111001 **1**	1111001 **1**	1111001 **1**
Block checksum calculated at Receiver		1111111 **1** (error detected)	1111001 **1** (error not detected)
Notes: Assume even parity for columns (BCC). Ignoring character parity (rows). Bold indicates errors in the received characters. PB means parity bit.			

Table 4.1
Vertical/longitudinal redundancy check using even parity

Arithmetic checksum

An extension of the VLRC is the arithmetic checksum, which is a simple sum of characters in the block. The arithmetic checksum provides better error checking capabilities than VLRC. The arithmetic checksum can be 1 byte (for messages up to 25 characters) or 2 bytes (for messages up to 50 characters in length).

Transmitted		Received		
Message MSB LSB		Double-bit Error in 1 Row	Double-bit Error in 2 Rows	Single-bit Error in 2 Columns
A	1000001	1000001	1000001	100000**0**
B	1000010	1000**100**	1000**100**	100001**1**
z	1111010	1111010	1111**100**	1111010
checksum	1111101	1111101	1111101	1111101
Block checksum calculated at receiver		1111111 (error detected)	0000001 (error detected)	1111101 (error not detected)
Notes: Assume even parity for columns (BCC). Ignoring character parity (rows). Bold indicates errors in the received characters. Parity Bit is left out for convenience.				

Table 4.2
Block redundancy: arithmetic checksum

Cyclic redundancy check (CRC)

For longer messages, an alternative approach has to be used. For example, an Ethernet frame has up to 1500 bytes or 12 000 bits in the message. A popular and very effective error checking mechanism is cyclic redundancy checking. The CRC is based upon a branch of mathematics called algebra theory, and is relatively simple to implement. Using a 16 bit check value, CRC promises detection of errors as shown in Table 4.3. [1]

Single bit errors	100%
Double bit errors	100%
Odd numbered errors	100%
Burst errors shorter than 16 bits	100%
Burst errors of exactly 16 bits	99.9969%
All other burst errors	99.9984%

Table 4.3
CRC error reduction [1]

The CRC error detection mechanism is obviously very effective at detecting errors, particularly difficult to handle 'burst errors', where an external noise source temporarily swamps the signal, corrupting an entire string of bits. The CRC is effective for messages of any length.

Polynomial notation

Before discussing the CRC error checking mechanisms, a few words need to be said about expressing the CRC in polynomial form. The binary divisor, which is the key to the successful implementation of the CRC, is:

10001000000100001

This can be expressed as:

$1 \times X^{16} + 0 \times X^{15} + 0 \times X^{14} + 0 \times X^{13} + 1 \times X^{12} ... + 1 \times X^5 + ... 1 \times X^0$
which when simplified equals:

$X^{16} + X^{12} + X^5 + 1$

The polynomial language is preferred for describing the various CRC error checking mechanisms because of the convenience of this notation.

There are two popular 16-bit CRC polynomials.

- CRC-CCITT
- CRC-16

[1] *Source*: Tanenbaum, Andrew S, *Computer Networks* (Prentice Hall, 1981)

CRC-CCITT

'The....information bits, taken in conjunction, correspond, to the coefficients of a message polynomial having terms from X^{n-1} (n = total number of bits in a block or sequence) down to X^{16}. This polynomial is divided, modulo 2, by the generating polynomial $X^{16} + X^{12} + X^5 + 1$. The check bits correspond to the coefficients of the terms from X^{15} to X^0 in the remainder polynomial found at the completion of this division.' [2]

CRC-CCITT was used by IBM for the first floppy disk controller (model 3770) and quickly became a standard for microcomputer disk controllers. This polynomial is also employed in IBM's popular synchronous protocols HDLC/SDLC (high-level data link control/synchronous data link control) and XMODEM – CRC file transfer protocols.

CRC-16

CRC-16 is another widely used polynomial, especially in industrial protocols:

$$X^{16} + X^{15} + X^2 + 1$$

CRC-16 is not quite as efficient at catching errors as CRC-CCITT, but is popular due to its long history in IBM's binary synchronous communications protocol (BISYNC) method of data transfer.

The CRC-16 method of error detection uses modulo-2 arithmetic, where addition and subtraction give the same result. The output is equivalent to the exclusive OR (XOR) logic function, as given in Table 4.4.

Transmitted	Received	
Message MSB LSB PB	Double-bit Error in 1 Row	Double-bit Error in 2 Rows
A 1000001 **0**	1000001 **0**	1000001 **0**
B 1000010 **0**	1000**100** **0**	1000**100** **0**
z 1111010 **1**	1111010 **1**	1111**100** **1**
checksum 1111001 **1**	1111001 **1**	1111001 **1**
Block checksum calculated at Receiver	1111111 **1** (error detected)	1111001 **1** (error not detected)

Notes:Assume even parity for columns (BCC).
    Ignoring character parity (rows).
    Bold indicates errors in the received characters.
    PB means parity bit.		

Table 4.4
Truth table for exclusive OR (XOR) or Modulo-2 addition and subtraction

[2] CRC-CCITT is specified in recommendation V.41, 'Code-Independent Error Control System', in the CCITT Red Book

Using this arithmetic as a basis, the following equation is true:

Equation 4.1:

(Message × 2^{16}) / Divisor = Quotient + Remainder

where:

Message – is a stream of bits, e.g., the ASCII sequence of H E L P with even parity:

2^{16} – in multiplying, effectively adds 16 zeros to the right side of the message.

Divisor – is a number which is divided into the (message × 2^{16}) number and is the generating polynomial.

Quotient – is the result of the division.

Remainder – is the value left over from the result of the division and is the CRC checksum.

Equation 4.1 then becomes:

Equation 4.2:

[(Message × 2^{16}) + Remainder] / Divisor = Quotient

This information is implemented in the transmitter, using Equation 4.1, as follows:

- Take the message which consists of a stream of bits
  [01001000] [11000101] [11001100] [0101000]
- Add 16 zeros to the right side of the message to get
  [01001000] [11000101] [11001100] [0101000] [00000000] [00000000]
- Divide modulo-2 by a second number, the divisor (or generating polynomial) e.g. 1100000000000101 (CRC-16) the resulting remainder is called the CRC checksum
- Add on the remainder as a 16 bit number to the original message stream (i.e. replace the 16 zeros with the 16 bit remainder) and transmit it to a receiver

At the receiver the following sequence of steps is followed, using Equation 4.1:

- Take the total message plus the CRC checksum bits and divide by the same divisor as used in the transmitter
- If no errors are present, the resulting remainder is all zeros (as per Equation 4.2)
- If errors are present then the remainder is non zero

The CRC mechanism is not perfect at detecting errors. Intuitively, the CRC checksum (consisting of 16 bits) can only take on one of 2^{16} (65 536) unique values. The CRC checksum, being a 'fingerprint' of the message data, has only 1 of 65 536 types. Logically it should be possible to have several different bit patterns in the message data, which is greater than 16 bits that can produce the same fingerprint. The likelihood that the original data and the corrupted data will both produce the same fingerprint is however negligible.

The error detection schemes examined only allow the receiver to detect when data has been corrupted. They do not provide a means for correcting the erroneous character or frame. This correction is normally accomplished by the receiver informing the transmitter that an error has been detected and requesting another copy of the message to be sent. This combined error detection/correction cycle is known as error control.

Forward error correction

Forward error correction is where the receiver can not only detect the presence of errors in a message, but also reconstruct the message into what it believes to be the correct form. It may be used where there are long delays in requesting retransmission of messages or where the originating transmitter has difficulty in retransmitting the message when the receiver discovers an error. Forward error correction is generally used in applications such as NASA space probes operating over long distances in space where the turn around time is too great to allow a retransmission of the message.

Hamming codes and hamming distance

In the late 1940s, Richard Hamming and Marcel Golay did pioneering work on error detecting and error correcting codes. They showed how to construct codes which were guaranteed to correct certain specified numbers of errors, by elegant, economic and sometimes optimal means.

Coding the data simply refers to adding redundant bits in order to create a codeword. The extra information in the codeword, allows the receiver to reconstruct the original data in the event of one or more bits being corrupted during transmission.

An effective method of forward error correction is the use of the Hamming codes. These codes detect and correct multiple bits in coded data. A key concept with these codes is that of the Hamming distance. For a binary code, this is just the number of bit positions at which two codewords vary. For instance, the Hamming distance between 0000 and 1001 is 2.

A good choice of code means that the codewords will be sufficiently spaced, in terms of the Hamming distance, to allow the original signal to be decoded even if some of the encoded message is transmitted incorrectly.

The following examples illustrate a Hamming code.

A code with a Hamming distance of 1 could represent the eight alphanumeric symbols in binary as follows:

000	A
001	B
010	C
011	D
100	E
101	F
110	G
111	H

If there is a change in 1 bit in the above codes, due to electrical noise for example, the receiver will read in a different character and has no way of detecting an error in the character. Consequently, the Hamming distance is 1, and the code has no error detection capabilities.

If the same three bit code is used to represent four characters, with the remaining bit combinations unused and therefore redundant, the following coding scheme could be devised.

000	A
011	B
110	C
101	D

This code has a Hamming distance of 2, as two bits at least have to be in error before the receiver reads an erroneous character.

It can be demonstrated that a Hamming distance of three requires three additional bits, if there are four information bits. This is referred to as a Hamming (7,4) code. For a 4-bit information code, a 7-bit code word is constructed in the following sequence:

$C_1 C_2 I_3 C_4 I_5 I_6 I_7$

where:

$I_3 I_5 I_6 I_7$ are the information, or useful bits

$C_1 C_2 C_4$ are the redundant bits calculated as follows:

- $C_1 = I_3$ XOR I_5 XOR I_7
- $C_2 = I_3$ XOR I_6 XOR I_7
- $C_4 = I_5$ XOR I_6 XOR I_7

For example:

If the information bits are 1101

$(I_3 = 1; I_5 = 1; I_6 = 0; I_7 = 1)$, the Hamming (7,4) codeword is:

- $C_1 = 1$ XOR 1 XOR 1 = 1
- $C_2 = 1$ XOR 0 XOR 1 = 0
- $C_4 = 1$ XOR 0 XOR 1 = 0

The codeword $(C_1 C_2 I_3 C_4 I_5 I_6 I_7)$ is then represented as 1010101.

If one bit is in error and the codeword 1010111 was received, the redundant bits would be calculated as:

- $C_1 = 1$ XOR 1 XOR 1 = 1

(and matches the 1 from the received codeword)

- $C_2 = 1$ XOR 1 XOR 1 = 1

(but does not match the 0 from the received codeword)

- $C_4 = 1$ XOR 1 XOR 1 = 1

(but does not match the 0 from the received codeword)

C2 and C4 indicate one bit out of place, which would be either I6 or I7 (as this is common to both). However C1 matches the check calculation, therefore I6 must be in error.

Hence, the code word should be: 1010101.

4.4   Other control mechanisms

There are obviously other control mechanisms in place between two communicating devices, which allow efficient and accurate transfer of messages. Error detection and control are not enough to ensure that data is transferred successfully from one point to another. An overall protocol framework is required to ensure that information is transferred correctly and any errors are handled appropriately. The subject of protocols is discussed in Chapter 8.

5

Cabling basics

To make sure that you get the best performance from communication cables, the type and size of cable should be chosen to suit the application.

Objectives

When you have completed studying this chapter you will be able to:

- List the four basic cable types
- Describe the general properties of copper-based cables
- Describe the properties and use of two-wire open lines
- Describe the properties and use of twisted pair cables
- Describe the properties and use of coaxial cables
- Describe the properties, use, theory of operation, handling and limitations of fiber-optic cables

5.1   Overview

The most common types of cables used in data communications systems are:

- Twisted pair
- Coaxial
- Fiber-optic

These offer differing:

- Signal and mechanical characteristics
- Installation convenience
- Costs

The noise susceptibility and data rate values of individual system components should also be considered when determining the type and specification of wiring to be used. Specific noise details of the various system components are normally available from the manufacturers or suppliers.

5.2   Copper-based cables

Two wire open lines, twisted pair and coaxial cables are all manufactured with copper conductors and extruded plastic insulation. This construction combines good electrical characteristics with mechanical flexibility, ease of installation and low cost. Fiber-optic cable is technologically different and is addressed later in this chapter. Aluminum conductors are seldom used for data communication cables because of their higher resistance and other physical limitations such as lack of flexibility.

The resistance of copper cables depends on the cross-sectional area of the conductor, usually expressed in mm², and the length of the cable. The thicker the conductor, the lower the resistance, the lower the signal voltage drop, and the higher the current it can carry without excessive heating.

The wire size must reflect the current carrying requirement of the application, while the voltage rating should be equal to or exceed the anticipated circuit rated voltage. Physical stresses imposed on the cable during installation and operation must also be considered to make sure that the mechanical strength of the cable is acceptable. It is possible to increase the cable strength by using multiconductor grouping within a single jacket. The signal voltage drop, which is expressed ($V_{drop} = I \times R$), depends on:

- The line current, which is dependent on the receiver input and transmitter output impedances
- The conductor resistance which is dependent on wire size and length

For dc voltages and low frequency signals, the resistance of the conductor is the only major concern. The voltage drop along the cable affects the magnitude of the signal voltage at the receiving end. In the presence of noise, the voltage drop affects the signal-to-noise ratio and the quality of the signal received.

As the frequency (or data transfer rate) increases, the other characteristics of the cable, such as capacitance and series inductance, become important. Inductance and capacitance are factors that are affected by:

- How the cable is made
- The number and thickness of shields used
- The number of conductors in the cable
- The insulation materials used

The resistance, inductance, and capacitance are distributed along the length of the cable and, at high frequencies, combine to present the effects of a low pass filter. The equivalent electrical circuit of a cable is illustrated in Figure 5.1 with these parameters shown distributed along the length of the cable.

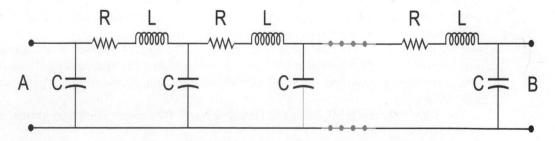

Figure 5.1
The main parameters of a data communications cable

To optimize data communications performance, the correct type and size of cable must be chosen for the application. The following information applies to most data applications:

Lower data transfer rates: Low frequency type cables (e.g. twisted pair cables)
High data transfer rates: High frequency type cables (e.g. coaxial cables, optic fiber or high quality twisted pair data cables)
High noise environment cables: Shielded copper or optic fiber

Note: There are some new types of twisted pair cables available that give good high frequency performance.

Another important consideration is the type of outer insulation and protection. For example, a cable may have the following options:

- A thin aluminum tape wound around the cable, under the plastic insulation, to provide a barrier against moisture ingress (moisture will permeate through plastic over a period of time) and noise
- A steel armored outer cover for areas where protection is required against excessive heat, fire, mechanical damage, and noise
- A filling of petroleum jelly between pairs to provide a good moisture barrier
- A nylon coated outer to provide a slippery jacket for ease of installation in conduits as well as protection from rodents
- Plenum cables, which are made with non toxic insulation, for installation in vented areas to ensure no poisonous fumes are produced in fires

5.3   Twisted pair cables

Twisted pair cables are the most economical solution for data transmission and allow for rates of up to 180 Mbps on communication links of up to 100 meters (330 feet). Longer distances are possible with lower data transfer rates. Twisted pairs are either shielded twisted pair (STP) or unshielded twisted pair (UTP).

100 Mbps Ethernet is widely used on twisted pair (Cat 5) cables over 100 m and 16 bps. Ethernet is now becoming available on copper cables.

Twisted pair cables are made from two identical insulated conductors, which are twisted together along their length at a specified number of twists per meter, typically forty, twists per meter (twelve twists per foot). The wires are twisted to reduce the effect of electromagnetic and electrostatic induction. An earth screen and/or shield is often placed around the wires to help reduce the electrostatically (capacitive) induced noise. An insulating PVC sheath is provided for overall mechanical protection. The cross-sectional area of the conductor will affect the voltage loss, so for long distances thicker conductor sizes are recommended. The capacitance of a twisted pair is fairly low at about 40 to 160 pF/m, allowing a reasonable bandwidth and an achievable slew rate.

For full-duplex digital systems using balanced transmission, two sets of screened twisted pairs are required in one cable; each set with individual and overall screens. The entire cable is then covered by a protective PVC sheath.

The 1970s and '80s saw a rapid increase in the use of twisted pair cables for data communications. This increase left the EIA to develop a set of rules and standards for the selection and installation of UTP cables in data communications applications up to 100 Mbps.

The EIA-568 standard divides UTP cables into five application categories, which are listed below:

- Category 1 UTPLow-speed data and analogue voice

- Category 2 UTPISDN data
- Category 3 UTPHigh-speed data and LAN (10 Mbps)
- Category 4 UTPExtended distance LAN
- Category 5 UTPExtended frequency LAN (100 Mbps)

The connection point of a landline into a building or equipment shelter is at the main distribution frame (MDF) or intermediate distribution frame (IDF).

In making data connections to modems, telemetry units or computer equipment, it is common to use withdrawable multiconductor connectors (e.g. 9 pin, 15 pin, 25 pin, 37 pin, 50 pin etc.). These connectors are usually classified as follows:

- The type, make or specification of the connector
- The number of associated pins or connections
- The gender (male or female)
- Mounting (socket or plug)

For example, the common connector DB-25 SM specifies a D-type, 25 pin socket, male (with pins).

There are many different types of connectors used by computer manufacturers such as IBM, Hewlett Packard, Wang, Apple etc. and the various manufacturers of printers, radio equipment, modems, instrumentation and actuators. The following is a selection of some of the more popular connectors:

- DB-9, DB-15, DB-25, DB-37, DB-50
- Amphenol 24-pin
- Centronics 36-pin
- Telco 50-pin
- Berg 50-pin
- RJ-11 4-wire
- RJ-12 6-wire
- RJ-45 8-wire
- DEC MMJ
- M/34 (ITU V.35)
- M/50

There is also a wide range of DIN-type connectors (German/Swiss), IEC-type connectors (French/European), BS-type connectors (British), and many others for audio, video, and computer applications. With all connectors, the main requirement is to ensure compatibility with the equipment being used. Suitable types of connectors are usually recommended in the manufacturer's specifications.

The DB-9, DB-25, and DB-37 connectors, used with the EIA standard interfaces such as RS-232, RS-422, and RS-485 have become very common in data communication applications. The interface standards for multidrop serial data communications RS-422 and RS-485 do not specify any particular physical connector. Manufacturers, who sell equipment, complying with these standards, can use any type of connector, but the DB-9, DB-25 (pin assignments to EIA-530) DB-37 (pin assignments to EIA-449) and sometimes screw terminals, have become common. Another connector commonly used for high speed data transmission is the ITU V.35 34-pin connector.

5.4　　Coaxial cables

Coaxial cables are used, almost without exception, for all antennas operating between the HF band of frequencies up to the SHF band around 2 GHz, where waveguides begin to take over.

The impedance of a cable is determined by the ratio of the surrounding shield and the diameter of the inner conductors. Although the characteristic impedance of a television antenna is 75 Ω, most communications antennas have an impedance of 50 Ω and care should always be taken to use the correct cable.

The size of a coaxial cable is determined by two factors – the transmitter power being fed to the antenna system and the frequency being used.

If a transmitter has an output power of 500 watt, the peak voltage across a 50 Ω cable will be 223 volt and the current will be about 3.3 amp. If the dielectric insulation is insufficient, the cable will breakdown and if the inner conductor is too small, there will be a high resistive loss in the cable.

Higher radio frequencies energy tends to travel on the surface of a conductor rather than through the center so a small diameter inner conductor will obviously have a small surface area and consequently a high resistance. It follows that as the frequency increases, so should the diameter of the inner conductor but the impedance of the cable is determined in part by the capacitance between the inner conductor and the screen. Therefore, in order to maintain the correct impedance, the size of the inner conductor and the spacing between the conductors, i.e. the dielectric, are the critical design elements.

Smaller types of coaxial cable, up to about 10 mm (0.4") diameter, use a copper braided sleeve as the outer conductor because this is efficient and cheap to manufacture. The largest coaxial cables are 200 mm (8") diameter, and as these cables may have to be curved around a bend any deformity badly affects the cable performance, a new type of shield conductor was developed.

The first outer conductor was an aluminum tube but this proved difficult to handle and was replaced with a copper tube with spiral corrugations. The corrugations ensure that the diameter, at any point along the length of the tube, is always constant. In this way, the average distance between the inner conductor and the outer is constant so the impedance remains stable even though the cable may be easily curved around quite a small radius without any damage.

Cable manufacturers publish accurate data on the characteristics of the cables they produce. The selection of a cable may seem to involve little more than finding the cheapest cable that will carry the power involved. However, in most cases where radio links are involved, this will be a minor consideration and the attenuation of the cable will be the major factor.

5.5　　Fiber-optic cables

Fiber-optic cables are normally used for the transmission of digital signals. The capabilities of fiber-optic cables will satisfy any future requirement in data communications, allowing transmission rates in the Gigabits per second (Gbps) range. There are many currently installed systems operating at around 10 Gbps.

Fiber-optic cables are generally cheaper than coaxial cables, especially when comparing data capacity per unit cost. However, the transmission and receiving equipment, together with more complicated methods of terminating and joining these cables, makes fiber-optic cable the most expensive medium for data communications. The cost of the cables has halved since the late 1980s and is becoming insignificant in an economic equation and it is worth noting that fiber-optic technology has become more affordable over the last decade and this trend will continue into the future.

The main benefits of fiber-optic cables are:

- Enormous bandwidth (greater information carrying capacity)
- Low signal attenuation (greater speed and distance characteristics)
- Inherent signal security
- Low error rates
- Noise immunity (impervious to EMI and RFI)
- Logistical considerations (light in weight, smaller in size)
- Total galvanic isolation between ends (no conductive path)
- Safe for use in hazardous areas
- No crosstalk

Theory of operation

The optical fiber forms a wave guide for light with the light being guided through the core of the fiber.

Communication over fiber-optic cables works on the principle that light propagates through different media at different speeds (in the same manner as radio waves). When light moves from one medium of a certain density to another of a different density, the light will change direction. This phenomenon is known as refraction.

The effectiveness of a medium to propagate light can be expressed as a ratio of an absolute reference; light traveling through a vacuum (3×10^8 m/s) i.e. speed of light in free space. This ratio is known as the 'Refractive Index' and is calculated as:

$$\text{Refractive Index } n = \frac{\text{Speed of light in vacuum}}{\text{Speed of light in the medium}}$$

In a typical fiber-optic medium, light travels at approximately 2×10^8 m/s. Therefore the refractive index is:

$$n_1 = \frac{3 \times 10^8}{2 \times 10^8}$$

$$n_1 = 1.5$$

The optical medium is said to have a refractive index of 1.5.

Fiber-optics follows Snell's law, which states that the ratio of the sine of the angle of incidence (Q_i) to the sine of the angle of refraction (Q_r) is equal to the ratio of the speed of light in the two respective media (C_1/C_2). This is equal to a constant (K), which has a ratio of the refractive index of medium 2 to medium 1 (n_2/n_1).

The formula is:

$$\frac{\text{Sin}^q i}{\text{Sin}^q r} = \frac{C_1}{C_2} = K = \frac{n_2}{n_1}$$

Fiber-optic cables are manufactured with a pure optical glass core, surrounded by a glass cladding. The core and cladding are treated with an impurity so that their refractive indices are different. Figure 5.2 and Figure 5.3 show the basic construction of the optical fiber. This construction allows the core to guide the light pulses to the receiver.

Because the refractive index of the core and cladding is different, light entering the core at an acceptable angle of entry will propagate the length of the fiber without losing light through the cladding. Light must enter the fiber within a 'cone of acceptance' angle. When light attempts to enter at an angle greater than the 'cone of acceptance', it will not reflect from the cladding and is lost.

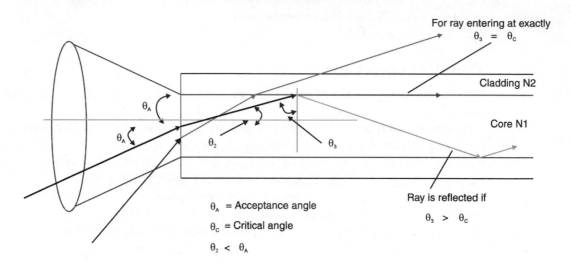

Figure 5.2
Optical fiber principles

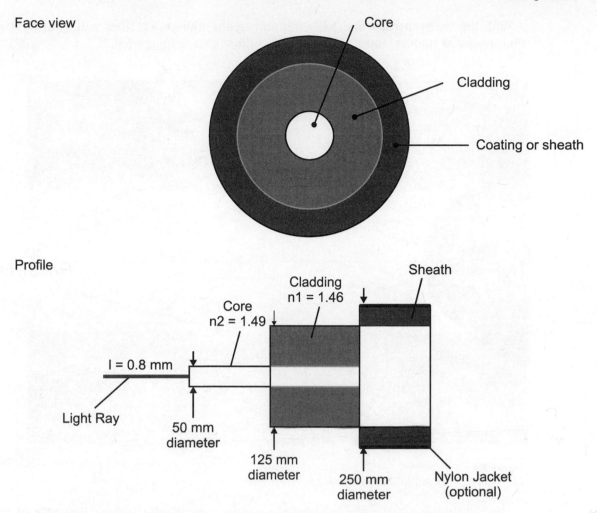

Figure 5.3
Typical optical fiber values (multimode fiber/step index)

The optical fiber acts as a conduit (or wave guide) for pulses of light generated by a light source. The light source is typically a laser diode or light emitting diode (LED) operating at wavelengths of 0.85, 1.3 or 1.55 micrometers.

The optical fiber is coated with a protective colored sheath to provide environmental protection and easy identification.

Modes of propagation

Fiber types are generally identified by the number of paths that the light follows inside the fiber core called 'modes' of propagation. There are two main modes of light propagation through an optic fiber, which give rise to two main constructions of fiber, 'multimode', and 'monomode' (also known as 'single mode').

Multimode fibers are easier and cheaper to manufacture than monomode fibers. Multimode cores are typically 50 times greater than the wavelength of the light signal they will propagate. With this type of fiber, an LED transmitter light source is normally used because it can be coupled with less precision than a laser diode.

With the wide aperture and LED transmitter, the multimode fiber will send light in multiple paths (modes) toward the receiver as illustrated in Figure 5.4.

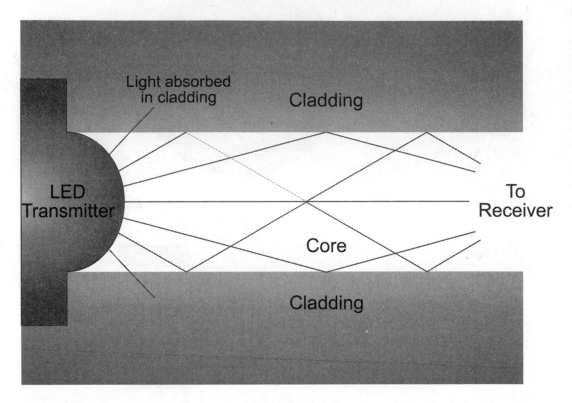

Figure 5.4
LED light source coupled to a multimode fiber (Step Index)

The light takes many paths between the two ends as it reflects from the sides of the fiber core. This causes the light paths to arrive both out of phase and at different times resulting in a spreading of the original pulse shape. As a result, the original sharp pulses sent from one end become distorted by the time they reach the receiving end.

The problem becomes worse as data rates increase. Multimode fibers, therefore, have a limited maximum data rate (bandwidth) as the receiver can only differentiate between the pulsed signals at a low data rate. The effect is known as 'modal dispersion' and its result referred to as 'intersymbol interference'. For slower data rates over short distances, multimode fibers are quite adequate and speeds of up to 300 Mbps are readily available.

A further consideration, with multimode fibers, is the 'index' of the fiber (how the impurities are applied in the core). The cable can be either 'graded index' (more expensive but better performance) or 'step index' (less expensive), refer to Figure 5.5. The type of index affects the way in which the light waves reflect or refract off the walls of the fiber. Graded index cores focus the modes as they arrive at the receiver, and consequently improve the permissible data rate of the fiber.

The core diameters of multimode fibers typically range between 50–100 mm. The two most common core diameters are 50 and 62.5 mm.

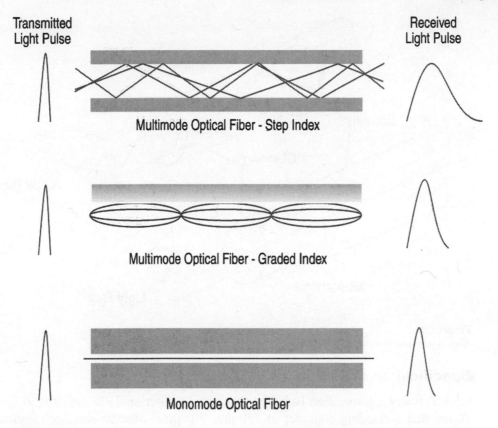

Figure 5.5
Optical fibers and their characteristics

'Monomode' or 'single mode' fibers are less expensive but more difficult to interface. They allow only a single path or mode for the light to travel down the fiber with minimal reflections. Monomode fibers typically use lasers as light sources.

Monomode fibers do not suffer from major dispersion or overlap problems and permit a very high rate of data transfer over much longer distances. The core of the fibers is much thinner than multimode fibers at approximately 5–10 mm. The cladding diameter is 125 mm, the same as for multimode fibers.

Source lighting must be powerful and aimed precisely into the fiber to overcome any misalignment (hence the use of laser diodes). The thin monomode fibers are difficult to work with when splicing, terminating, and are consequently expensive to install.

A typical application for a monomode installation would be a high capacity telephone link where the traffic volume makes a large bandwidth necessary.

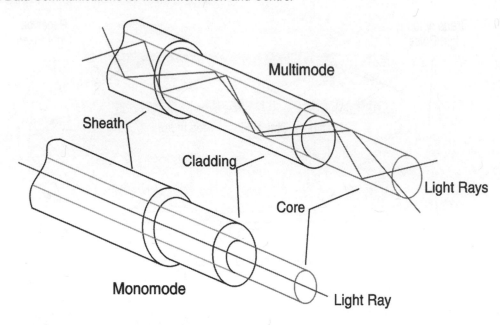

Figure 5.6
Monomode and multimode optic fibers

Specification of cables

Optical fibers are specified based on diameter. A fiber specified as 50/150 has a core of 50 μm and a cladding diameter of 150 μm. The most popular sizes of multimode fibers are 50/125, used mainly in Europe, and 62.5/125, used mainly in Australia and the USA.

Another outer layer provides an external protection against abrasion and shock. Outer coatings can range from 250–900 μm in diameter, and very often cable specifications include this diameter, for example: 50/150/250.

To provide additional mechanical protection, the fiber is often placed inside a loose, but stiffer, outer jacket which adds thickness and weight to the cable. Cables made with several fibers are most commonly used. The final sheath and protective coating on the outside of the cable depends on the application and where the cable will be used. A strengthening member is normally placed down the center of the cable to give it longitudinal strength. This allows the cable to be pulled through a conduit or hung from a power pole without causing damage to the fibers. The tensile members are made from steel or Kevlar, the latter being more common. In industrial and mining applications, fiber cores are often placed inside cables used for other purposes, such as trailing power cables for large mining, stacking, or reclaiming equipment.

Experience has shown that optic fibers will break two or three times in a 25 year period. In general, the incremental cost of extra fiber cores in cables is not very high when compared to overall costs (including installation and termination costs). Therefore, it is often worthwhile specifying extra cores as spares, or for future use.

Joining cables

In the early days of optic fibers, connections and terminations were a major problem. Largely, this has improved but connections still require a great deal of care to avoid signal losses that will affect the overall performance of the communications system.

There are three main methods of splicing optic fibers:

Mechanical: Where the fibers are fitted into mechanical alignment structures
Chemical: Where the two fibers are fitted into a barrel arrangement with epoxy
                     glue in it – they are then heated in an oven to set the glue
Fusion splicing: Where the two fibers are heat-welded together

To overcome the difficulties of termination, fiber-optic cables can be provided by a supplier in standard lengths such as 10 m, 100 m or 1000 m with the ends cut and finished with a mechanical termination ferrule that allows the end of the cable to slip into a closely matching female socket. This enables the optical fiber to be connected and disconnected as required. The mechanical design of the connector forces the fiber into a very accurate alignment with the socket and results in a relatively low loss. Similar connectors can be used for in-line splicing using a double-sided female connector.

Although the loss through this type of connector can be an order of magnitude greater than the loss of a fused splice, it is much quicker and requires no special tools or training. Unfortunately, mechanical damage or an unplanned break in a fiber requires special tools and training to repair and re-splice. One way around this problem is to keep spare standard lengths of pre-terminated fibers that can quickly and easily be plugged into the damaged section. The techniques for terminating fiber-optic cables are constantly being improved to simplify these activities.

Limitations of cables

On the negative side, the limitations of fiber-optic cables are as follows:

- The cost of source and receiving equipment is relatively high.
- It is difficult to 'switch' or 'tee-off' a fiber-optic cable so fiber-optic systems are most suitable for point-to-point communication links.
- Techniques for joining or terminating fibers (mechanical and chemical) are difficult and require precise physical alignment. Special equipment and specialized training are required.
- Equipment for testing fiber-optic cables is different and more expensive from traditional methods used for electronic signals.
- Fiber-optic systems are used almost exclusively for binary digital signals and are not suitable for long distance analog signals.

6

Electrical noise and interference

Sources of electrical noise and the ability of a cable to exclude them are important issues when selecting and installing data cables. This chapter examines the various categories of noise and where each of the various noise reduction techniques applies. In addition, a brief examination is included of noise suppression techniques and filtering of the noise that gets into the signal system.

Objectives

When you have completed studying this chapter you will be able to:

- Define the terms noise and SNR
- Explain the frequency spectrum analysis of the three noise groups
- Give examples of electrical noise sources
- Explain the four forms of electrical coupling of noise:
  - Impedance coupling
  - Electrostatic coupling
  - Magnetic/inductive coupling
  - Electromagnetic radiation
- Explain shielding techniques and shielding performance ratios, including cable ducting and cable spacing
- Describe earthing and grounding requirements
- Describe suppression techniques
- Describe the filtering of specific noise sources

6.1   Definition of noise

Noise, or interference, can be defined as undesirable electrical signals, which distort or interfere with an original (or desired) signal. In many cases, the noise will be un-predictable due to transients (or spikes) caused, for example, by lightning. In other cases, it may be due to the predictable 50 or 60 Hz ac 'hum' from power circuits close to the data communications cable. This unpredictability makes the design of a data communications system quite challenging.

Noise can be generated from within the system itself (internal noise) or from an outside source (external noise).

Examples of noise sources are:

- Internal noise
- Thermal noise (due to electron movement within the electrical circuits)
- Imperfections (in the electrical design)
- External noise
- Natural origins (electrostatic interference and electrical storms)
- Electromagnetic Interference (EMI) – from currents in cables
- Radio Frequency Interference (RFI) – from radio systems radiating signals
- Crosstalk (from other cables separated by a small distance)

It is commonly accepted that the main techniques used to reduce noise consist of

- Applying shielding around the signal wires
- Increasing the distance between the noise source and the signal
- Proper grounding of the shield and twisting of the signal wires

Noise is only important if it is measured in relation to the communication signal, which carries the data information. In previous chapters, it has been demonstrated that electronic receiving circuits for digital communications have a broad voltage range, which determines whether a signal is a binary bit '1' or '0'. The noise voltage has to be high enough to take the signal voltage outside these limits for errors to occur.

The ratio of the signal voltage to the noise voltage determines the strength of the signal in relation to the noise. This is called 'signal to noise ratio' (SNR) and is important in assessing how well the communication system will operate. In data communications, the signal voltage is relatively stable and is determined by the voltage at the source (transmitter) and the volt drop along the line due to the cable resistance (size and length). SNR is therefore a measure of the interference on the communication link.

The SNR is usually expressed in decibels (dB), which is the logarithmic ratio of the signal voltage (S) to noise voltage (N).

$$SNR = 10\log S/N \text{ dB}$$

An SNR of 20 dB is considered low (bad), while an SNR of 60 dB is considered high (good). The higher the SNR, the easier it is to provide acceptable performance with simpler circuitry and cheaper cabling.

In data communications, a more relevant performance measurement of the link is the bit error rate (BER). This is a measure of the number of successful bits received compared to bits that are in error. A BER of 10^{-6} means that one bit in a million will be in error and is considered poor performance on a bulk data communications system with high data rates. A BER of 10^{-12} (one error bit in a million, million) is considered very good. Over industrial systems, with low data requirements, a BER of 10^{-4} could be quite acceptable.

There is a relationship between SNR and BER. As the SNR increases, the error rate drops off rapidly as is shown in Figure 6.1. Most communications systems start to provide reasonably good BERs when the SNR is above 20 dB.

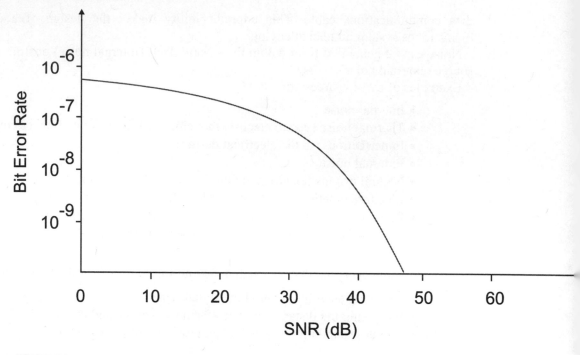

Figure 6.1
Relationship between the bit error rate and the signal to noise ratio

6.2   Frequency analysis of noise

Another useful way of evaluating the effects of noise is to examine its frequency spectrum. Noise can be seen to fall into three groups:

- Wideband noise
- Impulse noise
- Frequency specific noise

The three groups are shown in the simplified frequency domain as well as the conventional time domain. In this way, we can appreciate the signal's changing properties as well as viewing the amplitude in the customary time domain.

Wideband noise contains numerous frequency components and amplitude values. These are depicted in the time domain graph shown in Figure 6.2 and in the frequency domain graph shown in Figure 6.3.

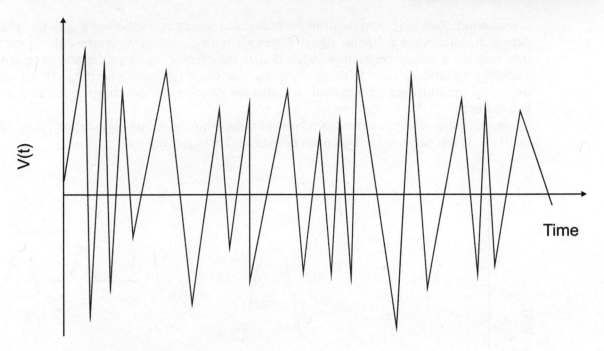

Figure 6.2
Time domain plot of wideband noise

In the frequency domain, the energy components of wideband noise extend over a wide range of frequencies (frequency spectrum).

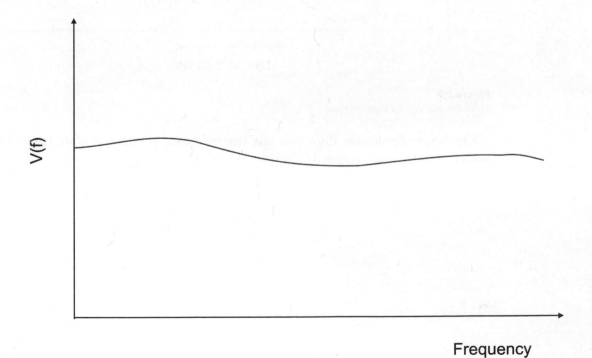

Figure 6.3
Frequency domain plot of wideband noise

Wideband noise will often result in the occasional loss or corruption of a data bit. This occurs at times when the noise signal amplitude is large enough to confuse the system into making a wrong decision on what digital information or character was received. Encoding techniques such as parity checking and block character checking (BCC) are important for wideband error detection so that the receiver can determine when an error has occurred.

Impulse noise is best described as a burst of noise which may last for duration of say up to 20 ms. It appears in the time domain as indicated in Figure 6.4.

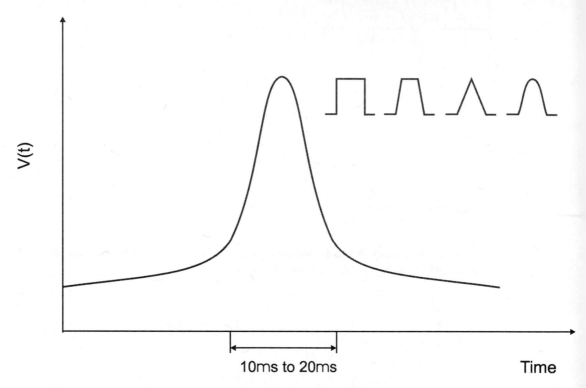

Figure 6.4
Time domain plot of impulse noise

The frequency domain illustrates this type of noise. It affects a wide bandwidth with decreasing amplitude versus frequency.

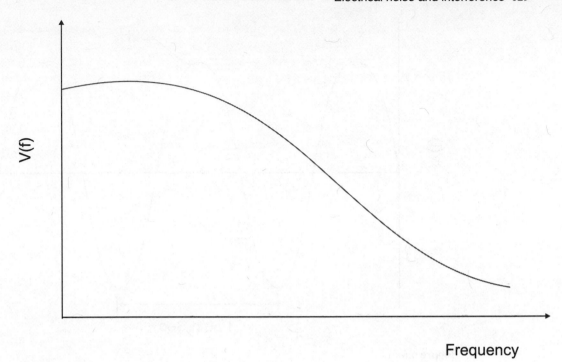

Figure 6.5
Frequency domain plot of impulse noise

Impulse noise is brought about by the transient disturbances in electrical activity such as when an electric motor starts up, or from switching elements within telephone exchanges. Impulse noise swamps the desired signal, thus corrupting a string of data bits. As a result of this effect, synchronization may be lost or the character framing may be disrupted. Noise of this nature usually results in garbled data making messages difficult to decipher. Cyclic redundancy checking (CRC) error detection techniques may be required to detect such corruption.

Although more damaging than wideband noise, impulse noise is generally less frequent. The time and frequency domain plots for impulse noise will vary depending on the actual shape of the pulse. Pulse shapes may be squared, trapezoidal, triangular or sine for example.

In general, the narrower and steeper a pulse, the more energy is placed in the higher frequency regions.

Frequency specific noise is characterized by a constant frequency, but its amplitude may vary depending on how far the communication system is from the noise source, the amplitude of the noise signal and the shielding techniques used.

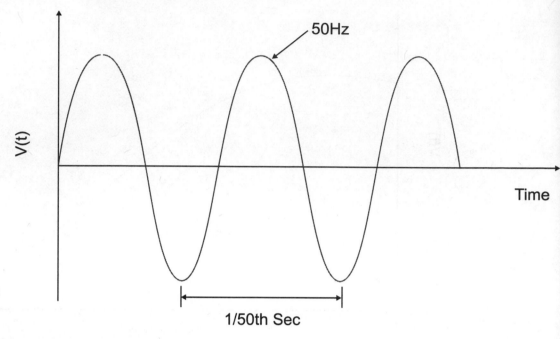

Figure 6.6
Time domain plot of constant frequency noise

This noise group is typical of ac power systems and can be reduced by separating the data communication system from the power source. Because this form of noise has a predictable frequency spectrum, noise resistance is easier to implement within the system design.

Filters are typically used to reduce this to an acceptable level.

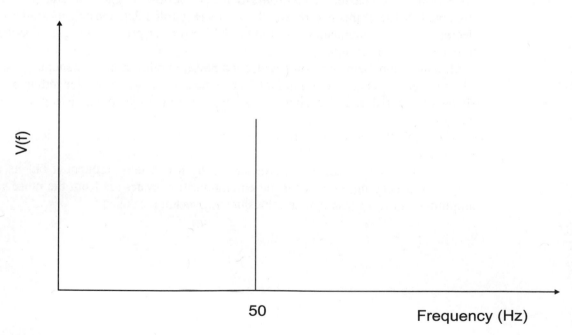

Figure 6.7
Frequency domain plot of constant frequency noise

6.3　Sources of electrical noise

Typical sources of noise are devices, which produce quick changes (or spikes) in voltage or current, such as:

- Large electrical motors being switched on
- Fluorescent lighting tubes
- Lightning strikes
- High voltage surges due to electrical faults
- Welding equipment

From a general point of view, there must be three contributing factors before an electrical noise problem can exist. These are:

- A source of electrical noise
- A mechanism coupling the source to the affected circuit
- A circuit conveying the sensitive communication signals

6.4　Electrical coupling of noise

There are four forms of coupling of electrical noise into the sensitive data communications circuits. These are:

- Impedance coupling (sometimes referred to as conductance coupling)
- Electrostatic coupling
- Magnetic or inductive coupling
- Radio frequency radiation (a combination of electrostatic and magnetic)

Each of these noise forms will be discussed in some detail in the following sections. Although the order of discussion is indicative of the frequency of problems, this will obviously depend on the specific application.

Impedance coupling (or common impedance coupling)

For situations where two or more electrical circuits share common conductors, there can be some coupling between the different circuits with deleterious effects on the connected circuits. Essentially, this means that the signal current from the one circuit proceeds back along the common conductor resulting in an error voltage along the return bus, which affects all the other signals. The error voltage is due to the capacitance, inductance, and resistance in the return wire. This situation is shown in the Figure 6.8.

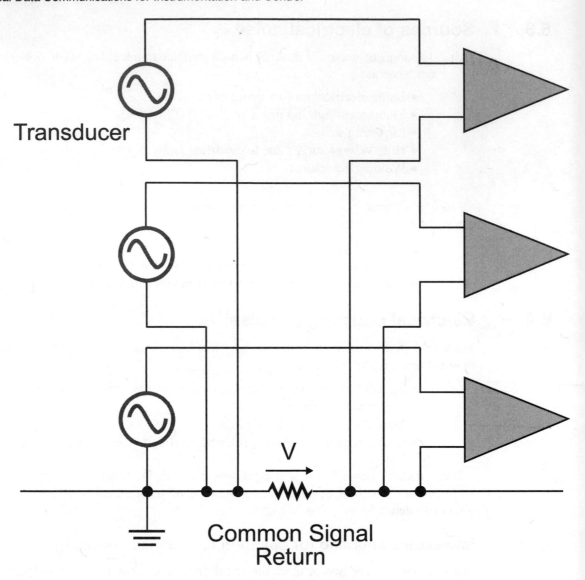

Transducer

Figure 6.8
Impedance coupling

Obviously, the quickest way to reduce the effects of impedance coupling is to minimize the impedance of the return wire. The best solution is to use a balanced circuit with separate returns for each individual signal.

044480

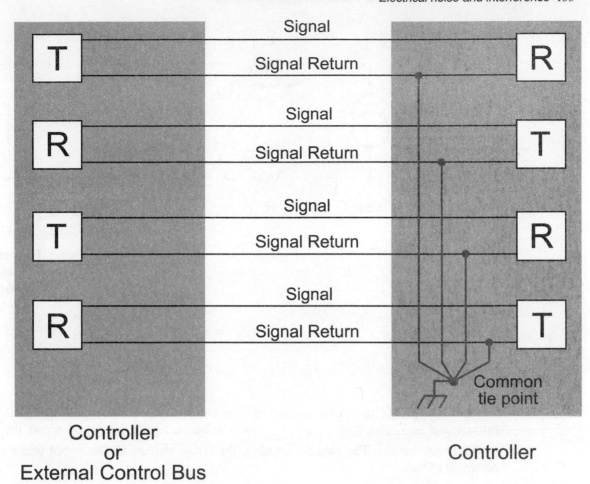

Figure 6.9
Impedance coupling eliminated with balanced circuits

Electrostatic or capacitive coupling

This form of coupling is proportional to the capacitance between the noise source and the signal wires. The magnitude of the interference depends on the rate of change of the noise voltage and the capacitance between the noise circuit and the signal circuit.

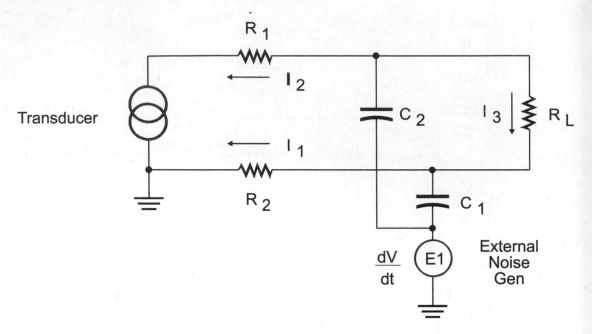

Figure 6.10
Electrostatic coupling

In the Figure 6.10, the noise voltage is coupled into the communication signal wires through the two capacitors C_1 and C_2, and a noise voltage is produced across the resistances in the circuit. The size of the noise (or error) voltage in the signal wires is proportional to the:

- Inverse of the distance of noise voltage from each of the signal wires
- Length (and hence impedance) of the signal wires into which the noise is induced
- Amplitude (or strength) of the noise voltage
- Frequency of the noise voltage

There are four methods for reducing noise induced by electrostatic coupling. They are:

- Shielding of the signal wires
- Separating from the source of the noise
- Reducing the amplitude of the noise voltage (and possibly the frequency)
- Twisting of the signal wires

Figure 6.11 indicates the situation that occurs when an electrostatic shield is installed around the signal wires. The currents generated by the noise voltages prefer to fl w down the lower impedance path of the shield rather than the signal wires. If one of the signal wires and the shield are tied to the earth at one point, which ensures that the shield and the signal wires are at an identical potential, then reduced signal current flows between the signal wires and the shield.

Note: The shield must be of a low resistance material such as aluminum or copper. For a loosely braided copper shield (85% braid coverage), the screening factor is about 100 times or 20 dB i.e. C_3 and C_4 are about 1/100 C_1 or C_2. For a low resistance multi-layered screen, this screening factor can be 35 dB or 3000 times.

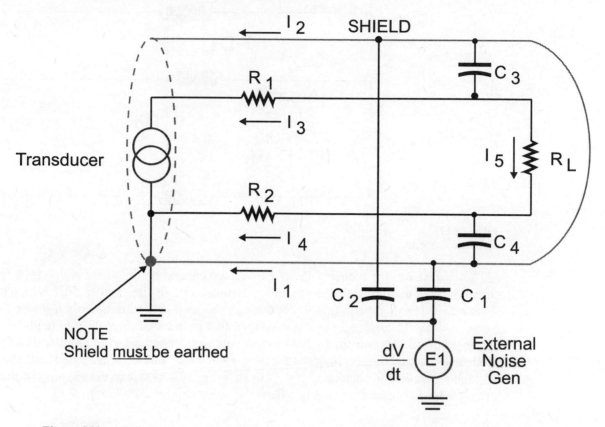

Figure 6.11
Shield to minimize electrostatic coupling

Twisting of the signal wires provides a slight improvement in the induced noise voltage by ensuring that C_1 and C_2 are closer together in value; thus ensuring that any noise voltages induced in the signal wires tend to cancel one another out.

Note: Provision of a shield by a cable manufacturer ensures that the capacitance between the shield and the wires are equal in value (thus eliminating any noise voltages by cancellation).

Magnetic or inductive coupling

This depends on the rate of change of the noise current and the mutual inductance between the noise system and the signal wires. Expressed slightly differently, the degree of noise induced by magnetic coupling will depend on the:

- Magnitude of the noise current
- Frequency of the noise current
- Area enclosed by the signal wires (through which the noise current magnetic flux cuts)
- Inverse of the distance from the disturbing noise source to the signal wires

The effect of magnetic coupling is shown in Figure 6.12.

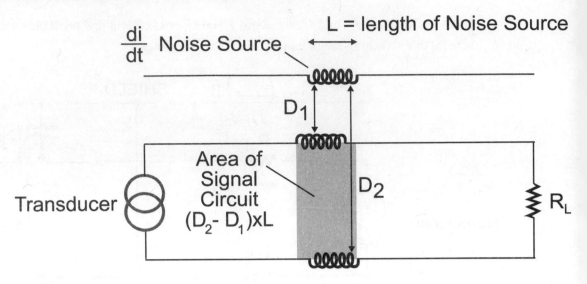

Figure 6.12
Magnetic coupling

The easiest way of reducing the noise voltage caused by magnetic coupling is to twist the signal conductors. This results in lower noise due to the smaller area for each loop. This means less magnetic flux to cut through the loop and consequently a lower induced noise voltage. In addition, the noise voltage that is induced in each loop tends to cancel out the noise voltages from the next sequential loop. Hence, an even number of loops will tend to have the noise voltages canceling each other out. It is assumed that the noise voltage is induced in equal magnitudes in each signal wire due to the twisting of the wires giving a similar separation distance from the noise voltage. See Figure 6.13.

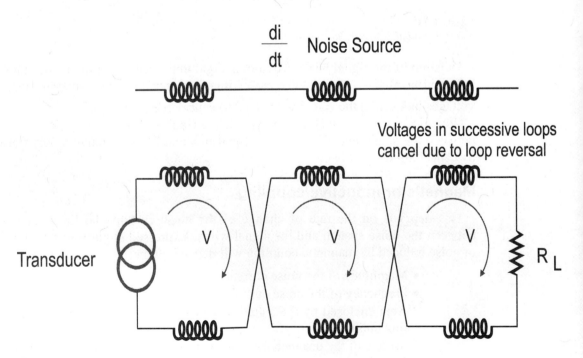

Figure 6.13
Twisting of wires to reduce magnetic coupling

The second approach is to use a magnetic shield around the signal wires. The magnetic flux generated from the noise currents induces small eddy currents in the magnetic shield. These eddy currents then create an opposing magnetic flux $\varnothing_1$ to the original flux $\varnothing_2$. This means a lesser flux $(\varnothing_2 - \varnothing_1)$ reaches our circuit!

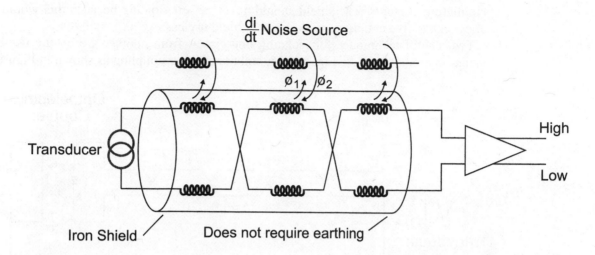

Figure 6.14
Use of magnetic shield to reduce magnetic coupling

Note: The magnetic shield does not require earthing. It works merely by being present. High permeability steel makes best magnetic shields for special applications. However, galvanized steel conduit makes a quite effective shield.

Radio frequency radiation

The noise voltages induced by electrostatic and inductive coupling (discussed above) are manifestations of the near field effect, which is electromagnetic radiation close to the source of the noise. This sort of interference is often difficult to eliminate and requires close attention of grounding of the adjacent electrical circuit and the earth connection is only effective for circuits in close proximity to the electromagnetic radiation. The effects of electromagnetic radiation can be neglected unless the field strength exceeds 1 volt/meter. This can be calculated by the formula:

$$Field\ strength = \frac{0.173\sqrt{\text{Power}}}{\text{Distance}}$$

where:

– Field strength is in volt/meter
– Power is in kilowatt
– Distance is in km

The two most commonly used mechanisms to minimize electromagnetic radiation are:

• Proper shielding (iron)
• Capacitors to shunt the noise voltages to earth

Any incompletely shielded conductors will perform as a receiving aerial for the radio signal and hence care should be taken to ensure good shielding of any exposed wiring.

6.5 Shielding

It is important that electrostatic shielding is only earthed at one point. More than one earth point will cause circulating currents. The shield should be insulated to prevent inadvertent contact with multiple points, which behave as earth points resulting in circulating currents. The shield should never be left floating because this would tend to allow capacitive coupling, rendering the shield useless.

Two useful techniques for isolating one circuit from another are by the use of opto-isolation as shown in the Figure 6.15, and transformer coupling as shown in Figure 6.16.

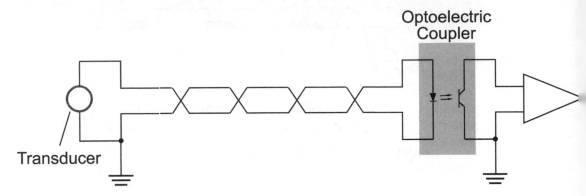

Figure 6.15
Opto-isolation of two circuits

Although opto-isolation does isolate one circuit from another, it does not prevent noise or interference being transmitted from one circuit to another.

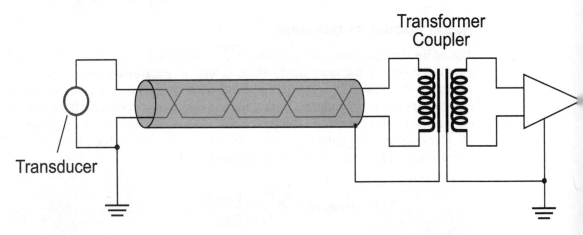

Figure 6.16
Transformer coupling

Transformer coupling can be preferable to optical isolation when there are very high speed transients in the one circuit. There is some capacitive coupling between the LED and the base of the transistor, which in the opto-coupler can allow these types of transients to penetrate one circuit from another. This is not the case with transformer coupling.

6.6 Good shielding performance ratios

The use of some form of low resistance material covering the signal conductors is considered good shielding practice for reducing electrostatic coupling. When comparing shielding with no protection, this reduction can vary from copper braid (85% coverage) which returns a noise reduction ratio of 100:1 to aluminum Mylar tape, with drain wire, with a ratio of 6000:1.

Twisting the wires to reduce inductive coupling reduces the noise (in comparison to no twisting) by ratios varying from 14:1 (for four inch lay) to 141:1 (for one inch lay). In comparison, putting parallel (untwisted) wires into steel conduit only gives a noise reduction of 22:1.

On very sensitive circuits with high levels of magnetic and electrostatic coupling the approach is to use coaxial cables. Double shielded cable can give good results for very sensitive circuits.

Note: With double shielding, the outer shield could be earthed at multiple points to minimize radio frequency circulating loops. This distance should be set at intervals of less than 1/8th the wavelength of the radio frequency noise.

6.7 Cable ducting or raceways

These are useful in providing a level of attenuation of electric and magnetic fields. These figures are 60 Hz for magnetic fields and 100 kHz for electric fields.

Typical screening factors are:

- 5 cm (0.2 inch) aluminum conduit with 0.154 inch thickness
  - Magnetic fields 1.5:1
  - Electric fields 8000:1
- Galvanized Steel conduit 5 cm (0.2 inch), wall thickness 0.154 inch with
  - Magnetic fields 40:1
  - Electric fields 2000:1

6.8 Cable spacing

In situations where there are a large number of cables varying in voltage and current levels, the IEEE 518-1982 standard has developed a useful set of tables indicating separation distances for the various classes of cables. There are four classification levels of susceptibility for cables. Susceptibility, in this context, is understood to be an indication of how well the signal circuit can differentiate between the undesirable noise and required signal. It follows that a data communication physical standard such as RS-232E would have a high susceptibility and a 1000 volt, 200 amp ac cable has a low susceptibility.

The four susceptibility levels defined by the IEEE 518-1982 standard are briefly:

- Level 1 – High
  This is defined as analog signals less than 50 volt and digital signals less than 15 volt. This would include digital logic buses and telephone circuits. Data communication cables fall into this category.

- Level 2 – Medium
  This category includes analog signals greater than 50 volt and switching circuits.
- Level 3 – Low
  This includes switching signals greater than 50 volt and analog signals greater than 50 volt. Currents less than 20 amp are also included in this category.
- Level 4 – Power
  This includes voltages in the range 0–1000 Volt and currents in the range 20–800 amp. This applies to both ac and dc circuits.

The IEEE 518 also provides for three different situations when calculating the separation distance required between the various levels of susceptibilities.

In considering the specific case where one cable is a high susceptibility cable and the other cable has a varying susceptibility the required separation distance would vary as follows:

- Both cables contained in a separate tray
    - Level 1 to Level 2 – 30 mm
    - Level 1 to Level 3 – 160 mm
    - Level 1 to Level 4 – 670 mm
- One cable contained in a tray and the other in conduit
    - Level 1 to Level 2 – 30 mm
    - Level 1 to Level 3 – 110 mm
    - Level 1 to Level 4 – 460 mm
- Both cables contained in separate conduit
    - Level 1 to Level 2 – 30 mm
    - Level 1 to Level 3 – 80 mm
    - Level 1 to Level 4 – 310 mm

The figures are approximate as the original standard is quoted in inches.

A few words need to be said about the construction of the trays and conduits. It is expected that the trays are manufactured from metal and be firmly earthed with complete continuity throughout the length of the tray. The trays should also be fully covered preventing the possibility of any area being without shielding.

6.9 Earthing and grounding requirements

This is a contentious issue and a detailed discussion laying out all the theory and practice is possibly the only way to minimize the areas of disagreement. The picture is further complicated by the different national codes, which whilst not actively disagreeing with the basic precepts of other countries tend to lay down different practical techniques in the implementation of a good earthing system.

A typical design should be based around two separate electrically insulated earth systems. The two earth systems are:

- The equipment earth
- The instrumentation (and data communications) earth

The aims of these two earthing systems are as follows:

- To minimize the electrical noise in the system
- To reduce the effects of fault or earth loop currents on the instrumentation system
- To minimize the hazardous voltages on equipment due to electrical faults

Earth (or ground) is defined as a common reference point for all signals in equipment situated at zero potential. Below 10 MHz, the principle of a single point earthing system is the optimum solution. Two key concepts to be considered when setting up an effective earthing system are:

- To minimize the effects of impedance coupling between different circuits (i.e. when three different currents for example flow through a common impedance)
- To ensure that earth or ground loops are not created (e.g. by mistakenly tying the screen of a cable at two points to earth)

There are three types of earthing systems possible as shown in Figure 6.17. The series single point is perhaps the more common, while the parallel single point is the preferred approach with a separate earthing system for groups of signals:

- Safety or power earth
- Low level signal (or instrumentation) earth
- High level signal (motor controls) earth
- Building earth

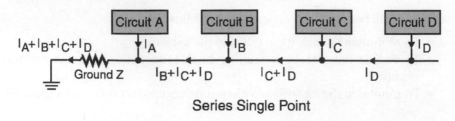

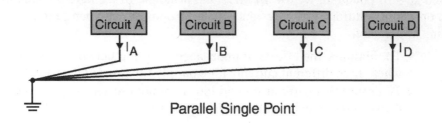

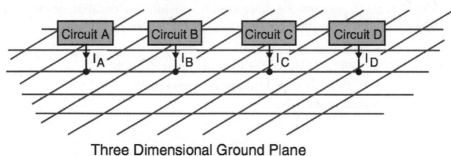

Figure 6.17
Various earthing configurations

6.10 Suppression techniques

It is often appropriate to approach the problem of electrical noise proactively by limiting the noise at the source. This requires knowledge of the electrical apparatus that is causing the noise and then attempting to reduce the noise caused here. The two main approaches are shown in Figure 6.18.

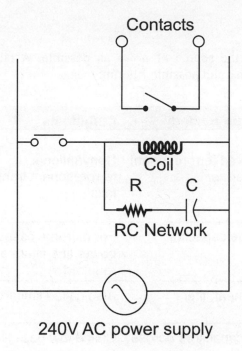

Suppression Network for AC

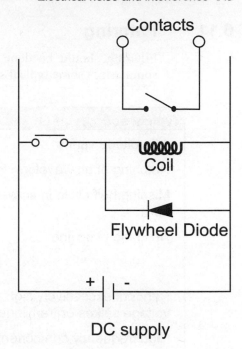

Suppression Network for DC

Figure 6.18
RC network in parallel across coil

In Figure 6.18, the inductance will generate a back emf across the contacts when the voltage source applied to it is switched off. This RC network then takes this back emf and thus reduces damage to the contacts.

The voltage can be limited by various combinations of devices (depending on whether the circuit is ac or dc).

The user of these techniques should be aware that the response time of the coil can be reduced by a significant time e.g. the dropout time of a coil can be increased by a factor of ten. Hence, this should be approached with caution where quick response is required from regular switched circuits (apart from the obvious deleterious impact on safety due to slowness of operation).

Two other areas to consider are:

Silicon controlled rectifiers (or SCRs) and triacs

Generate considerable electrical noise due to the switching of large currents. A possible solution is to place a correctly sized inductor placed in series with the switching device.

Lightning protection

Can be affected by the use of voltage limiters (suitably rated for the high level of current and voltage) connected across the power lines.

6.11 Filtering

Filtering should be done as close to the source of noise as possible. A table below summarizes some typical sources of noise and possible filtering means.

Typical sources of noise	Filtering Remedy	Comments
· AC voltage varies · Notching of ac waveform form · Missing half cycle in ac wave-form	Improved ferroresonant transformer	Conventional ferroresonant transformer fails
· Notching in dc line	Storage capacitor	For extreme cases, active power line filters are required
· Random excessively high voltage spikes or transients	Non-linear filters	Also called limiters
· High frequency components	Filter capacitors across the line	Called low pass filtering. Great care should be taken with high frequency vs performance of "capacitors" at this frequency
· Ringing of filters	Use T filters	From switching transients or high level of harmonics
· 60 Hz or 50 Hz interference	Twin-T RC notch filter networks	Sometimes low pass filters can be suitable
· Common mode voltages	Avoid filtering (isolation transformers or common-mode filters)	Opto isolation is preferred - eliminates ground loop
· Excessive Noise	Auto or cross correlation techniques	Extracts the signal spectrum from the closely overlapping noise spectrum

Table 6.1
Typical noise sources and some possible means of filtering

7

Modems and multiplexers

This chapter reviews the concepts of modems and multiplexers, their practical use, position and importance in the operation of a data communications system.

Objectives

When you have completed studying this chapter you will be able to:

- Explain the modes of operation of a modem
- Describe the role of interchange circuits
- Describe three methods of flow control
- Explain the causes of signal distortion
- Describe the different methods of modulation:
    - ASK
    - FSK
    - PSK
    - QAM
    - TCM
- List and describe the components of a modem
- Describe the properties of different modem types:
    - Dumb modems
    - Smart modems, including their states and commands
- Describe radio modems:
    - Terms
    - Modes
    - Features
    - Spread spectrum modems
- Describe the error detection protocols
- Describe data compression techniques
- List and describe the CCITT and Bell modem standards
- Explain troubleshooting
- Describe modem selection
- Describe three multiplexing concepts

- Describe terminal multiplexers
- Describe statistical multiplexers

7.1 Introduction

Communications systems, whether they are telephone, landline, or radio, cannot directly transport digital information without some distortion of the signal. This is due to the bandwidth limitation inherent in any of the connecting mediums. A conversion device, called a modem (modulator/demodulator), is required to convert the digital signals generated by the transmitting computer, into an analogue form suitable for long distance transmission. The demodulator in the modem receives analogue information and converts it back to the original digital information. Figure 7.1 gives a schematic view of the place of the modem in the communications hierarchy.

Figure 7.1
The modem as a component in a typical communication system

The bandwidth in a telephone network, for example, is limited by cable capacitance and inductance. The bandwidth is defined as the difference between the upper and lower allowable frequency and is typically 300 Hz to 3400 Hz for a telephone cable. This is illustrated in Figure 7.2.

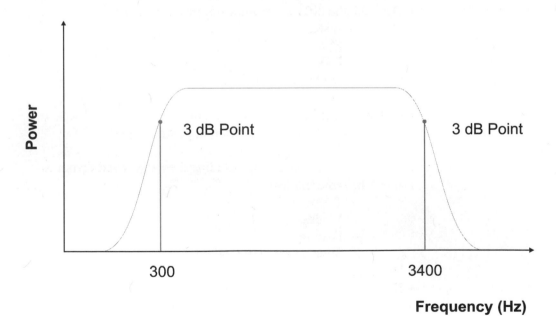

Figure 7.2
The bandwidth limitation problem

An example of what a digital signal would look like at the far end of a cable without conversion to an analog signal is given in Figure 7.3.

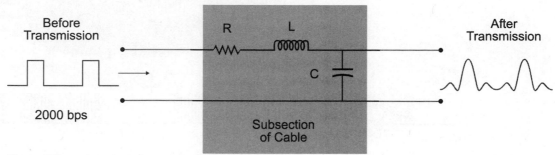

Figure 7.3
Injection of a digital signal down a cable

7.2 Modes of operation

Modems can operate in two modes:

- Half duplex
- Full duplex

A full duplex system is more efficient than a half duplex system, as data can flow in both directions simultaneously. A full duplex system requires a communication capacity of at least twice that of a half duplex system, where data can flow in both directions, but in only one direction at a time, as discussed in Chapter 2.

7.3 Synchronous or asynchronous

Modems can operate in either of two modes.

- Asynchronous
- Synchronous

Asynchronous

In asynchronous communication each character is encoded with a start bit at the beginning of the character bit stream and a parity and stop bit at the end of the character bit stream. The start bit allows the receiver to synchronize with the transmitter so that the receiver looks for each character as it is sent. Once the character has been received the communications link returns to the idle state and the receiver waits for the next start bit indicating the arrival of the next character. This is illustrated in Figure 7.4.

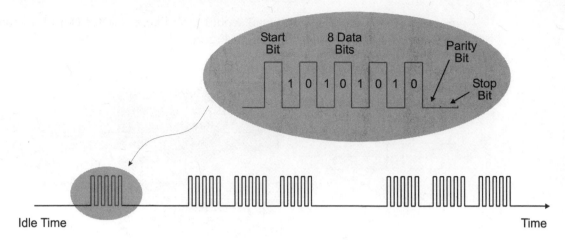

Figure 7.4
Asynchronous transmission of a few characters

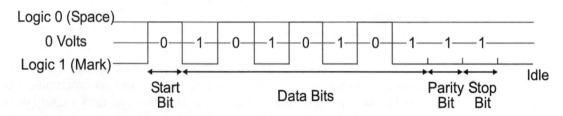

Figure 7.5
Format of a typical serial asynchronous data message

Synchronous communications

Synchronous communication relies on all characters being sent in a continuous bit stream. The first few bytes in the message contain synchronization data allowing the receiver to synchronize to the incoming bit stream. Synchronization is then maintained by a timing signal or clock. The receiver follows the incoming bit stream and maintains a close synchronization between the transmitter clock and receiver clock. Synchronous communication provides for far higher speeds of transmission of data, but is avoided in many systems because of the greater technical complexity of the communications' hardware.

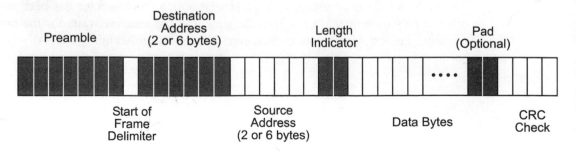

Figure 7.6
Synchronous communication protocol frame

The major difference between asynchronous and synchronous communications with modems is the need for timing signals.

A synchronous modem outputs a square wave on Pin 15 of the RS-232 DB-25 connector. Pin 15 is called the transmit clock pin or more formally the DCE transmitter signal element timing pin. The square wave is set to the frequency of the modem's bit rate. The attached personal computer, the DTE, synchronizes its transmission of data from Pin 2 to the modem.

7.4   Interchange circuits

The interchange circuits that can be employed to change the operation of the attached communications devices are:

- Signal quality detector
- Data signal rate selector

Signal quality detector (CG, Pin 21)

If there is high probability of error in the received data to the modem because of poor signal quality this line is set to OFF.

Data signal rate selector (CH/CI, Pin 23)

If the signal quality detector pin indicates that the quality of the signal is unacceptable, that is, it is set to OFF, the terminal may set Pin 23 to ON to select a higher data rate; or OFF to select a lower data rate. This is called the CH circuit. If, however, the modem selects the data rate and advises the terminal on Pin 23 (ON or OFF), the circuit is known as circuit CI.

7.5   Flow control

Flow control techniques are widely used to ensure that there will be no overflow of data by the device receiving a stream of characters, which it is temporarily unable to process or store. The receiving device needs a facility to signal the transmitter to temporarily cease sending characters down the line. Flow control between the PC and modem can be achieved either through hardware or software handshaking.

There are three mechanisms of flow control:

- XON/XOFF signaling, software based
- ENQ/ACK, software based
- RTS/CTS signaling, hardware based

When the modem decides that it has too much data arriving, it sends an XOFF character to the connected terminal to tell it to stop transmitting characters. This typically occurs when the modem memory buffer is approximately 66% full. The delay in transmission of characters by the terminal allows the modem to process the data in its memory buffer. Once the data has been processed and the memory buffer has emptied to typically 33% full, the modem sends an XON character to the terminal and transmission of data to the modem resumes. XON and XOFF are two defined ASCII characters DC1 and DC3 respectively.

XON/XOFF signaling works well unless there are flow control characters (XON/XOFF) in the normal data stream. These characters can cause problems and should be removed from the standard stream of transmitted information and reserved for control purposes.

ENQ/ACK

The terminal sends an ENQ control character to the modem when it wants to transmit a finite block of data. When the modem is ready to receive characters, it transmits an ACK, which then allows the terminal to commence transmission of this block of data. The process is repeated for subsequent blocks of data.

RTS/CTS signaling

This technique of hardware flow control is a simplified version of the full hardware handshaking sequence discussed. When the terminal wants to transmit data to the modem, it asserts the request to send (RTS) line and waits for the modem to assert the clear to send (CTS) line before transmitting. When the modem is unable to process any further characters it switches off, or inhibits, the CTS control line. The terminal device then stops transmitting characters until the CTS line is again asserted.

7.6   Distortion

There are two significant causes of distortion in the signal during communications (as discussed in Chapter 4). These are:

- Attenuation distortion
- Envelope delay distortion

Both forms of distortion are illustrated in Figure 7.7

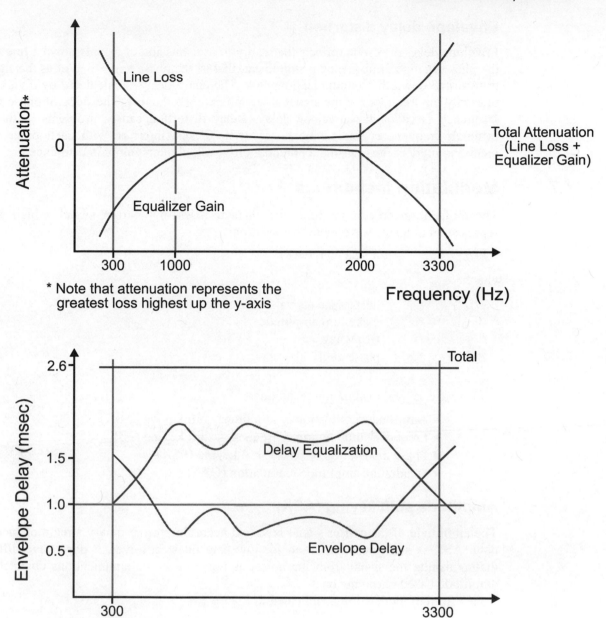

Figure 7.7
Attenuation distortion and envelope delay

Attenuation distortion

Attenuation distortion indicates that the theoretical smooth, horizontal plot of power transmitted versus frequency is not realized in practice. Higher frequencies tend to attenuate more easily and attenuation becomes more non-linear at the edges of the operating bandwidth, or 'passband'. Hence, the 'equalizer' compensates with an equal and opposite effect, giving a constant total loss throughout the passband.

Envelope delay distortion

Envelope delay distortion reflects the reality of transmission of signals down a line where the phase change to frequency is non linear, that is, the phase tends to alter as the signal is transmitted down the communications link. The phase delay is calculated by dividing the phase by the frequency of the signal at any point along the line. The slope of phase versus frequency is called the envelope delay. Delay distortion causes problems in that two different frequencies (indicating a, '1' or a '0' bit) interfere with each other at the receiving modem thus causing a potential error, called intersymbol interference.

7.7   Modulation techniques

The modulation process modifies the characteristics of a carrier signal, which can be represented as a sine wave, with the equation:

$$F(t) \quad = \quad A \sin (2 \, /ft + \text{Æ})$$

where:

F(t)	=	instantaneous value of voltage at time t
A	=	maximum amplitude
f	=	frequency
Æ	=	phase angle

There are several modulation techniques:

- Amplitude modulation or amplitude shift keying (ASK)
- Frequency modulation or frequency shift keying (FSK)
- Phase modulation or phase shift keying (PSK)
- Quadrature amplitude modulation (QAM)

Amplitude shift keying (ASK)

The amplitude of the carrier signal is varied according to the binary stream of incoming data. ASK is sometimes still used for low data rates, however, it does have difficulty distinguishing the signal from the noise, as noise in the communications channel is an amplitude based phenomenon.

This form of modulation is indicated in Figure 7.8.

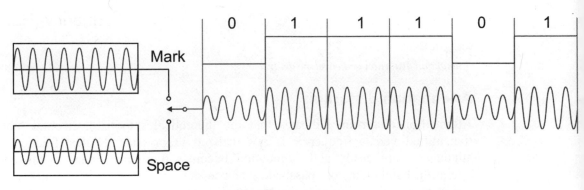

Figure 7.8
Operation of amplitude shift keying

Frequency shift keying (FSK)

Frequency modulation allocates different frequencies to logic 1 and logic 0 of the binary data message. FSK is primarily used by modems operating at data rates of up to 300 bps in full duplex mode and 1200 bps in half duplex mode.

The Bell 103/113 and the compatible ITU V.21 standards are indicated in Table 7.1.

Specification	Originate (Mark)	Originate (Space)	Answer (Mark)	Answer (Space)
CCITT V.21	1270 Hz	1070 Hz	2225 Hz	2025 Hz
Bell 103	980 Hz	1180 Hz	1650 Hz	1850 Hz

Table 7.1
CCITT V.21 and Bell System 103/113 modems frequency allocation

The Bell 103/113 modems are setup in either originate or answer mode. Typically, terminals are connected to originate modems and main frame computers are connected to answer type modems. It is easy to communicate when originate modems are connected to answer mode modems, but similar modems, for example, two originate modems connected together, cannot communicate with each other as they expect different frequencies.

Because of the two different bands of frequencies in which the sets of signals operate, full duplex operation is possible with these modems. Note that they fit into the allowable bandwidth of the communications channel.

Phase shift keying (PSK)

PSK is the process of varying the carrier signal by phase. There are two forms of phase modulation:

- Quadrature phase shift keying (QPSK)
- Differential PSK

QPSK

In QPSK four phase angles are used for encoding:

$$0°, \ 90°, \ 180° \ \text{and} \ 270°$$

There are four phase angles possible at any one time, allowing the basic unit of data to be a 2-bit pair, or dibit. The weakness of this approach is that a reference signal is required as indicated in Figure 7.9.

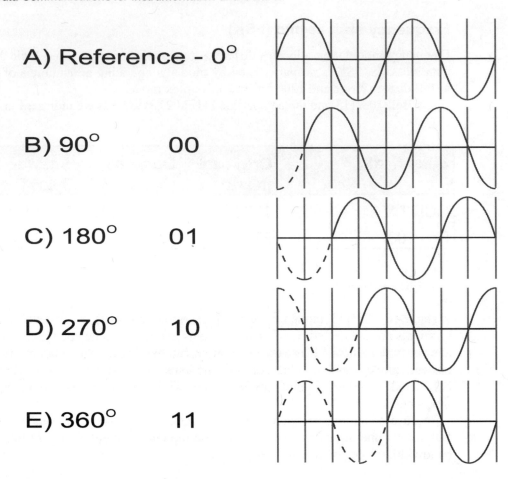

A) Reference - 0°

B) 90°   00

C) 180°   01

D) 270°   10

E) 360°   11

Figure 7.9
Quadrature phase shift keying

Differential PSK

The preferred option is to use differential PSK where the phase angle for each cycle is calculated relative to the previous cycle as shown in Figure 7.10.

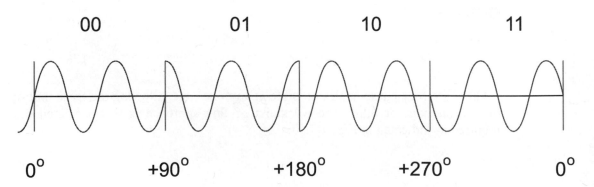

| 00 | 01 | 10 | 11 |

| 0° | +90° | +180° | +270° | 0° |

Figure 7.10
Differential PSK

A modulation rate of 600 baud results in a data rate of 1200 bps using two bits for each phase shift.

A typical allocation of dibits, or two bit codes, for each phase shift is as follows:

Dibit	Phase Shift (degrees)
00	0°
01	90°
10	180°
11	270°

Table 7.2
Allocation of dibits for differential PSK

Quadrature amplitude modulation (QAM)

Two parameters of a sinusoidal signal, amplitude and phase, can be combined to give QAM. QAM allows for 4 bits to be used to encode every amplitude and phase change. Hence, a signal at 2400 baud would provide a data rate of 9600 bps. The first implementation of QAM provided for 12 values of phase angle and 3 values of amplitude.

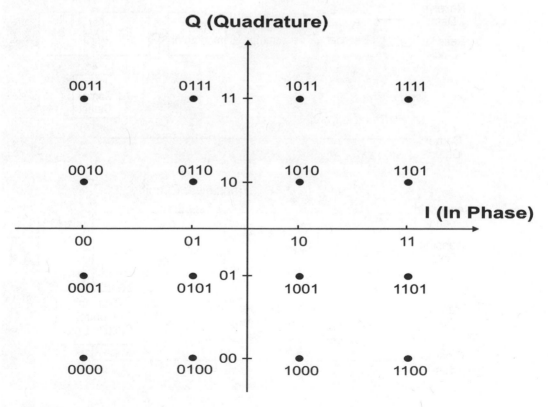

Figure 7.11
CCITT V.22bis quadrature amplitude modulation

QAM also uses two carrier signals. The encoder operates on 4 bits for the serial data stream and causes both an in-phase (IP) cosine carrier and a sine wave that serves as the quadrature component (QC) of the signal to be modulated. The transmitted signal is then changed in amplitude and phase resulting in the constellation pattern illustrated above.

Trellis coding

QAM modems are susceptible to noise; hence, a new technique called trellis coding was introduced. Trellis coding allows 9600 to 14 400 bps transmission over normal telecommunication lines and 14 400 bps and higher over good quality leased lines. In order to minimize the errors that occur when noise is evident on the line, an encoder adds a redundant code bit to each symbol interval.

Only certain sequences are valid. If there is noise on the line, which causes the sequence to differ from an accepted sequence, the receiver will select the valid signal point closest to the observed signal without needing a retransmission of the affected data.

A conventional QAM modem, which might require 1 out of every 10 data blocks to be retransmitted, could be replaced by a modem using trellis coding where only one in every 10 000 data blocks might be in error.

7.8   Components of a modem

The components of a modem are indicated in Figure 7.12.

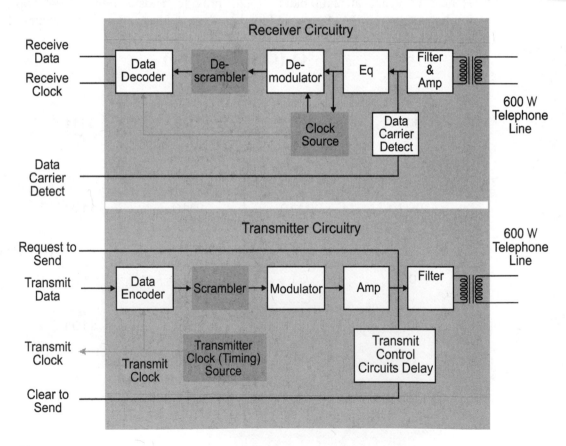

Figure 7.12
Basic components of a modem

The components of a modem can be divided into two areas:

- Modem transmitter
- Modem receiver

Modem transmitter

The modem transmitter contains the following:

- Data encoder
- Scrambler
- Modulator
- Amplifier

Data encoder

The data encoder takes the serial bit stream and uses multilevel encoding, where each signal change represents more than one bit of data, to encode the data. Depending on the modulation technique used the bit rate can be two, four, or more times the baud rate.

Scrambler

The scrambler is used for synchronous operation only. It modifies the bit stream so that long sequences of 1s and 0s do not occur. Long sequences of 1s and 0s are difficult to use in synchronous circuits because of the difficulties they cause in extracting clocking information.

Modulator

The bit stream is converted into the appropriate analogue form using the selected modulation technique. Where initial contact is established with the receiving modem, a carrier is put on the line.

Amplifier

The amplifier increases the level of the signal to the appropriate level for the telephone line and matches the impedance of the line.

Modem receiver

The modem receiver contains the following:

- Filter and amplifier
- Equalizer
- Demodulator
- Descrambler
- Data decoder

Filter and amplifier

Noise is removed from the signal and the resultant signal is amplified.

Equalizer

The equalizer minimizes the effect of attenuation and delay on the various components of the transmitted signal. A predefined modulated signal, called a training signal, is sent down the line by the transmitting modem. The receiving modem knows the ideal characteristics of the training signal and the equalizer will adjust its parameters to correct for the attenuation and delay characteristics of the signal.

Demodulator

The demodulator retrieves the bit stream from the analogue signal.

Descrambler

The descrambler is used in synchronous operation only. The descrambler restores the data to its original serial form after it has been encoded in the scrambler circuit, ensuring that long sequences of 1s and 0s do not occur.

Data decoder

The final bit stream is produced in the data decoder in true RS-232 format.

7.9 Types of modem

There are two types of wire modems available today:

- Dumb, or non-intelligent modems
- Smart modems (Hayes compatible)

Dumb modems

Dumb, or non-intelligent, modems depend on the computer to which they are connected to instruct the modem when to perform most of its tasks such as answering the telephone.

Smart modems

Smart modems have an on-board microprocessor enabling them to perform such functions as automatic dialing and selection of the appropriate method of modulation.

As defined by RS-232, any interaction between a traditional dumb modem and the computer equipment occurs by exchanging signal voltages across wires. For example, without pin 20 (DTR) asserted, modems are disabled. However, the smart modem interacts with peripheral equipment by exchanging ASCII character sequences. The smart modem also handles such normally complex tasks as answering the phone automatically, and is capable of answering on a particular ring. A *de facto* standard has been established based on the Hayes Smartmodem.

The Hayes Smartmodem employs the minimum number of RS-232E functions necessary for full duplex control. RS-232E connections are made through the DB-25S (female) connector.

The smart modem has three states:

- On-line state
- Command state
- Comatose state

On-line state

The on-line state occurs whenever the smart modem is engaged in a carrier link with another modem. In this state, it behaves as a conventional modem transferring all RS-232 input directly to its transmitter.

Command state

When not on-line the smart modem is said to be in the command state and all RS-232 data is treated as a potential command. Usually, modems power up into the command state.

Comatose state

The comatose state is where the DTR pin is inhibited and the modem does not acknowledge commands, participate in dialing activity, or exhibit modem behavior.

Smart modems typically do not use DIP switches to select options because all options and commands are implemented in software. In the command state, the smart modem monitors the bytes incoming from the RS-232 port and watches for a particular sequence of characters referred to as the command sequence introducer. After the smart modem has executed the commands in its command buffer, the smart modem responds with its own sequence of ASCII characters.

There are two general classes of commands:

- Mode commands
- Numeric register commands

Mode commands

There are four basic sets of mode commands:

- User interface group
- Primary answer/dial group
- Answer/dial group
- Miscellaneous

The user interface group commands alter the way in which the smart modem interacts with the user and includes commands, which alter speaker level setting, for example.

The primary answer/dial group commands control the dialing process with commands such as answer, dial, and hook.

The answer/dial group commands affect the characteristics of the primary dial command, the second group with, for example, a pause setting.

The miscellaneous group commands handle such things as managing its own carrier and resetting the modem.

Numeric register commands

The second class of commands is the numeric class which sets up the thirteen status registers (S0 to S12). Other more modern modems use a larger number of registers. There are also three bit mapped registers (S13, S14 and S15). These enable the programmer to query the smart modem about the state of its internal variables, command flags, and current data format.

Status registers

Smart modems use the ATS Command to set and read status registers. A typical selection of S-registers for a smart modem is listed below:

S0	Number of rings before answering	Defines the number of ring tones before answering.
S1	Number of rings detected	Used with S0 to count incoming ring tones.
S2	Escape code character	Used for switching to the on-line command state and for terminating test sequences and can be changed to any ASCII character.
S3	Carriage return character	Contains the value of the carriage return character.
S4	Line feed character	Contains the value of the line feed character used in command mode.
S5	Backspace character	Contains the value of the backspace character used in command mode. Some systems may require other values such as the DEL code (7F) to be used.
S6	Time limit to wait for dial tone	Sets the time delay used to wait for dial tone when initiating a call.
S7	Time to wait for carrier	Sets the time the modem waits for carrier after any connection request. If no carrier is received within this period the modem hangs up and returns to command mode.
S8	Length of dial pause	Sets the time delay used by the pause dial modifier when initiating a call.
S9	Carrier detect response time	Sets the carrier OFF to ON detection time. Can be used to delay control signals in CCITT mode to ensure data is not lost when the modem connects to a remote modem.
S10	Time limit for disconnection on loss of carrier signal	Sets the maximum period the carrier can be dropped before the modem disconnects the line. If S10 = 0, the modem will hold the line indefinitely.
S11	Not user configurable	
S12	Escape code guard time	The pause time required before and after the escape sequence.
S13	Not user configurable	
S14	Not user configurable	

S15	Status register	Stores the condition of modem signals. This register is used by technicians and network managers for remote interrogation of the modem and is a read only register.

Bit	Status
0	RTS
1	CTS
2	DSR
3	DCD
4	DTR
5	OH
6	TEST
7	RLSD (carrier)

S16	Not user configurable	
S17	Data inactivity timeout	If no data is sent or received by the modem for the period set in register S17, the modem will disconnect the line and return to command mode. Setting S17 to 0 disables this feature.
S18	Test timer	Sets the period that loopback testing will be enabled. If set to 0, these tests will continue until a terminate command is issued by the operator.
S19	Error test bit error count	Stores the number of bit errors, when performing local and remote loopback with self-test. The maximum number returned is 255, even if the error count exceeds this.
S20	Error test bit count	Stores the total number of bits transmitted (x10,000), when performing local and remote loopback with self-test. The diagnostics section shows how to use S19 and S20 to calculate the Bit Error Rate (BER).
S21-S24	Not user configurable	
S25	Delay to DTR	Sets the time the modem waits before disconnecting after loss of DTR (see AT&D2).
S26	Not user configurable	
S27	Successful connection count	Records the number of successful connections since S27 was last reset. Along with S28 and S29, this information is used by network managers to compile operating statistics. The maximum number returned is 255, even if the count exceeds this.
S28	Failed connection count	Records the number of unsuccessful connections since S28 was last reset. The maximum number returned is 255, even if the count exceeds this.

S29	Failed security access count	Records the number of unsuccessful connections due to invalid security logons since S29 was last reset. The maximum number returned is 255, even if the count exceeds this.
S30S31	Not user configurable	
S32	ATIO response value	Contains the modem's model identification Code. This register determines the response returned to the inquiry command (ATI).
S33	Remote configuration escape character	Used by network managers and service technicians to remotely configure the modem. The modem monitors the received data for the escape sequence (+++) and for the escape code guard time (S12). If detected, remote configuration access is given. If S33 is greater than 127, this disables remote configuration.
S34	Not user configurable	
S35	Security dial exchange time out	When executing a security dial back, S35 sets the period after hanging up before dialing the caller back, which overcomes delays in some telephone exchanges.

7.10   Radio modems

Radio modems are suitable for replacing wire lines to remote sites or as a backup to wire or fiber-optic circuits, and are designed to ensure that computers and PLCs, for example, can communicate transparently over a radio link without any specific modifications required.

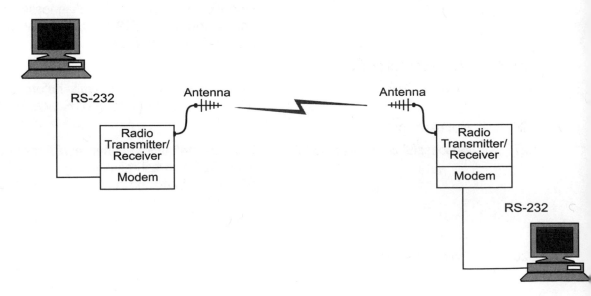

Figure 7.13
Radio modem configuration

Modern radio modems operate in the 400 to 900 MHz band. Propagation in this band requires a free line of sight between transmitting and receiving antennae for reliable communications. Radio modems can be operated in a network, but require a network management software system (protocols) to manage network access and error detection. Often, a master station with hot change-over, communicates with multiple radio field stations. The protocol for these applications can use a simple poll/response technique.

The more sophisticated peer-to-peer network communications applications require a protocol based on carrier sensing multiple access with collision detection (CSMA/CD). A variation on the standard approach is to use one of the radio modems as a network watchdog to periodically poll all the radio modems on the network and to check their integrity. The radio modem can also be used as a relay station to communicate with other systems, which are out of the range of the master station.

The interface to the radio modem is typically RS-232 but RS-422, RS-485, and fiber-optics are also options. Typical speeds of operation are up to 9600 bps. A buffer is required in the modem and is typically a minimum of 32 kilobytes. Hardware and software flow control techniques are normally provided in the radio modem firmware, ensuring that there is no loss of data between the radio modem and the connecting terminal.

Typical modulation techniques are two level direct FM (1200 to 4800 bps) to three level direct FM (9600 bps).

A typical schematic of a radio modem is given in Figure 7.14.

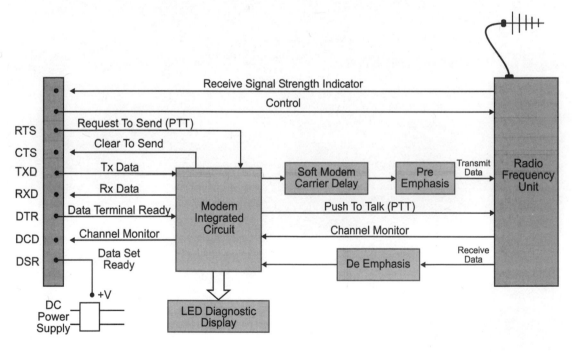

Figure 7.14
Typical block diagram of a radio modem

The following terms are used in relation to radio modems:

PTT   Push to talk signal
RSSI   Receive signal strength indicator – indicates the received signal strength with a proportionally varying dc voltage.

noise squelch	Attempts to minimize the reception of any noise signal at the discriminator output.
RSSI squelch	Opens the 'receive audio path' when the signal strength of the RF carrier is of a sufficiently high level.
channel monitor	Indicates if the squelch is open.
soft carrier delay	Allows the RF transmission to be extended slightly after the actual end of the data message which avoids the end of transmission bursts that occur when the carrier stops and the squelch almost simultaneously disconnects the studio path.
RTS, CTS, DCD, clock, transmit data, receive data	All relate to RS-232.

The radio modem has a basic timing system for communications between a terminal and the radio modem, indicated in Figure 7.15.

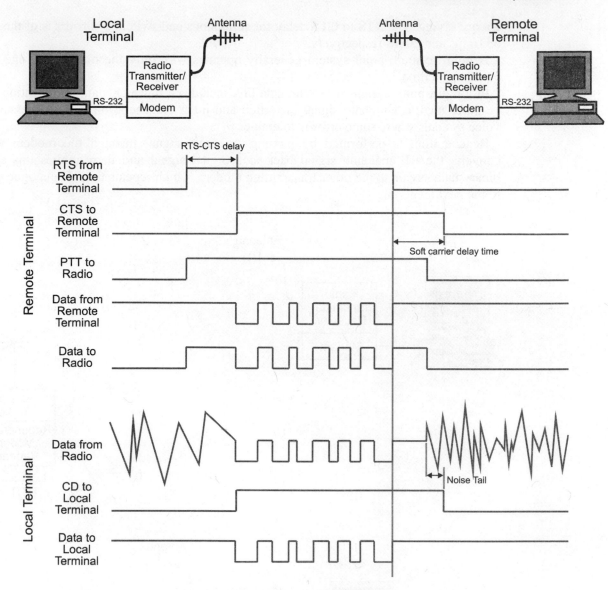

Figure 7.15
Radio modem timing diagram

Data transmission begins with the RTS line becoming active at the remote terminal side. The radio modem then raises the CTS line to indicate that transmission can proceed. At the end of the transmission, the PTT is kept active to ensure that the receiving side detects the remaining useful data before the RF carrier is removed.

Modes of radio modems

Radio modems can be used in two modes:

- Point to point
- Point to multi-point

A point to point system can operate in continuous RF mode, which has a minimal turn on delay in transmission of data, and non-continuous mode where there is a considerable energy saving. The RTS to CTS delay for continuous and switched carriers is of the order of 10 ms and 20 ms respectively.

A point to multi-point system generally operates with only the master and one radio modem at a time.

In a multi-point system when the data link includes a repeater, data regeneration must be performed to eliminate signal distortion and jitter. Regeneration is not necessary for voice systems where some error is tolerable.

Regeneration is performed by passing the radio signal through the modem which converts the RF analogue signal back to a digital signal and then applies this output binary data stream to the other transmitting modem, which repeats the RF analogue signal to the next location.

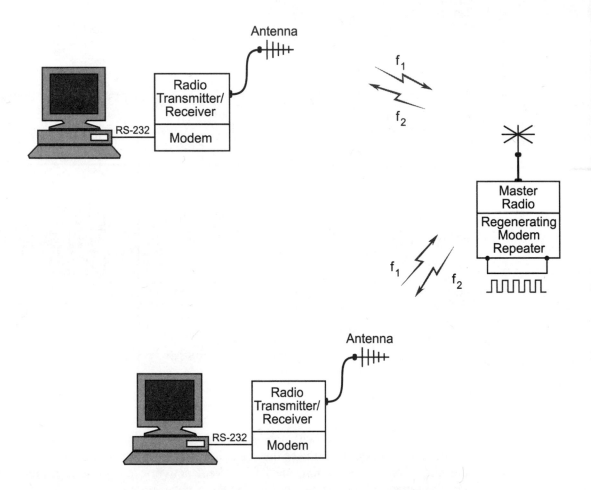

Figure 7.16
Regeneration of a signal with a radio modem

Features of a radio modem

Typical features that have to be configured in the radio modem are:

- Transmit/receive radio channel frequency.
  In a point to point configuration running in a dual frequency/split channel assignment, two radios will operate on opposing channel sets.

- Host data rate and format
  Data rate/character size/parity type and number of stop bits for RS-232 communications.
- Radio channel data rate
  Data rate across the radio channel defined by the radio and bandwidth capabilities. Note that these specifications are generally set at the time of manufacture.
- Minimum radio frequency signal level
  Should not be set too low on the receiver otherwise noise data will also be read.
- Supervisory data channel rate
  Used for flow control and should therefore not be set too low otherwise the buffer on the receiver will overflow. Typically one flow control bit to 32 bits of serial data is standard.
- Transmitter key up delay
  The time for the transmitter to energize and stabilize before useful data is sent over the radio link. Transmitter key up delay should be kept as low as possible to minimize overheads.

Spread spectrum radio modems

Several countries around the world have allocated a section of bandwidth for use with spread spectrum radio modems. In Australia and America, this is in the 900 MHz area.

In brief, a very wide band channel is allocated to the modem, for example, approximately 3.6 MHz wide. The transmitter uses a pseudo random code to place individual bits, or groups of bits, broadly across the bandwidth and the receiver uses the same random code to receive them. Because they are random, a number of transceivers can operate on the same channel and a collision of bits will be received as noise by a receiver in close proximity.

The advantage of 'spread spectrum' radio modems is very high data security and data speeds of up to 19.2 kbps. The disadvantage is the very inefficient use of the radio spectrum.

7.11 Error detection/correction

The most popular form of error detection was initially cyclic redundancy check (CRC), especially CRC-16. CRC is discussed in detail in Chapter 4. Unfortunately different manufacturers implemented minor variations on the CRC approach, which resulted in incompatibilities between different products. The advent of the Microcom networking protocol (MNP), licensed by Microcom to numerous other manufacturers resulted in a *de facto* standard developing.

MNP protocol classes

MNP defines a system for the detection and correction of errors by retransmission between modems.

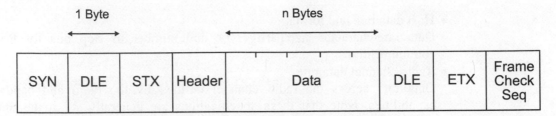

Asynchronous Data Frame

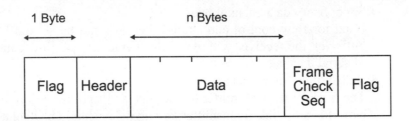

Synchronous Data Frame

Figure 7.17
Asynchronous and synchronous MNP frame formats

There are nine MNP Protocol classes defined in Table 7.3, which cover the transmission alternatives. Smart modems are programmed to attempt an MNP connection at the highest class that both modems can support. An initial frame called the link request is used to establish the standards to be followed in transferring the data. If MNP connection fails, the normal mode is used without error detection, correction, or data compression.

MNP Class	Async/ Synchronous	Half or Full Duplex	Efficiency	Description
1	Asynchronous	Half	70%	Byte oriented protocol
2	Asynchronous	Full	84%	Byte oriented protocol
3	Synchronous	Full	108%	Bit oriented protocol – communications between (PC) terminal and modem is still asynchronous.
4	Synchronous	Full	120%	Adaptive Packet Assembly (large data packets used if possible). Data phase optimization (elimination of protocol administrative overheads)
5	Synchronous	Full	200%	Data compression ratio of 1.3 to 2.0
6	Synchronous	Full	-	9600 bps V.29 modulation universal link negotiation allows modems to locate the highest operating speed and use statistical multiplexing
7	Synchronous	Full	-	Huffman encoding (enhanced data compression) reduces data by 42%
8	Synchronous	Full	-	CCITT V.29 fast Train Modem technology added to class 7
9	Synchronous	Half Duplex emulates Full Duplex	-	CCITT V.32 modulation + Class 7 enhanced data compression. Selective retransmission in which error packets are retransmitted.

Table 7.3
MNP protocol classes

Link access protocol modem (LAP-M)

This is recognized as the primary method for error detection and correction under the ITU-T V.42 recommendation. MNP error detection and correction is considered the secondary mechanism.

7.12   Data compression techniques

Data compression is used to achieve higher effective speeds in the transmission of the data and a reduction in transmission time.

Two of the most popular data compression methods are Adaptive Computer Technology's (ACT) compressor technology and Microcom's MNP class 5 and class 7 compression procedures. In 1990, the CCITT promulgated the V.42 bis standard which defines a new data compression method known as Lempel-Ziv.

The data compression standards that will be discussed here are:

- MNP class 5
- MNP class 7, Huffman
- V.42bis, Lempel-Ziv

MNP class 5 compression

MNP-5 involves a two stage process:

- Run length encoding
- Adaptive frequency encoding

Run length encoding

The first three bytes indicate the beginning of a run length encoded sequence. The next byte is the repetition count of bits, with a maximum number of 250 bits. For runs of similar bits, this can reduce the total size of the data bytes dramatically.

Essentially, the number of successive bits, which are the same, are counted and then coded into an eight bit symbol, for example. The eight-bit symbol is then transmitted.

Data compression is used extensively in the fax machine. For example in Group 3 machines, a regular 11 inch sheet of paper can be vertically digitized into 100 lines per inch to produce 1100 lines and horizontally each line is further digitized into 1700 bits/line.

Total size of the file   = 1700 bits/line × 1100 lines   = 1.87 Mbits

Assuming this file is sent on a 2400 baud modem, the transmission time for one page of text would be 779 seconds, as calculated below. However, in practice, the transmission time of a page is about 30 to 60 seconds. Data compression is used to achieve these results.

The microprocessor on the facsimile machine can process the data bits before sending them and uses a compression algorithm for compressing the data into fewer bits.

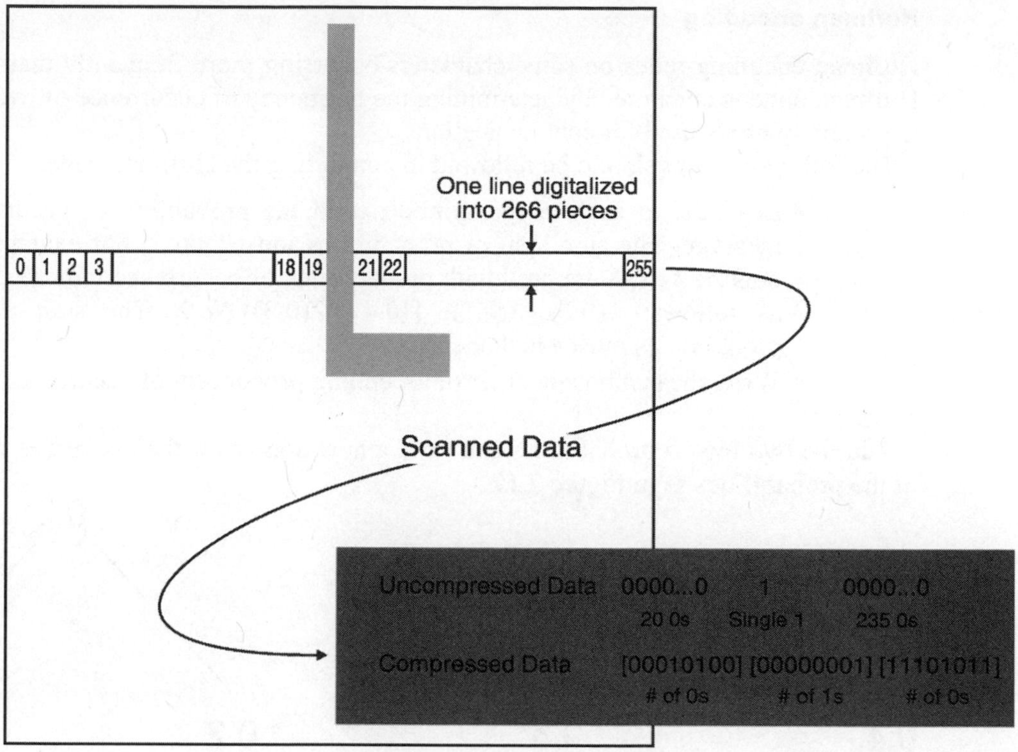

Figure 7.18
Data compression techniques applied to a scanned line

Adaptive frequency coding

In adaptive frequency coding a compression token is substituted for the actual byte transferred. Shorter tokens are substituted for more frequently occurring data bytes. A compression token consists of two parts:

- A fixed length header, 3 bits long, which indicates the length of the body
- A variable length body

At compression initialization, a table is set up for each byte from 0 to 255. To encode a data byte, the token to which it is mapped is substituted for the actual data byte in the data stream. The frequency of occurrence of the current data byte is increased incrementally by one. If the frequency of occurrence of the current data byte is greater than the frequency of the next most frequently occurring data byte, the two tokens are swapped. This comparison process is repeated for the next most frequently occurring data byte and the tokens are again swapped.

MNP class 7: enhanced data compression

MNP class 7 combines run length encoding with an adaptive encoding table. The table is used to predict the probability of a character occurring, based on the value of the previous character. Up to 256 (28) coding tables are kept for each 8-bit pattern. All characters are organized according to the rules of Huffman coding.

Huffman encoding

Huffman encoding relies on some characters occurring more frequently than others. The Huffman code is computed by determining the frequency of occurrence of each symbol in the set of symbols used for communications.

The following steps should be followed in computing the Huffman codes:

- List next to each of the symbols used the probability of occurrence in the message. The sum total of probabilities must total 1. For example, the symbols *A, X, Y, S* are used with probabilities of occurrence indicated in brackets as follows: *A*(0.2); *X*(0.1); *Y*(0.4); *Z*(0.3) *Note:* The sum totals of the probabilities must equal one.
- Write the symbols in order of ascending probability of occurrence.

Add the two lowest probabilities and form a new node over the two nodes with the sum of the probabilities as in Figure 7.19.

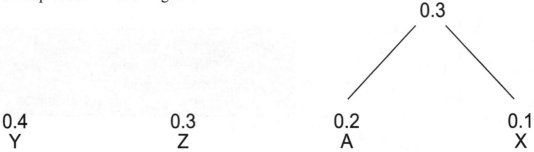

Figure 7.19
First node generation

- Repeat the process with the new node created and the next node to the left in order of probability.
- Repeat this process until completed, with a result as indicated in Figure 7.20.

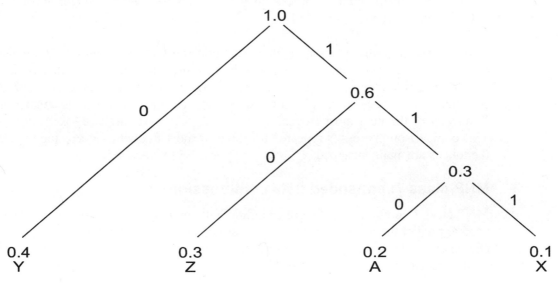

Figure 7.20
Second and third nodes

- Assign a 1 to the branches leaning in one direction as indicated above and 0 to the remaining branches.
- Compute the Huffman code for each symbol by tracing the path from the apex of the pyramid to each base.

Hence:

Y = 0
Z = 10
A = 110
X = 111

In order to compute the compression ratio, as compared to the standard 7-bit ASCII code, assume there are 1000 symbols (that is, *Y, Z, A* and *X*) transmitted.

Total bits using Huffman encoding =
(Probability of occurrence of symbol of 0.4 * 1000 symbols) * 1 bit/symbol *Y* +
(Probability of occurrence of symbol of 0.3 * 1000 symbols) * 2 bits/symbol *Z* +
(Probability of occurrence of symbol of 0.2 * 1000 symbols) * 3 bits/symbol *A* +
(Probability of occurrence of symbol of 0.1 * 1000 symbols) * 3 bits/symbol *X*

$$= 400 + 600 + 600 + 300$$
$$= 1900 \text{ bits}$$

If the ASCII code had been used this would have resulted in –

1000 symbols * 7 bits/symbol   = 7000 bits
Hence the compression ratio   = 7000/1900   = 3.68.

Once the Huffman code has been computed, the software converts each symbol into its equivalent code and includes the table used for translating the code back to symbols in the original transmission. The receiver software will then decompress the stream of bits into the original stream of symbols.

Run length encoding is used if there are four or more identical characters in a specified sequence of characters. The first three characters are encoded (as for the rules of Huffman encoding) and the number of remaining, identical characters is encoded in a 4-bit nibble.

Decoding the data stream is achieved quite simply because the receiving modem keeps the same compression table as the transmitting modem.

V.42bis

V.42bis relies on the construction of a dictionary, which is continually modified as data is transferred between two modems. The dictionary consists of a set of trees in which each root corresponds to a character in the alphabet. When communications is established each tree comprises a root node with a unique code word assigned to each node. The sequence of characters received by the modem from its attached terminal is compared with and matched against the dictionary.

The maximum string length can vary from 6 to 250 characters and is defined by the two connecting modems. The number of code words has a minimum of 512, but any value above this default minimum value can be agreed between the two connecting modems.

V.42bis data compression, in substituting a code word for a string, is between twenty and thirty per cent more efficient than MNP class 5 compression. V.42bis is effective for large file transfers, but not for short strings of data.

7.13 Modem standards

Table 7.4 summarizes the ITU-T modem standards.

Modem Type	Data Rate	Asynch/Synch	Mode	Modulation	Line Use Switched/Leased
V.21	300	Async	Half/Full	FSK	Switched
V.22	600	Async	Half/Full	DPSK	Switched/Leased
	1200	Async/Sync	Half/Full	DPSK	Switched/Leased
V.22 bis	2400	Async	Half/Full	QAM	Switched
V.23	600	Async/Sync	Half/Full	FSK	Switched
	1200	Async/Sync	Half/Full	FSK	Switched
V.26	2400	Sync	Half/Full	DPSK	Leased
	1200	Sync	Half	DPSK	Switched
V.26 bis	2400	Sync	Half	DPSK	Switched
V.26 ter	2400	Sync	Half/Full	DPSK	Switched
V.27	4800	Sync	Full	DPSK	Leased
V.27 bis	4800	Sync	Full	DPSK	Leased
	2400	Sync	Full	DPSK	Leased
V.27 ter	4800	Sync	Half	DPSK	Switched
	2400	Sync	Half	DPSK	Switched
V.29	9600	Sync	Half/Full	QAM	Leased
V.32	9600	Async	Half/Full	TCM/QAM	Switched
V.33	14400	Sync	Half/full	TCM	Leased

Table 7.4
ITU-T Modem standards

ITU V.34 and V.90 are high-speed dialup modem standards that are commonly used to connect to the Internet. Many data communication systems use modems to connect to the Internet. The V.34 and V.90 standard use a modulation scheme very similar to V.22bis. It has a symbol rate or baud rate of 3429 symbols per second. It can transmit up to 10 bits per symbol. With overheads, this averages out to approximately 33.6 k.

V.34 and V.90 use a modified QAM system called 'super constellation' that has 1664 possible symbol combinations. Not all symbols are used in every conversation. At the beginning of the conversation, the modems transmit special test strings that are used to formulate the best possible connection. The V.34 and V.90 modems will accept asynchronous data from the modem. They then change the asynchronous data to synchronous before sending it down the telephone line. Both standards also use a scrambler and Trellis coding to increase the quality of the signal.

V.90 modems like the V.34 modems check the telephone line when they first connect and can modify their parameters for optimal data communications. They also can change parameters midstream if either modem sees the need.

y \ x	-43	-39	-35	-31	-27	-23	-19	-15	-11	-7	-3	1	5	9	13	17	21	25	29	33	37	41	45
45										408	396	394	400	414									
41							398	375	349	339	329	326	335	347	359	386							
37					412	371	340	314	290	279	269	265	273	281	302	322	353	390					
33				401	357	318	282	257	236	224	216	212	218	228	247	270	298	337	378				
29			406	350	306	266	234	206	185	173	164	162	170	181	197	220	253	288	327	379			
25			360	310	263	226	193	165	146	133	123	121	125	137	154	179	207	242	289	338	391		
21		384	324	277	229	189	156	131	110	96	87	83	92	100	117	140	172	208	254	299	354		
17		355	294	243	201	160	126	98	79	64	58	54	62	71	90	112	141	180	221	271	323	387	
13	392	330	274	222	177	135	102	77	55	41	35	31	37	48	65	91	118	155	198	248	303	361	
9	380	316	255	203	158	119	84	60	39	24	17	15	20	30	49	72	101	138	182	230	283	348	415
5	367	304	244	194	148	108	75	50	28	13	6	4	8	21	38	63	93	127	171	219	275	336	402
1	362	296	238	186	142	103	69	43	22	9	1	0	5	16	32	56	85	122	163	213	267	328	395
-3	365	300	240	190	144	106	73	45	25	11	3	2	7	18	36	59	88	124	166	217	272	331	397
-7	372	307	251	199	152	113	80	52	33	19	12	10	14	26	42	66	97	134	174	225	280	341	409
-11	388	320	261	210	167	128	94	67	47	34	27	23	29	40	57	81	111	147	187	237	291	351	
-15	410	343	284	232	183	149	115	89	68	53	46	44	51	61	78	99	132	168	209	258	315	376	
-19		369	311	259	214	175	139	116	95	82	74	70	76	86	104	129	157	195	235	285	342	399	
-23		403	345	292	249	205	176	150	130	114	107	105	109	120	136	161	191	227	268	319	373		
-27			382	332	287	250	215	184	169	153	145	143	151	159	178	202	231	264	308	358	413		
-31				377	333	293	260	233	211	200	192	188	196	204	223	245	278	312	352	404			
-35					383	346	313	286	262	252	241	239	246	256	276	295	325	363	407				
-39						405	370	344	321	309	301	297	305	317	334	356	385						
-43								411	389	374	366	364	368	381	393								

Figure 7.21
One quarter of the points (in decimal) in a super-constellation

Motorola and Rockwell originally designed 56 k modems for Internet communications. Both the Motorola and Rockwell company standards would 'talk' at the same speed but they were slightly different. For a while, Internet Service Providers (ISP) used both standards. Then the ITU standardized the 56 k-modem communication system under ITU V.90.

The 56 k ISP to modem communication system transfers data from the user to the ISP at 33.6 k and from the ISP to the user at 56 k. This happens because the ISP uses a digital modem connected to the telephone exchange, whereas the user is connected using an analog modem. V.90 modems only connect at 56 k when they are connected to a digital modem. The 33.6 k part of the V.90 standard is the same as the V.34 standard.

The following table summarizes the various Bell modem standards.

Modem Type Bell System	Data Rate	Transmission Technique	Modulation Technique	Transmission Mode
103 A,E,F	300	Asynchronous	FSK	Half/Full
201 B,C	2400	Synchronous	PSK	Half/Full
202 C	1200	Asynchronous	FSK	Half
202 D/R	1800	Asynchronous	FSK	Half/Full
202 T	1800	Asynchronous	FSK	Half/Full
208 A	4800	Synchronous	PSK	Half/Full
208 B	4800	Synchronous	PSK	Half
209 A	9600	Synchronous	QAM	Full
21	0 - 300	Asynchronous	FSK	Half/Full
	1200	Asynchronous / Synchronous	PSK	Half/Full

Table 7.5
Bell modems

7.14 Troubleshooting a system using modems

There are two aspects to troubleshooting a system, which uses modems. These relate to:

- Satisfactory operation of the RS-232 system
- Specifics of the modem

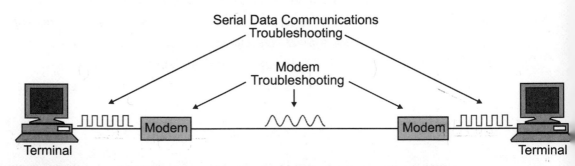

Figure 7.22
Troubleshooting a system using modems

Troubleshooting the modem

There are various tests available for troubleshooting operational problems associated with a modem, which fall into two categories:

- Self test
- Loop back test

Self Test

The self test is where the modem connects its transmitter to its receiver. The connection with the communications line is broken and a specified sequence of bits is transmitted to the receiving parts of the modem where this is then compared with a defined pattern. An error will be indicated on the modem front panel if the transmitted sequence does not match the expected pattern.

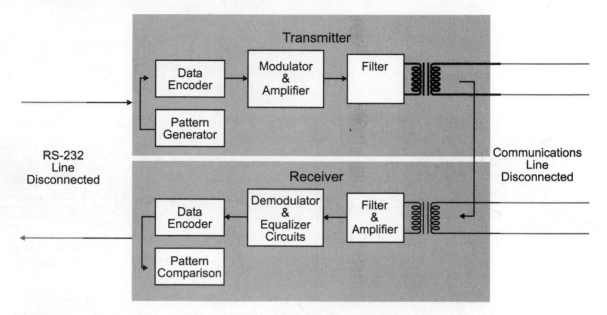

Figure 7.23
Modem internal self test

Loop back tests

The second set of tests is the loop back tests. There are four forms of loop back tests:

- Local digital loop to test the terminal or computer and connecting RS-232 line
- Local analog loop to test the modem's modulator and demodulator circuitry
- Remote analog loop to test the connecting cable and local modem
- Remote digital loop to test the local and remote modem and connecting cable

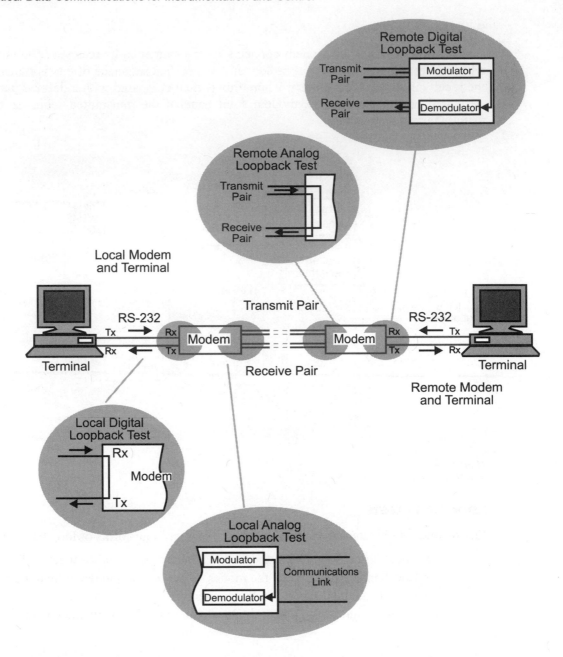

Figure 7.24
Loop back tests for modems

7.15   **Selection considerations**

There are certain features you should especially consider when selecting a modem for use in an industrial or telemetry application. Some of the more important are listed here:

Automatic smart features   Most asynchronous modems are compatible with the Hayes AT command set, which automates most modem features.

Data rate   Usually, the data rate of a modem is one of the first features considered. It is important to distinguish between the data rate and the baud rate, and the difference between the nominal data rate before compression and the effective data rate when compression has been performed should be noted.

Asynchronous/synchronous	The ability to switch between both modes allows more modes flexibility for future applications, and is sometimes provided as a dip switch configuration.
Transmission modes	The most efficient and preferable method of data transfer operation is full duplex rather than half duplex where the line turnaround time introduces a considerable amount of data transfer inefficiency.
Modulation techniques	The two most popular modulation techniques are V.22bis, which supports 1200 and 2400 bps transmission, and V.34+, which has V.22bis as a subset and supports the other almost universal transmission capability.
Data compression	The modem should have compatibility with the four main techniques compression standards used for telecommunication switched lines:

- ACT
- MNP class 5
- MNP class 7
- CCITT V.42bis

Error correction/detection	The most popular error detection and correction mechanism is MNP-4, which ITU have incorporated into the V.42 standard, which also allows LAP-M.
Flow control	Useful in controlling the flow of data from an attached terminal so that it does not overload the modem. You should ensure that the existing terminals and hardware support the necessary flow control protocols such as ENQ/ACK, RTS/CTS, or XON/XOFF.
Optimal blocking of data	Before transfer of data occurs, two modems negotiate with (protocol spoofing) each other for the specific file transfer protocol that should be used. This avoids unnecessary acknowledgments from the terminal device connected to the modem. If two modems can transfer 500 character blocks between them but the terminal to modem only supports 100 character blocks, the modem would accumulate 5 sets of 100 character blocks and transfers this in one hit to the receiving modem. The receiving modem would transfer 5 sets of 100 character blocks to the receiving terminal, which would acknowledge each 100 character blocks in turn.
Rack	Selection must be made based on the application. Many mounted/internal/stand industrial systems use rack mounted modems for space alone modems saving and the ease of providing appropriate power supplies.
Power supply	The latest modems have a separate power supply or derive power from the telephone lines.

| Self testing features | Ensure that the modem can perform a self test and the standard local and remote loop back tests. |

7.16 Multiplexing concepts

Multiplexing allows an existing link or channel to be used for more than one message at a time and has the potential to dramatically expand line utilization. It should be noted that multiple stages of multiplexing are possible.

Demultiplexing is the process of extracting the individual channel messages from the multiplexed data.

There are three possible multiplexing techniques:

- Space division multiplexing (SDM)
- Frequency division multiplexing (FDM)
- Time division multiplexing (TDM)

Space division multiplexing (SDM)

SDM is where multiple paths are created by running new physical channels next to the existing ones to connect a receiver and transmitter as shown in Figure 7.24. Some authorities feel that SDM is not a true multiplexing method. The technique is generally considered unattractive because additional cables, transmitters, and receivers are required.

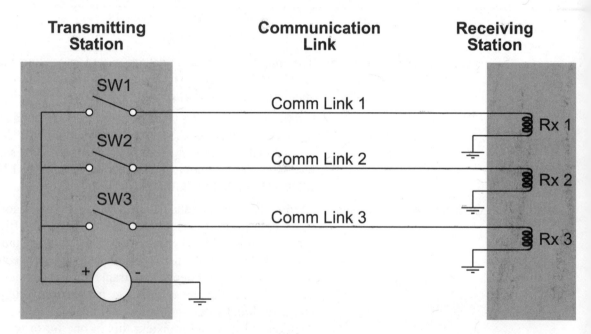

Figure 7.25
Space division multiplexing

The best example of space division multiplexing is the local telephone system. Each telephone is connected to the central office by a local loop not shared by other subscribers.

Demultiplexing SDM systems is virtually unnecessary as each signal has its own independent link and receiver/transmitter equipment.

Frequency division multiplexing (FDM)

FDM is where different, unique frequencies are used by each channel enabling several channels to use the same medium electrical cable. FDM occurs, therefore, when the bandwidth of the link is greater than the bandwidth of the messages sent over the link.

FDM is used extensively in telemetry and radio/TV broadcast applications where each sensor signal, representing for example, temperature, pressure and speed, is within a 0–1 V range, suitable for narrow band FM, and each has a final bandwidth of 4000 Hz. A basic signal of 0–4 kHz is called the 'baseband' signal. All of these baseband signals are multiplexed with various sub-carriers spaced 4 kHz apart with a bandwidth extending from 0 Hz to 4 kHz times the number of telemetry channels. Figure 7.26 gives an example of the division of a frequency spectrum.

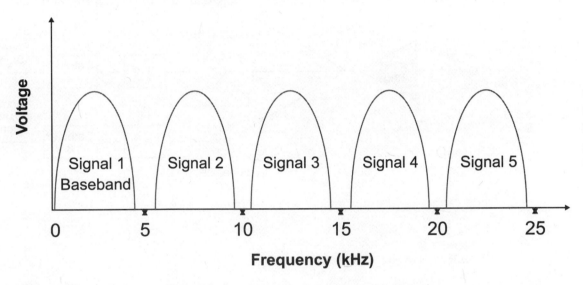

Figure 7.26
FDM spectrum containing five signals

Care should be taken to maximize the available total bandwidth. For example, using a 4 MHz bandwidth spacing to send 4 kHz bandwidth signals wastes (4000–4) kHz = 3996 kHz, using only 7% of the spectrum.

At the receiver, demultiplexing begins with translation of the multiplexed signal down to an intermediate frequency, with a local oscillator and mixer for each channel. The intermediate frequency bandwidth is set equal to the bandwidth of an individual baseband signal after modulation. A receiver that demultiplexes ten signals requires ten separate local oscillators and mixers, with each oscillator operating at the frequency appropriate for its intended signal.

Figure 7.27 is a block diagram of a basic FDM transmitter and receiver circuit.

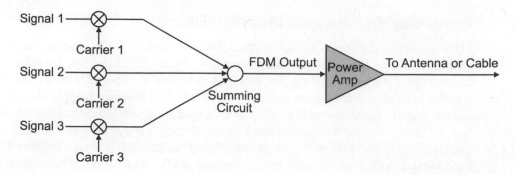

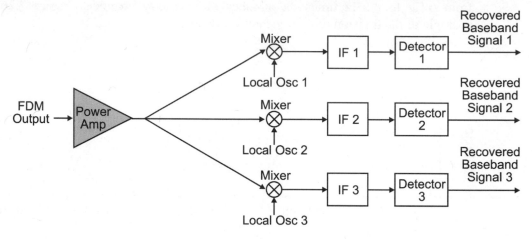

Figure 7.27
Basic FDM receiver and transmitter circuit

Time division multiplexing (TDM)

In a communications system, it is possible for the many users to time share the physical links by switching each signal for a short period of time, as in Figure 7.28. As the scanning rate rises, the system will eventually become ineffective because of an increase in the following:

- Propagation delays
- Noise
- Errors
- Retransmission

Unlike analog signals, digital TDM has greater latitude in sampling each bit of each channel. As long as samples occur within the bit period, even though it may be late or early in the specific bit period, no data will be lost. The greatest limitation of TDM lies in the bandwidth of the communications medium. As the bit rate increases, the frequency requirements of that medium also increase.

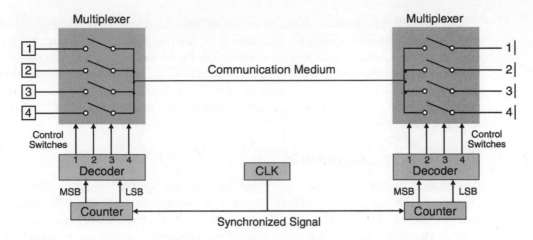

Figure 7.28
A basic TDM circuit

Practical example

The Modbus Plus bridge/multiplexer is a practical example of an alternative use of a multiplexer in TDM mode. This permits the connection of Modbus based devices to Modbus Plus networks. The example illustrates the use of a multiplexer in a broader sense, rather than maximizing the use of a communication channel. Figure 7.29 shows the NW-BM85-000 Modbus Plus bridge/multiplexer from AEG Modicon.

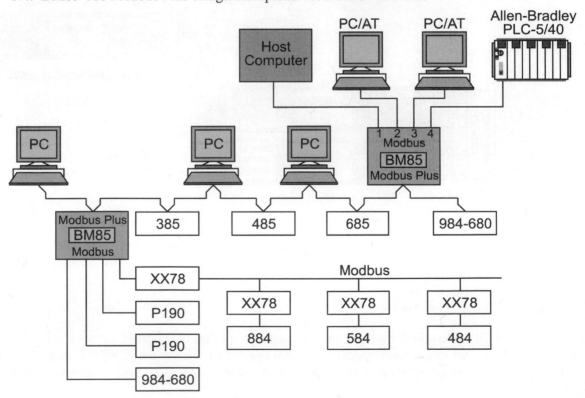

Figure 7.29
Example of use of BM85 multiplexer

This multiplexer permits interconnection of four Modbus ports and one Modbus Plus port. Configuration is possible by setting a few switches and doing some software configuration, which is then stored in an EEPROM. A buffer in the multiplexer allows up to 512 transactions to be queued. The software automatically connects the relevant port to its destination. Alternatively, it can be used as a four channel modbus port expander permitting connection of master devices to either individual slaves or several networks of modbus slave devices.

7.17   Terminal multiplexers

A terminal multiplexer employs TDM to connect groups of terminals to a central computer. The same multiplexer is used at both ends of the link with operation transparent to the users of the terminals. The microprocessor in each multiplexer continually polls each connected terminal for incoming characters. In using TDM, each terminal or UART in the multiplexer, gets a time slot. Each sequence of time slots is preceded by a synchronizing control character, such as SYN, so that the receiving multiplexer can determine the beginning of the next sequence of time slots. When no characters are received at the UART from a terminal, the multiplexer inserts a NULL character into the time slot for that terminal. With the additional overhead of this synchronizing character, it is important that the bit rate of the common data link connecting both terminal multiplexers is greater than the sum total of the bit rates of the terminals. Figure 7.30 is a block diagram of a terminal multiplexer.

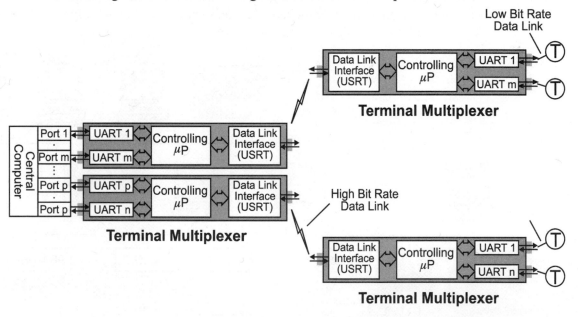

Figure 7.30
Terminal multiplexer

Terminal multiplexers can vary in performance from devices servicing up to four terminals each operating at 1200 bps with a throughput of 9600 bps, to 32 terminals each operating at up to 9600 bps with the common data link operating at well over 310 kbps.

7.18 Statistical multiplexers

A terminal multiplexer can be inefficient because time slots are allocated irrespective of whether or not a particular terminal requires them. A statistical multiplexer operates on the principle that not all terminals are active at a particular time and that inactive channels are 'skipped' until such time as they become active again. Each terminal has an identification tag, which is transmitted before a set of characters.

There are two methods of operation:

- The statistical multiplexer buffers a stream of characters from a particular terminal up to a pre-defined limit. It then prefixes this block of characters with a unique terminal identifier and appends an error checking frame check sequence before dispatching the message to a central computer, for example.
- The statistical multiplexer creates a block of characters consisting of groups of characters from all the active terminals. A unique +-terminal identifier is prefixed to each group of characters. An overall frame check sequence is appended to the block before it is dispatched to the central computer.

Figure 7.31 shows a block diagram of a statistical multiplexer.

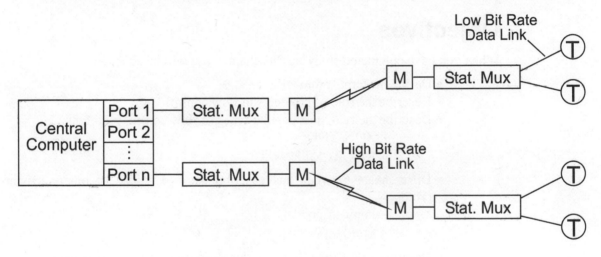

Figure 7.31
Statistical multiplexer

8

Introduction to protocols

A protocol can be defined as a set of rules governing the exchange of data between a transmitter and a receiver over a communications link or network.

Objectives

When you have completed studying this chapter you will be able to:

- Define the term 'protocol'
- Describe the role of protocols in flow control
- Describe the two most popular types of flow control protocols:
  - XON/XOFF
  - ETX/ACK
- Differentiate between and describe the two modes of binary synchronous protocols:
  - Point-to-point
  - Multipoint
- Describe HDLC and SDLC protocols in terms of:
  - Frame format
  - Frame contents
  - Operation
  - Error/flow control
- Describe file transfer protocols
- Describe ARQ protocols

As we will see in Chapter 9, data communications systems, which follow the OSI reference model, are made up of several hierarchical layers. Each of these layers contains working software or hardware elements referred to as entities. One of the elements in each layer is a protocol entity, which has its own specification, and broadly speaking is a protocol in its own right. The purpose of the protocol entity is to determine how messages are transferred across a network to a peer entity in another node.

Actual transfers across the physical link are determined by the data link layer protocol, and it is these types of protocols that are discussed in this chapter. Other more complex protocols are discussed in later chapters.

A protocol is concerned with some or all of the following:

- Initialization, to start the transmission of data.
- Framing and frame synchronization – this defines the beginning and end of a frame and ensures that the receiver can synchronize with the frame.
- Flow control, to ensure that the receiver is not swamped.
- Line control, applies to half duplex links in which the transmitter tells the receiver to start transmitting.
- Error control.
- Timeout control, so action can be taken if no acknowledgment is received within a certain period.

8.1 Flow control protocols

The most elementary protocols are only concerned with flow control and were introduced as an improvement on simple techniques such as the insertion of delays between characters or echoing of received characters to the transmitter. The two most popular types of flow control protocols are XON/XOFF and ETX/ACK.

8.2 XON/OFF

This is a character based flow-control protocol, which uses two special characters. Typically, these are the ASCII characters DC1 for XON and DC3 for XOFF. The transmitter sends data until it receives an XOFF from the receiver; it then waits for an XON before resuming transmission. A typical example can be found in a printer buffer. When the buffer reaches a certain point (say 66%), the printer sends an XOFF to the PC, then sends an XON when the buffer is emptied to another point (say 33%).

One disadvantage of XON/XOFF is that the data stream being sent may contain one of the control characters, although this is not a problem in applications such as printer control.

8.3 Binary synchronous protocol

The binary synchronous control (BSC) protocol was designed by IBM in 1966 for computer-to-terminal and computer-to-computer communications. It can be used in point-to-point or in multipoint mode. BSC is a character-based protocol as opposed to the high-level data link control (HDLC) protocol, which is bit-based. HDLC is discussed under 'HDLC and SDLC protocols' below.

The XON/XOFF flow control mechanism can easily handle short interactive messages between a terminal and a computer. They are less adequate in 'block mode', that is when passing complete messages, with hundreds or even thousands of characters, between terminals. The BSC protocol, on the other hand, was specifically designed to handle large blocks of data.

Control characters are used to separate the different fields in a BSC message and for the exchange of acknowledgment information.

Table 8.1 lists the control characters involved in BSC.

Abbreviation	Meaning	Description
ACK	Acknowledgment	Received block is OK
ACK1	Acknowledgment 1	Received odd block is OK
ACK2	Acknowledgment 2	Received even block is OK
DLE	Data Link Escape	Control character follows
ENQ	Enquiry	Please respond with data
EOT	End of Transmitted Data	The transmission is over
ETB	End of Transmission Block	End of block
ETX	End of Text	End of data text
ITB	End of Intermediate Block	More blocks coming
NAK	No Acknowledgment	Problems with received blocks
SOH	Start of Header	Routing information
STX	Start of Text	Message data text commences
SYN	Synchronous Idle	Allows receiver to synchronize

Table 8.1
BSC control characters

There are various types of BSC messages, as given in Figure 8.1.

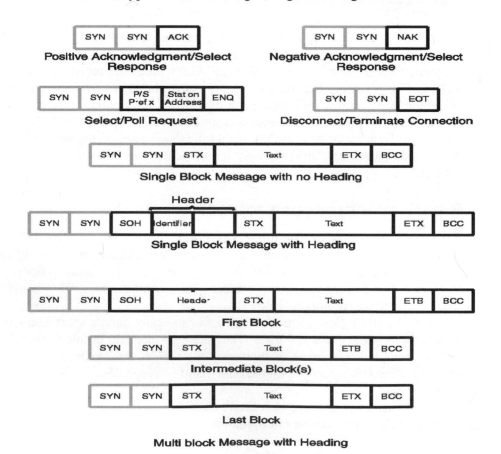

Figure 8.1
BSC message format

The receiver of the message uses the two SYN characters, (bit pattern 0010110) to synchronize with the start of the message. Note that the SYN characters are not considered to be part of the message itself and are therefore not used in the calculation of the block check character (BCC). In order to maintain synchronization, the transmitter inserts SYN characters into text messages once every second; however, as before, these are not used in BCC calculations.

A text field starts with the STX character and ends with ETX, ETB, EOT or ITB, as appropriate. The BCC field consists of a vertical/longitudinal check or a cyclic redundancy check such as the CRC-16 used for transparent mode (see Data Transparency later in this chapter).

If a message is received without error, the receiver responds with ACK1 for the first response, ACK2 for the next, and so on, alternating at each response so that odd numbered messages return ACK1 and even numbered messages return ACK2. This enables the transmitter to track the responses and detect any messages not acknowledged. ACK1 is represented by the sequence DLE00, and ACK2 is represented by the sequence DLE01.

If the receiver detects an error in a message, it responds with a NAK.

Point-to-point mode full sequence transmission is as follows:

Station 1	Station 2	
SYN SYN *		
	Three possible responses:	
	*	Ready to receive SYN SYN DLE 00
	*	Not ready to receive SYN SYN NAK
	*	Try again later SYN SYN NAK
SYN SYN STX [data chars] ETX BCC		
	*	Positive ackn/ACK0 SYN SYN DLE 00
Station 1 terminates the exchange * SYN SYN EOT .		

Table 8.2
Sequence of transmission in point-to-point mode

Multipoint mode

In this mode there is one primary station and one or more secondary stations on the same line.

All exchanges are initiated by the primary station as one of two types of transaction:

- The primary station issues a poll to determine if a secondary station has any data to transmit.
- The primary station selects a secondary station in order to transmit data to it.

Time out functions

- The transmitting station times out if there is no response within three seconds.

- A station on a switched network (the normal telephone system) disconnects itself if there is no activity for 20 seconds.
- A transmitter receiving a TTD or WAK character waits for two seconds before retrying.

Data transparency

- As an ASCII based protocol, BSC cannot usually operate transparently. That is, it is unable to handle binary data because the 7-bit data field limits the range of numbers to 0 to 127. Binary (or hexadecimal) data needs the range 0 to 225 (8 bits).
- It is possible to make BSC handle binary data by using eight data bits (and thus no parity bit) and preceding each BSC control character with a DLE. Thus ETX becomes DLE ETX, and so forth. In this way, control character sequences in the binary data will not be erroneously interpreted as BSC control characters. To avoid the possibility of problems with a DLE in the binary data, the transmitter inserts an additional DLE into the data when it detects a DLE. This is removed by the receiver before it passes on the data.
- Error checking uses a cyclic redundancy check (CRC) polynomial code in transparent operation because the eighth bit is not available for the parity calculation.

BSC limitations

- BSC is a half duplex protocol in which each message must be acknowledged by the receiver. This is very slow compared to more efficient protocols that number each message and send them out in multiples, requiring only an acknowledgment for the group. In transparent mode, the extra number of DLE characters that may be required is wasteful.

8.4   HDLC and SDLC protocols

HDLC has been defined by the International Standards Organization for use on both multipoint and point-to-point links. Other descriptions of it include SDLC (synchronous data link control used by IBM) and ADCCP (advanced data communication control procedure used by ANSI). HDLC will be the reference used throughout the following text. In contrast to the BSC protocol, HDLC is a bit-based protocol. It is interesting to note that it is a predecessor to the local area network datalink protocols.

The two most common modes of operation for HDLC are:

- Unbalanced normal response mode (NRM). This is used with only one primary (or master) station initiating all transactions.
- Asynchronous balanced mode (ABM). In this mode each node has equal status and can act as either a primary or secondary node.

Frame format

The standard format is indicated Figure 8.2. The three different classes of frames used are as follows:

Unnumbered frames: Used for setting up the link or connection and to define whether NRM or ABM is to be used. They are called unnumbered frames because no sequence numbers are included.

Information frames: Used to convey the actual data from one node to another.

Supervisory frames: Used for flow control and error control purposes. They indicate whether the secondary station is available to receive the information frames; they are also used to acknowledge the frames. There are two forms of error control used: a selective retransmission procedure because of an error, or a request to transmit a number of previous frames.

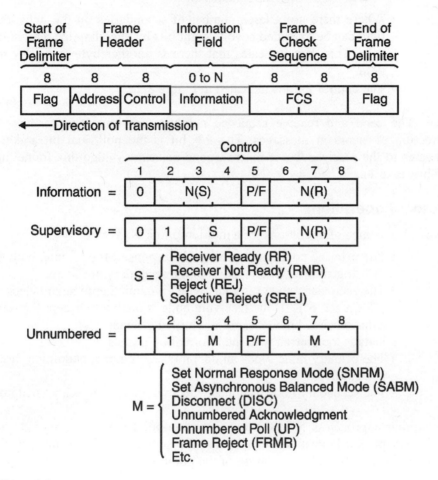

Figure 8.2
HDLC frame format and types

Frame contents

The frame contents are as follows:

- The flag character is a byte with the value 01111110. To ensure that the receiver always knows that the character it receives is unique (rather than merely some other character in the sequence); a procedure called zero insertion is adopted. This requires the transmitter to insert a 0 after a sequence of five 1s in the text, so that the flag character can never appear in the message text. The receiver removes the inserted zeros.

- The frame check sequence (FCS) uses the CRC-CCITT methodology, with sixteen 1s to the tail of the message before the CRC calculation proceeds, and the remainder is inverted.
- The address field can contain one of three types of address for the request or response messages to or from the secondary node:
  - Standard secondary address
  - Group addresses for groups of nodes on the network
  - Broadcast addresses for all nodes on the network (here the address contains all 1s)
- Where there are a large number of secondaries on the network, the address field can be extended beyond eight bits by encoding the least significant bit as a 1. This then indicates that there is another byte to follow in the address field.
- The control field is indicated in Figure 8.2.

Note: The send and receive sequence numbers are important for the detection and correction of errors in messages. The P/F bit is the poll/final bit and when set to 1 indicates to the receiver that it must respond or acknowledge this frame (again with the P/F bit set to 1).

Protocol operation

A typical sequence of operations for a multidrop link is given below:

1 The primary node sends a Normal Response Mode frame, with the P/F bit set to 1, together with the address of the secondary node.
2 The secondary node responds with an unnumbered acknowledgment with the P/F bit set to 1. If the receiving node is unable to accept the setup command, a disconnected mode frame is returned instead.
3 Data is transferred with the information frames.
4 The primary node sends an unnumbered frame containing a disconnect in the control field.
5 The secondary node responds with an unnumbered acknowledgment.

A similar approach is followed for a point-to-point link using asynchronous balanced mode, except that both nodes can initiate the setting up of the link and the transfer of information frames, and the clearing of the point-to-point link. The following differences also apply:

- When the secondary node transfers the data, it transmits the data as a sequence of information frames with the P/F bit set to 1 in the final frame of the sequence.
- In NRM mode, if the secondary node has no further data to transfer, it responds with a Receiver Not Ready frame with the P/F bit set to 1.

Error control/flow control

For a half duplex exchange of information frames, error control is by means of sequence numbers. Each end maintains a transmit sequence number and a receive sequence number. When a node successfully receives a frame, it responds with a supervisory frame containing a receiver ready (RR) indication and a receive sequence number. The number is that of the next frame expected, thus acknowledging all previous frames.

If the receiving node responds with a negative acknowledgment (REJ) frame, the transmitter must transmit all frames from the receive sequence number in the REJ frame. This happens when the receiver detects an out-of-sequence frame.

It is also possible for selective retransmission to be used. In this case the receiver would return a selection rejection frame containing only the sequence number of the missing frame.

A slightly more complex approach is required for a point-to-point link using asynchronous balanced mode with full duplex operation, where information frames are transmitted in two directions at the same time. The same philosophy is followed as for half duplex operation except that checks for correct sequences of frame numbers must be maintained at both ends of the link.

Flow control operates on the principle that the maximum number of information frames awaiting acknowledgment at any time is seven. If seven acknowledgments are outstanding, the transmitting node will suspend transmission until an acknowledgment is received. This can be either in the form of a receiver ready supervisory frame, or piggybacked in an information frame being returned from the receiver.

If the sequence numbers at both ends of the link become so out of sequence that the number of frames awaiting acknowledgment exceeds seven, the secondary node transmits a frame reject or a command reject frame to the primary node. The primary node then sets up the link again, and on an acknowledgment from the secondary node, both sides reset all the sequence numbers and commence the transfer of information frames.

It is possible for the receiver to run out of buffer space to store messages. When this happens it will transmit a receiver not ready (RNR) supervisory frame to the primary node to instruct it to stop sending any more information frames.

8.5   File transfer protocols

In most asynchronous file transfers used on PCs, the basic structure is the packet (or frame) consisting of a group of fields. Only one of these fields contains the actual data. The remaining fields, known as service fields, contain the information required for the receiver to verify that the packet is error free.

Automatic repeat request (ARQ) protocols

The most common type of packet protocol is the automatic repeat request (ARQ) protocol in which an error detected in a received packet and an unacknowledged packet automatically results in the retransmission of the packet.

Send and wait ARQ

Here the receiver inputs the packets and after verifying that the packet is in the correct sequence relative to the previous packet, computes a local check value on the data portion of the packet. On a successful match with the one in the packet the receiver acknowledges with an ACK; or sends a NAK. When the transmitter receives the ACK it then sends the next packet.

Continuous ARQ

Here the transmitter sends several packets in a row with no delay between packets. The receiver sends a NAK or ACK (as per the send and wait ARQ) together with the packet's number. The transmitter continually examines the stream of acknowledgments returning and keeps track of the packets with errors. At the end of the transmission the packets with errors are retransmitted.

The send and wait ARQ remains the most popular file transfer protocol found on PCs.

Packet design

There are three approaches for design of packets as shown in the figures below.

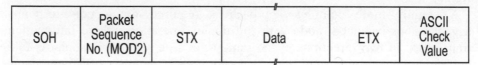

SOH	Packet Sequence No. (MOD2)	STX	Data	ETX	ASCII Check Value

Figure 8.3
Packet for transfer of text files (64 to 512 bytes)

SOH	Number Sequence	LEN	Data	Check Value

Figure 8.4
Packet for transfer of binary files

SOH	Number Sequence	LEN	Fixed Data Field	Check Value

Figure 8.5
Packet for fixed length data field packet

The two types of protocols that are used for file transfer are XMODEM and Kermit.

XMODEM

XMODEM is a simple send and wait ARQ protocol that uses a fixed length data field. The check value is a single byte arithmetic checksum.

SOH	Packet Number Sequence	1's Complement of Packet Sequence Number	Data (128 Bytes)	Arithmetic Checksum

Figure 8.6
Layout of XMODEM protocol

SOH	Start of header byte
Packet sequence No	Current packet number (up to 256) 1's complement of the current packet
Packet sequence	That 2's complement of the current packet number in the previous field
Data	Length of data (binary or text) is fixed at 128 bytes
Arithmetic checksum	A 1 byte arithmetic sum of the data field only – modulo 256

Table 8.3
Bytes in the XMODEM protocol

The mode of operation of the XMODEM protocol is briefly as follows:

- The receiver sends a NAK to transmitter to initiate transmission.
- The transmitter then sends a packetized 128 byte block of data. An ACK received by the transmitter commences transmission of the next packet. A NAK receipt means retransmit the packet; whilst a CAN aborts the transfer.
- When all the data has been sent, a sender transmits a solitary EOT, which the receiver acknowledges with an ACK.

Other versions of XMODEM use a 1 byte CRC instead of the single byte arithmetic checksum (XMODEM-CRC). This is a non-reversed CRC algorithm with the CCITT divisor polynomial $X^{16} + X^{12} + X^5 + 1$.

The problems with XMODEM

- The XMODEM protocol is designed to operate at lower error rates on communication links. Unless an in-built error-correction algorithm is used within the modem the XMODEM protocol will fail. Higher baud rates and lower quality telecommunication services (that provide a cheaper service) make the problem worse.
- Line noise with multiple bit changes often pass through XMODEM's simple error detection system undetected.
- The simple response control characters (ACK/NAK/EOT) are often corrupted to other codes resulting in erroneous actions taken with a resulting waste of time. If the characters are corrupted to the control-X character, file transfer is aborted.
- Sliding windows giving more efficiency in the use of the protocol cannot be implemented, as each NAK or ACK response character does not have an accompanying sequence number. This makes for inefficient operation over communication systems with long time delays.
- 8 bits are required in the construction of the characters in the packet. This makes it inappropriate for communication systems, which only allow 7 bits for each character.

The big advantage with XMODEM is that it is provided on all popular communication packages (such as the Windows Terminal package). This *de facto* standard is useful for transferring files between different incompatible computer systems, which have

XMODEM as the only common denominator. It should be noted however that the XMODEM is often merely then used to transfer the more efficient actual file transfer protocol (such as ZMODEM) from one machine to the other. This is then used to affect the actual file transfer. The result can be a huge saving in time and thus costs of communication time.

YMODEM

This was introduced as an improvement to the XMODEM protocol. This protocol has enjoyed wide spread acceptance as it was released into the public domain and it was written in C.

Features include:

- YMODEM has less overhead than XMODEM in its frame structure with 1024 bytes allowed to be transferred per block. YMODEM still has the flexibility to reduce the data block size to 128 bytes if the error rate is too high (causing repeated retransmissions).
- YMODEM is more reliable than XMODEM. In order to terminate a file transfer, the character sequence: control-X, control-X (or two CAN characters) must be received. This prevents unnecessary file transfer terminations due to errors on the line.
- CRC error detection is used to ensure a high level of error detection.
- File related information (file name; file time; file date; size) is also transferred to the receiving computer.
- Multiple files can be transferred using the batch file capability of the protocol.
- 8 data bits are the basis of file transfer. This causes some problems as discussed earlier with XMODEM.
- The stop and wait technique is used to send data packets. This lack of an adequate sliding window is the cause of inefficiencies on systems with significant time delays.

ZMODEM

This was developed to overcome the limitations in the previous file transfer protocols.
Typical features of the ZMODEM protocols are as follows:

- In addition to batch file transfer, which is a standard feature, a useful feature is that ZMODEM can recommence the file transfer at any point at which the communication link failed (to the precise byte at which communications were lost).
- ZMODEM will not perform a file transfer if the same file already exists on the receiver's hard disk. This check is performed automatically by ZMODEM. This saves on time.
- Better error detection and correction is performed by negotiating the use of the CRC-32 error detection mechanism.
- Data compression can be performed; thus speeding up file transfer.
- The sliding windows technique is used to improve performance over communication links with time delays. This size of the packet can be modified automatically by the software, downwards from 1024 bytes, to handle increasing noise on the communications link.

Kermit

The basic Kermit is a send and wait ARQ packet protocol. The sender transmits a packet and then awaits the receiver's acknowledgment of the packet. The receiver can then either request the next packet (ACK) or retransmission of the previous packet (NAK).

Although there are a number of similarities with XMODEM the main differences are:

- Kermit can transfer a number of files in one file transfer session.
- Packets may be of variable length.
- The I/O channels only have to transfer printable ASCII characters.
- Several types of packets are defined.
- The receiver's responses must consist of entire packets.
- Transmitter and receivers negotiate important operating parameters such as device padding, etc.
- Name of file is included in protocol.
- The herald packets make the protocol extensible.

The operating sequence of Kermit commences in a similar way to XMODEM where the receiver sends repeated NAK packets until the transmitter responds by sending a herald packet, which in Kermit is called a send initiate packet. The receiver makes its own preference known by including them in its ACK packet. When the entire file is transferred the sender transmits a special end of file packet. If there are more files to transmit, it sends the file header packet for the next file. When all files have been sent, the transmitter sends an end of transmission packet to signal the end of the session.

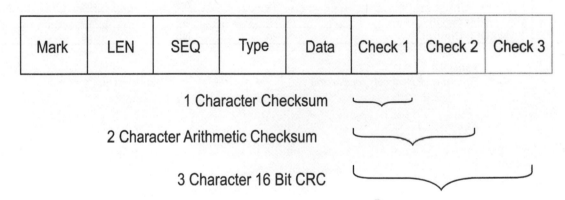

Figure 8.7
Kermit packet

Where:

Mark	This is the start of packet signature byte, SOH. This is the only canonical control bit permitted in the packet.
LEN	Number of bytes in the packet following this field.
SEQ	Characterized packet sequence number, Modulo-64.
Type	A single literal ASCII character identifying the type of packet (e.g. 'D' for data).
Data	The actual packet type determines the contents of this field.
Check 1	Packet check value.
Check 2	
Check 3	

Note: Kermit converts 'high risk' characters to printable characters before transmission. Control characters are thus moved to printable form before transmission and then transformed back after reception. High-risk characters are those that the computer system may elect to modify or throw away because of its system philosophy.

Description	XMODEM	YMODEM	ZMODEM	KERMIT
Types of File Transfer				
Single File	Yes	Yes	Yes	Yes
Batch of Files	No	Yes	Yes	Yes
Text/Data/Binary	Yes	Yes	Yes	Yes
Number of Data Bits	8	8	8	7 or 8
Data Compression	No	No	Yes	Yes
Error Checking	Checksum	Checksum/ CRC-CCITT	CRC-CCITT/ CRC-32	Checksum/ CRC-CCITT
Error Checking Response	Single char.	Single char.	Single char.	Error checked pkt
Data Packet Size (bytes)	128	1024	1024	0-95
File Attribute Transfer				
File Size/Time/ Date Stamp	No	Yes	Yes	Yes
Sliding Window Support	Single pkt	Single pkt	1 - more pkts	1 - more pkts
Negotiated Parameters				
Packet Size	No	Yes	Yes	Yes
Packets in Window	No	No	Yes	Yes
Error Check Type	No	Yes	Yes	Yes
Data Compression	No	No	Yes	Yes
XON/XOFF Control	No	No	Yes	Yes

Table 8.4
File transfer protocols comparison

9

Open systems interconnection model

The purpose of the open systems interconnection reference model is to provide a common basis for the development of systems interconnection standards. An open system is a system that conforms to specifications and guidelines, which are 'open' to all.

Objectives

When you have completed studying this chapter you will be able to:

- Describe data communications for instrumentation and control
- List and describe the seven layers of the OSI reference model
- Give an example of an OSI model application

9.1   Data communications for instrumentation and control

In digital data communications wiring together two or more devices is one of the first steps in establishing a network. As well as this hardware requirement, software must also be addressed. Where all the devices on a network are from the same manufacturer, any hardware and software problems are usually overcome easily because the system is generally designed within the same guidelines and specifications. However, it is where devices from different manufacturers are used on the same network that the problems seem to begin.

Networks which use devices from one manufacturer and that work with specific hardware connections and protocols are called 'closed systems'. Most of these networks were developed before standardization or when it was thought unlikely that equipment from other manufacturers would be included in the network.

In contrast, 'open systems' are those that conform to specifications and guidelines, which are 'open' to all. This enables equipment from various manufacturers, who claim compliance to a standard, to be used interchangeably on a network using that standard.

In 1978, faced with a proliferation of closed network systems, the International Standards Organization (ISO) defined a 'reference model for communication between open systems'. This is known as the open systems interconnection (OSI) reference model, or more correctly as the ISO/OSI Model (ISO 7498). The model can be applied to all communication systems, from personal computers to satellite systems.

The ISO defines the purpose of the OSI reference model as:

'.... providing a common basis for the coordination of standards development for the purpose of systems interconnection, while allowing existing standards to be placed into perspective within the overall reference model.'

OSI is essentially a 'management structure', which simplifies data communications into a hierarchy of seven layers. Each layer has a defined purpose, which is dependent on, and interfaces with, the layer above and below it. Standards are defined for each layer in a way, which allows some flexibility enabling system designers to develop independent protocol layers. Any two or more of these layers together are referred to as a 'protocol stack'.

It is important to realize that the OSI reference model is not a protocol or set of rules dictating how a protocol should be written but an overall framework in which to define protocols. The OSI model framework specifically and clearly defines the functions or services that have to be provided at each of the seven layers (or levels).

The OSI reference model consists of the following seven layers:

• Layer 1, physical layer	Electrical and mechanical definition of the system
• Layer 2, data link layer	Framing and error correction format of data
• Layer 3, network layer	Optimum routing of messages from one network to another
• Layer 4, transport layer	Channel for transfer of messages of one application process to another
• Layer 5, session layer	Organization and synchronization of the data exchange
• Layer 6, presentation layer	Data format or representation
• Layer 7, application layer	File transfer, message exchange

The OSI model can be visualized as a collection of entities, such as software programs or hardware integrated circuits, situated at each of the seven layers. Data in a network is exchanged in packet form, each packet originating at a source node and addressed to a destination node. In effect, a packet starts at an upper layer, and passes down through each of the layers. As a packet moves down from one layer to another, it is enclosed in a 'protocol envelope'. Each envelope, therefore, encloses the message data and any protocol envelope from the layer above. The protocol envelope carries addressing and control information that advises the next layer down what to do with the packet, and the lower layer only reads that information.

When a packet reaches layer 1, that is the physical layer, it is sent across a physical communications link to the next node en route to its destination. At the receiving node, the packet moves up the protocol stack, losing the outer protocol envelope at each layer. In effect, the receiving node's network layer, for example, sees the packet as coming directly from the transmitting node's network layer because it has the appropriate envelope. For this reason, equivalent layers in different nodes are called 'peer entities' and there is said to be a 'virtual link' between them.

Figure 9.1 shows that a packet needs to go only as far as Layer 3, the network layer, on an intermediate node on its route. This is because the network layer protocol envelope contains all the information required for routing the packet along its journey.

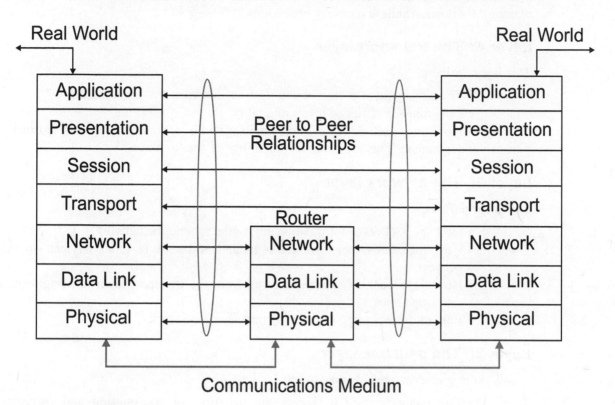

Figure 9.1
The layers of the OSI model

9.2   Individual OSI layers

Layer 7: The application layer

The application layer is the top most layer in the OSI/RM. It is responsible for giving applications access to the network. Examples of application layer tasks include file transfer, e-mail services and network management.

   To accomplish the tasks, the application layer passes program requests and data to the presentation layer, which is responsible for encoding the application layer's data in the appropriate form.

Layer 6: The presentation layer

This layer maps the data representations into a suitable format for the recipient. It translates the format and syntax of the data produced by applications, and handles encryption or compression of the data.

Layer 5: The session layer

This layer manages interactive sessions between devices. It defines the use of software that allows reference to other devices by name rather than binary address (logical naming). It allows seamless recovery of sessions.

Layer 4: The transport layer

This layer defines:

- The management of the communications between two end systems.
- Data transfer at agreed levels of quality.
- How the packets in a large message are accounted for and correctly segmented. This improves the reliability of data transfer.

Layer 3: The network layer

This layer defines:

- How the packets of information are routed around the network.
- How the status messages are regulated and sent to other devices on the network.
- How large packets of data received from the transport layer are fragmented into small ones.
- Enough frames to go through the underlying network.

Layer 2: The data link layer

This layer is always needed and defines:

- The methods used to access the network for transmitting and receiving messages.
- The handling of the information received and its acknowledgment.
- The procedures for managing flow control between users.

Layer 1: The physical layer

This layer is always needed and defines the physical connections between the computer and the network. It provides Layer 2, the Data Link Layer, with the physical means of transmitting data onto the network.

This layer is concerned with the following:

- The network topology
- The electrical aspects of voltages and currents
- The signal modulation techniques
- The mechanical aspects of the connection, i.e. the cables and connectors

Note: It stipulates, but does not include the actual medium.

9.3   OSI analogy

The following is a simple analogy to better explain the OSI model.

A French-speaking manager in her Paris office wants to send a letter containing a request to an English-speaking colleague in an office in Chicago. The French manager merely dictates the letter to her secretary who translates it into English. The French manager is like the application layer, while the secretary represents the presentation layer.

The letter is passed to a clerk who records its details in a file or correspondence with the Chicago company, making sure that details of title and address for the Chicago manager are correct. The clerk represents the session layer.

The shipping manager then receives the letter(s) and makes arrangements for the delivery and quality of service promised for the letter(s). He also makes a copy of the letter in case it has to be sent out again. He represents the transport layer.

Next in line is a shipping clerk who establishes the route for the letter. She decides that the best route for the letter is via the company's Boston office. She represents the network layer. The letter is then passed to the mailroom where it is weighed together with other letters going to Chicago by courier. The weight is labeled on the courier bag to ensure that there are no discrepancies at the other end. The mailroom is the data link layer.

The mailbag is then dropped off at the shipping dock, which acts as the physical layer.

Once the letter is received in Chicago the above process is repeated, but in reverse order.

9.4   An example of an industrial control application

The following is an example of the sequence of steps necessary to transfer a command, such as 'change the set point' on a remote programmable logic controller output:

- The user selects a command via the menu of a user program located on an operator station.
  For example, 'change the set point for a remote valve from 20% to 95%'.
- The menu-driven system software then instructs a command within the application layer program to make the necessary change. This layer can be visualized as being similar to the DOS interface on a computer, with a group of high level commands, acting as a high level interface.
- The application layer passes the message to the presentation layer, which translates it into a system usable form. For example, it could translate from the type coding used by the application layer into an ASCII based format suitable for the system.
- The session layer allows the two application protocol entities (at the operator station and the remote programmable logic controller) to synchronize, setup and manage the data interchange between them. This allows different requests to be queued and transmitted in an orderly manner.
- The transport layer forms the interface between the higher-level OSI layers and the lower network and data link layers. It shields the higher layers from the detailed operation of the lower layers. The transport layer provides various classes of service, for example adding the routing address onto the message received from the session layer.
- The message is passed onto the network layer, which arranges the detailed optimum routing, which the message is to follow across the network to arrive at the remote programmable logic controller.
- The message is then passed onto the data link layer, which calculates the frame check sequence for error checking and adds in source and destination addresses.
- The physical layer then modulates the physical connection with the appropriate bit pattern received from the data link layer.

The whole process is then repeated in reverse at the programmable logic controller.

9.5   Simplified OSI model

For many industrial protocols the use of the full seven layers of the OSI model is inappropriate as the application may require a high-speed response. Hence a simplified OSI model is often preferred for industrial applications where time critical communications is more important than full communications functionality provided by the seven-layer model.

Generally, most industrial protocols are written around three layers:

- The physical layer
- The data link layer
- The application layer

When the reduced OSI model is implemented the following limitations exist:

- As there is no transport layer, the maximum size of the application messages is limited by the maximum size allowed on the channel
- As there is no network layer, no routing of messages is possible   between different networks
- As there is no session layer, no full duplex communications are possible
- As there is no presentation layer, message formats must be the same for all nodes

The MiniMap and Fieldbus protocol standards use the reduced three layer OSI model. Similarly, other industrial protocols such as the Allen Bradley Data Highway Plus protocol, Modbus Plus and the HART smart instrumentation protocols have all standardized on the three layers only.

One of the challenges when using the OSI model is the concept of interoperability and the need to define another layer called the 'user' Layer. This topic is examined in Chapter 12.

10

Industrial protocols

The industrial protocols discussed in this chapter vary from a straightforward ASCII type protocol to the industry standard Modbus protocol. A fairly sophisticated Allen Bradley Data Highway Plus protocol is also reviewed. This chapter focuses on the software aspects of the protocols (as opposed to the physical aspects which are covered in separate chapters).

Objectives

When you have completed study of this chapter you will be able to:

- Describe the features of industrial protocols
- Describe the use of ASCII based protocols
- Describe the read and write commands of ANSI-X3.28-2.5-A4
- List and describe the three Modbus structures
- Describe the Modbus protocol:
    - Message format
    - Synchronization
    - Memory location
    - Function codes
    - Exception responses
- Describe the Allen Bradley Data Highway protocol
- Describe the Allen Bradley Data Highway Plus protocol
- Describe the OSI model layers used by the Allen Bradley Data Highway Plus protocol
- Describe the application of MAP/TOP protocols

10.1   Introduction

In some respects the distinction between an industrial and commercial (or data processing) protocol is somewhat artificial. There are, however, a few features contained in an industrial protocol, which can make it useful to an engineer on a plant.

These are:

- Ease of troubleshooting systems
  Where the level of understanding of industrial communication systems on a plant may be fairly low it makes sense to select a simple protocol such as one of the ASCII protocols.

- High level of integrity of data transfer
  In an industrial environment where there is electrical noise and no errors in the transfer of the data are acceptable (due to the communications link controlling critical equipment for example), a protocol should be selected with a high degree of error checking such as cyclic redundancy checks.

- Standardization of protocol
  There may be a requirement to interface to other manufacturers PLCs or industrial systems. In this case a commonly accepted industrial protocol such as Modbus would be appropriate.

- High speed update of parameters
  There may be a necessity to update a number of setpoints to a series of controlling devices virtually simultaneously. Here one of the new FieldBus protocols may be appropriate to ensure that there is no skew (or delays) between transferring the setpoint to the first and last devices on the data highway.

10.2   ASCII based protocols

ASCII based protocols are popular because of their simplicity. Their main disadvantage is that they are slow and become unwieldy for larger systems with the requirement for multiple nodes on a network that need to communicate with one another (rather than in a simple one master, multiple slave setup).

Consequently, ASCII based protocols are normally only used for slow systems with one master talking to a limited number of slaves (preferably only one).

ASCII based protocols are also popular for stand-alone instruments where a serial interface has been added on with no major design changes to the existing design. Essentially this means that the additional serial port is treated like another keypad by the instrument.

Although ASCII based protocols would appear to be the simplest; in the author's experience they have proved problematic in their implementation because of the lack of a tight definition by the particular manufacturer.

Two ASCII based protocol implementations are given below. The first is for the implementation of smart transmitters and the other for a variable speed drive. The smart transmitter is a fairly simple protocol structure whilst the ANSI-X3.28-2.5-A4 is a slightly more complex approach.

ASCII based protocols for digital transmitters

A variety of digital signal transmitters has recently appeared on the market, which accept a variety of sensor and process inputs and communicate the data back to the serial port of a computer or other processor based device in a digital format. Data is also sent down from the computer to the signal transmitter for controlling devices (via a digital or analog output from the conditioner unit). The RS-232 or RS-485 standards are used for communications between the signal transmitter and computer.

Each digital transmitter is a complete single channel interface system with analog signal conditioning electronics optimized for a specific input type. The analog input signals are digitized by an analog to digital (A/D) converter whilst the analog output signals are converted from their digital form by a digital to analog converter. All data is stored in an ASCII format in a buffer where contents can be updated about eight times per second. The host computer may transmit or request data from or to the transmitter by sending simple ASCII commands to it.

There are many variations on the standard transmitter available such as for high frequency inputs, digital inputs and outputs, thermocouple and RTD (resistance, temperature, dependent) inputs and analog outputs.

Communications hardware

The RS-232 standard is used as a point-to-point communication system but these transmitters can be setup with multiple units 'hanging off' the same RS-232 communications port. However, as the RS-232 standard does not allow a multi-node system, the units are daisy chained as indicated in the Figure 10.1. In this network any characters transmitted by the host computer are received by each transmitter in the chain and passed on to the next until a transmitter recognizes its address and then transmits a response which is rippled back through other transmitters in the chain.

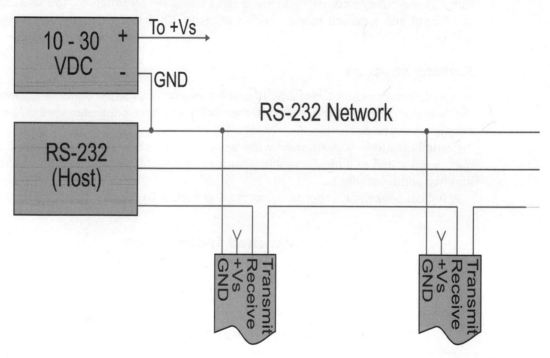

Figure 10.1
RS-232 Communications for smart transmitters

The RS-485 standard is used in a half duplex fashion for multidropped systems. If more than 32 modules are required to hang off the same RS-485 port, then an RS-485 repeater module is required to boost the signal and supply power to the additional modules.

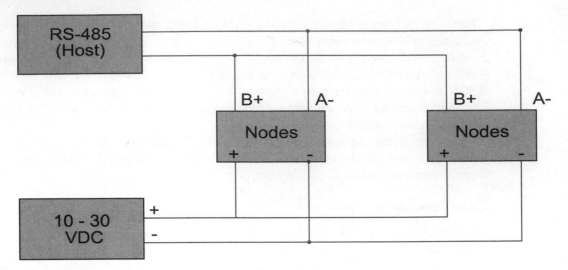

Figure 10.2
RS-485 Communications for smart transmitters

The transmitter module contains an EEPROM (electrically erasable programmable read only memory) to store setup information and calibration constants. As the communications parameters (such as baud rate) could be forgotten by the user, the module can be put into a default mode where it resets to 300 baud, no parity and recognizes any address.

Protocol structure

A simple command/response ASCII based protocol is used for communications between the host computer and the transmitter module. The host computer always generates the command sequence.

Communications is performed with two character ASCII command codes. All analog data is requested as a nine-character string consisting of a sign, five digits, decimal point and two additional digits.

A typical command/response is indicated in Figure 10.3.

Command from Host

$	1	R	D	[CR]

Response from the Transmitter Module

*	+	0	0	2	7	5	.	0	0	[CR]

Figure 10.3
Short form command and response messages

This command reads from transmitter at address 1 and receives a value of 275.00 in the response message.

The maximum length of the command and response messages is 20 printable characters (i.e. non ASCII control characters).

A variation on the short form command and response messages are the long form which is used to ensure greater response message integrity and echoes the command message and appends a block checksum at the end of the message. The long form command is initiated using a # in place of the $ signifying the commencement of the command message. Note that the two-character checksum can be optionally added to all command messages at the host computer's discretion.

A checksum is simply the sum of the hexadecimal values of all the ASCII characters in the message.

Command from Host

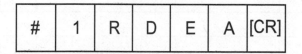

Response from the Transmitter Module

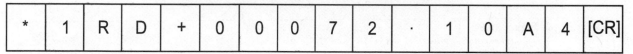

Figure 10.4
Long form command and response messages

Note: The calculation of the checksum for the response is performed as follows:

ASCII Character	Hex Value	Binary Value
*	2A	0101010
1	31	0110001
R	52	1010010
D	44	1000100
+	2B	0101011
0	30	0110000
0	30	0110000
0	30	0110000
7	37	0110110
2	32	0110001
.	2E	0101110
1	31	0110001
0	30	0110000
SUM	2A4	

Table 10.1
Calculation of checksum

Discard the 2 and add the A4 onto the end of the message.

Note that the A and 4 being hexadecimal characters have to be sent as their ASCII equivalents.

Errors

If the transmitter module indicates that it has received a message with an error in it, it will respond with '?' character. Alternatively, there may be no response at all if an incorrect address or command prompt has been used.

Typical error responses are illustrated in Figure 10.5.

?	1	[SP]	B	A	D	[SP]	C	H	E	C	K	S	U	M	[CR]

?	1	[SP]	S	Y	N	T	A	X	[SP]	E	R	R	O	R	[CR]

Note: [SP] is an ASCII space character

Figure 10.5
Typical error responses

10.3   ASCII based protocol ANSI-X3.28-2.5-A4

ANSI-X3.28-2.5-A4 is an example of an ASCII based protocol. One particular manufacturer of variable speed drives, Control Techniques, uses this protocol to communicate from a programmable controller or personal computer to up to 32 drives. Generally the RS-485 standard is preferred in the implementation of this protocol.

Overall approach

The ANSI-X3.28-2.5-A4 standard defines the character format and sequence of characters in a message. The typical structure adopted in the RS-485 standard is:

- A 10 bit character consisting of 1 start bit, 7 bit ASCII Code, even or no parity, 1 stop bit
- Baud rate selectable between 300 and 19 200 bps
- Up to 32 drives or slaves permitted on the network

There are two types of commands:

- READ command from the PC to the specific drive requesting information on a specific parameter
- A WRITE command from the PC to the specific drive to change READ/WRITE parameters

The read command

The read command and its response have the format shown in Figure 10.6.

Read Request Frame (Pass 1)

EOT (^D) Reset Data Link	ADD Address Field	PAR Parameter Field	ENQ (^E) Enquiry
1 Byte	4 Bytes	3 Bytes	1 Byte

Read Response Frame (Pass 1)

STX (^B) Start of Text	PAR Parameter Field	DATA Data Field	ETX (^C) End of Text	BCC Block Checksum
1 Byte	3 Bytes	6 Bytes	1 Byte	1 Byte

Figure 10.6
Format of READ request and its response (Pass 1)

The ASCII characters used in the READ request message are:

- EOT (1 character)
  This resets all the devices connected to the serial link

- ADD (4 characters)
  The protocol allows addressing up to 32 devices. Each character in the address is transmitted twice, to ensure data integrity. For example if drive address is 14, ADD = 1144

- PAR (3 characters)
  This consists of the parameter information, which is in the range of 0–999. For bit parameters, add 100 to the bit number

- ENQ (1 character)
  Termination of message

The ASCII characters used in the READ response message are:

- STX (1 character)
  This indicates to the master the start of the reply

- PAR (3 characters)
  As above

- DATA (6 characters)
  The first character consists of the polarity (or a space if this is irrelevant). A maximum of four digits and a decimal point form the remaining characters.

- ETX (1 character)
  This indicates to the master that the data is finished.

- BCC (1 character)
  The block checksum forms the error checking mechanism. This is calculated by doing an exclusive OR calculation on the parameter number, the data and the ASCII character EOT. Sometimes the BCC is disabled on the equipment and a CR is only returned.

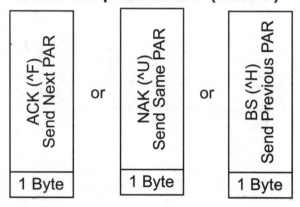

Figure 10.7
Format of READ command and the responses (Pass 2)

The message interchange between the master and slave can be continued after the first READ request and response frames by:

- Stepping forward to the next (sequential) parameter by sending an ACK

or

- Stepping backwards to the previous (sequential) parameter by sending a BS *or*

- Requesting the same parameter again by sending a NAK to the slave

This can be repeated indefinitely (until the end of the parameter list). Sending of EOT resets all the devices on the network.

The write command

Write Request Frame (Pass 1)

EOT (^D)	ADD	STX (^B)	PAR	DATA	ETX (^C)	BCC
Reset Data Link	Address Field	Start of Text	Parameter Field	Data Field	End of Text	Block Checksum
1 Byte	4 Bytes	1 Byte	3 Bytes	6 Bytes	1 Byte	1 Byte

Write Response Frame (Pass 1)

ACK (^F)		NAK (^U)
Data Received	or	Invalid Message
1 Byte		1 Byte

Figure 10.8
Format of the write command request and its response

The same descriptions as for the READ command apply. In addition the slave will return the NAK character if the drive parameter, data or BCC is in error.

If it is requested to write further data to the drive, the sequence outlined in Figure 10.9 should be followed.

Write Request Frame (Pass 2)

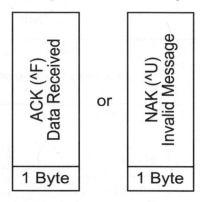

STX (^B) Start of Text	PAR Parameter Field	DATA Data Field	ETX (^C) End of Text	BCC Block Checksum
1 Byte	3 Bytes	6 Bytes	1 Byte	1 Byte

Write Response Frame (Pass 2)

ACK (^F) Data Received	or	NAK (^U) Invalid Message
1 Byte		1 Byte

Figure 10.9
Format of WRITE

10.4   Modbus protocol

General overview

Modbus transmission protocol was developed by Gould Modicon (now AEG) for process control systems. In contrast to the many other buses discussed, no interface is defined.

The user can therefore choose between RS-422, RS-485 or 20 mA current loops, all of which are suitable for the transmission rates, which the protocol defines.

Although the Modbus is relatively slow in comparison with other buses, it has the advantage of wide acceptance among instrument manufacturers and users. About 20 to 30 manufacturers produce equipment with the Modbus protocol and many systems are in industrial operation. It can therefore be regarded as a *de facto* industrial standard with proven capabilities. A recent survey in the well-known American *Control Engineering* magazine indicated that over 40% of industrial communication applications use the Modbus protocol for interfacing.

Besides the standard Modbus protocol, there are two other Modbus protocol structures:

- Modbus Plus
- Modbus II

The most popular one is Modbus Plus. It is not an open standard as the classical Modbus has become. Modbus II is not used much due to additional cabling requirements and other difficulties.

The Modbus is accessed on the master/slave principle, the protocol providing for one master and up to 247 slaves. Only the master initiates a transaction.

Transactions are a query/response type where only a single slave is addressed, or a broadcast/no response type where all slaves are addressed. A transaction comprises a single query and single response frame or a single broadcast frame.

Certain characteristics of the Modbus protocol are fixed, such as frame format, frame sequences, handling of communications errors and exception conditions and the functions performed. Other characteristics are selectable. These are transmission medium, transmission characteristics and transmission mode, RTU or ASCII. The user characteristics are set at each device and cannot be changed when the system is running.

The Modbus protocol provides frames for the transmission of messages between master and slaves. The information in the message is the address of the intended receiver, what the receiver must do, the data needed to perform the action and a means of checking errors. The slave reads the messages, and if there is no error it performs the task and sends a response back to the master. The information in the response message is the slave address, the action performed, the result of the action and a means of checking errors. If the initial message was of broadcast type, there is no response from the slaves.

Normally, the master can send another query as soon as it has received the response message. A timeout function ensures that the system still functions when the query is not received correctly.

Data can be exchanged in two transmission modes:

- ASCII – readable, used e.g. for testing
- RTU – compact and faster; used for normal operation (Hex)

The RTU mode (sometime also referred to as Modbus-B for Modbus binary) is the preferred Modbus mode and will be discussed in this section. The ASCII transmission mode has a typical message, which is about twice the length of the equivalent RTU message.

The Modbus also provides an error check for transmission and communication errors. Communication errors are detected by character framing, a parity check, a redundancy check or CRC. The latter varies depending on whether the RTU or ASCII transmission mode is being used.

Modbus functions

All functions supported by the Modbus Protocol are identified by an index number. They are designed as control commands for field instrumentation and actuators and are as follows:

- Coil control commands for reading and setting a single coil or a group of coils
- Input control commands for reading input status of a group of inputs
- Register control commands for reading and setting one or more holding registers
- Diagnostics test and report functions
- Program functions
- Polling control functions
- Reset

Protocol specifics

This section reviews the Modbus protocol in detail and is broken down into the following sections:

- Message format
- Synchronization
- Memory location
- Function codes
- Exception responses

Message format

A transaction consists of a single request from the host to a specific secondary device and a single response from that device back to the host. Both of these messages are formatted as Modbus message frames. Each such message frame consists of a series of bytes grouped into four fields as described in the following paragraphs. Note that each of these bytes indicated here are in Hex format (not ASCII).

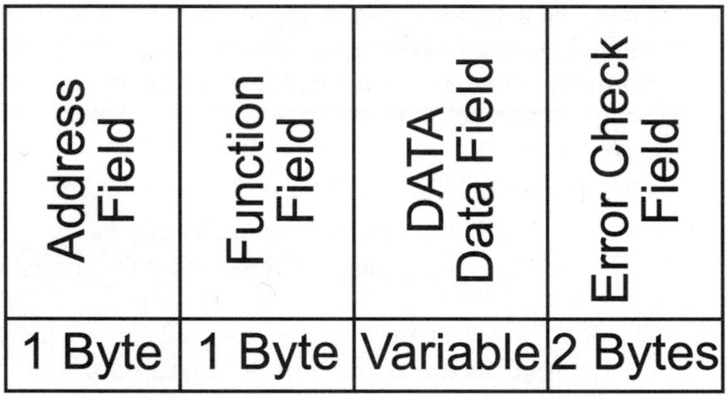

Address Field	Function Field	DATA Data Field	Error Check Field
1 Byte	1 Byte	Variable	2 Bytes

Figure 10.10
Format of Modbus message frame

The first field in each message frame is the address field, which consists of a single byte of information. In request frames, this byte identifies the controller to which the request is being directed. The resulting response frame begins with the address of the responding device. Each slave can have an address field between 1 and 247, although practical limitations will limit the maximum number of slaves. A typical Modbus installation will have one master and two or three slaves.

The second field in each message is the function field, which also consists of a single byte of information. In a host request, this byte identifies the function which the target PLC is to perform.

If the target PLC is able to perform the requested function, the function field of its response will echo that of the original request. Otherwise, the function field of the request will be echoed with its most-significant bit set to one, thus signaling an exception response. Table 10.2 summarizes the typical functions used.

The third field in a message frame is the data field, which varies in length according to which function is specified in the function field. In a host request, this field contains information the PLC may need to complete the requested function. In a PLC response, this field contains any data requested by that host.

The last two bytes in a message frame comprise the error-check field. The numeric value of this field is calculated by performing a cyclic redundancy check (CRC-16) on the message frame. This error checking ensures that devices do not react to messages that may have been changed during transmission.

Synchronization

In order to achieve reliable communication, the reception of a message must be synchronized with its transmission. In other words, the receiving device must be able to identify the start of a new message frame. Under the Modbus RTU protocol, frame synchronization is established by limiting the idle time between successive characters within a message frame. If three character times (approximately three milliseconds) elapse without the receiving device detecting a new character, the pending message will be flushed. The next byte will then be interpreted as the address field of a new message line.

Memory notation

The memory notation allows for four different data types: coils, discrete inputs, input registers and holding registers. Register variables consist of two bytes, while coils and discrete inputs are single bytes.

Each function references only one type of data. This allows message-frame memory references to be expressed as hexadecimal offsets relative to the lowest possible address for that data type. For example, the first holding register (40001) is referenced as 0000.

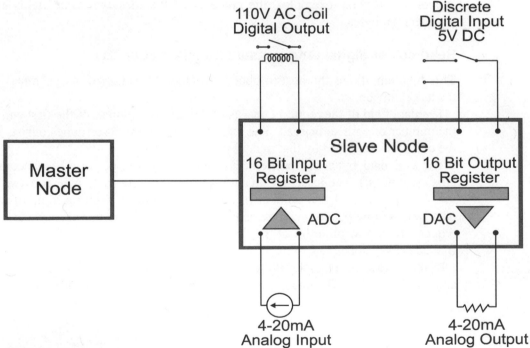

Figure 10.11
Diagram illustrating Modbus PLC notation

Table 10.2 lists the address range and offsets for these four data types, as well as the function codes, which apply to each. The diagram above also gives an easy reference to the Modbus data types.

Data Type	Absolute Addresses	Relative Addresses	Function Codes	Description
Coils	00001 to 09999	0 to 9998	01	Read Coil Status
Coils	00001 to 09999	0 to 9998	05	Force Single Coil
Coils	00001 to 09999	0 to 9998	15	Force Multiple Coils
Discrete Inputs	10001 to 19999	0 to 9998	02	Read Input Status
Input Registers	30001 to 39999	0 to 9998	04	Read Input Registers
Holding Registers	40001 to 49999	0 to 9998	03	Read Holding Register
Holding Registers	40001 to 49999	0 to 9998	06	Preset Single Register
Holding Registers	40001 to 49999	0 to 9998	16	Preset Multiple Registers
-	-	-	07	Read Exception Status
-	-	-	08	Loopback Diagnostic Test

Table 10.2
Modicon addresses and function codes

Function codes

Each request frame contains a function code that defines the action expected for the target controller. The meaning of the request data fields is dependent on the function code specified.

The following paragraphs define and illustrate most of the popular function codes supported. In these examples, the contents of the message-frame fields are shown as hexadecimal bytes.

Read coil or digital output status (function code 01)

This function allows the host to obtain the ON/OFF status of one or more logic coils in the target device.

The data field of the request consists of the relative address of the first coil followed by the number of coils to be read. The data field of the response frame consists of a count of the coil bytes followed by that many bytes of coil data.

The coil data bytes are packed with one bit for the status of each consecutive coil (1=ON, 0=OFF). The least significant bit of the first coil data byte conveys the status of the first coil read. IF the number of coils read is not an even multiple of eight, the last data byte will be padded with zeros on the high end. Note that if multiple data bytes are requested, the low order bit of the first data byte in the response of the slave contains the first addressed coil.

In the following example, the host requests the status of coils 000A (decimal 00011) and 000B (00012). The target device's response indicates both coils are ON.

Request Message

Address	Function Code	Initial Coil Offset		Number of Points		CRC
		Hi	Lo	Hi	Lo	
01	01	00	0A	00	02	9D C9

Response Frame

Address	Function Code	Byte Count	Coil Data	CRC
01	01	01	03	11 89

Figure 10.12
Example of read coil status read digital input status (Function Code 02)

This function enables the host to read one or more discrete inputs in the target device.

The data field of the request frame consists of the relative address of the first discrete input followed by the number of discrete inputs to be read. The data field of the response frame consists of a count of the discrete input data bytes followed by that many bytes of discrete input data.

The discrete-input data bytes are packed with one bit for the status of each consecutive discrete input (1=ON, 0=OFF). The least significant bit of the first discrete input data byte conveys the status of the first input read. If the number of discrete inputs read is not an even multiple of eight, the last data byte will be padded with zeros on the high end. The low order bit of the first byte of the response from the slave contains the first addressed digital input.

In the following example, the host requests the status of discrete inputs hexadecimal offsets 0000 and 0001 (i.e. decimal 10001 and 10002). The target device's response indicates that discrete input 10001 is OFF and 10002 is ON.

Request Message

Address	Function Code	Initial Coil Offset		Number of Points		CRC
		Hi	Lo	Hi	Lo	
01	02	00	00	00	02	F9 CB

Response Frame

Address	Function Code	Byte Count	Input Data	CRC
01	02	01	02	20 49

Figure 10.13
Example of read input status

Read holding registers (function code 03)

This function allows the host to obtain the contents of one or more holding registers in the target device.

The data field of the request frame consists of the relative address of the first holding register followed by the number of registers to be read. The data field of the response time consists of a count of the register data bytes followed by that many bytes of holding register data.

The contents, of each requested register, are returned in two consecutive register-data bytes (most significant byte first).

In the following example, the host requests the contents of holding register hexadecimal offset 0002 or decimal 40003. The controller's response indicates that the numerical value of the register's contents is hexadecimal 07FF or decimal 2047. The first byte of the response register data is the high order byte of the first addressed register.

Request Message

Address	Function Code	Starting Register		Register Count		CRC	
		Hi	Lo	Hi	Lo		
01	03	00	02	00	01	25	CA

Response Frame

Address	Function Code	Byte Count	Register Data		CRC	
			Hi	Lo		
01	03	02	07	FF	FA	34

Figure 10.14
Example of reading holding register

Reading input registers (function code 04)

This function allows the host to obtain the contents of one or more input registers in the target device.

The data field of the request frame consists of the relative address of the first input register followed by the number of registers to be read. The data field of the response frame consists of a count of the register-data bytes followed by that many bytes of input-register data.

The contents, of each requested register, are returned in two consecutive register-data bytes (most-significant byte first). The range for register variables is 0 to 4095.

In the following example, the host requests the contents of input register hexadecimal offset 000 or decimal 30001. The PLC's response indicates that the numerical value of that register's contents is 03FFH, which would correspond to a data value of 25 per cent (if the scaling of 0 to 100 per cent is adopted) and a 12-bit A/D converter with a numerical range of 0 to 4095 (0FFFH) is used.

Request Message

Address	Function Code	Starting Register		Register Count		CRC
		Hi	Lo	Hi	Lo	
01	04	00	00	00	01	31 CA

Response Frame

Address	Function Code	Byte Count	Register Data		CRC
			Hi	Lo	
01	04	02	03	FF	F9 80

Figure 10.15
Example of reading input register

Force single coil (function code 05)

This function allows the host to alter the ON/OFF status of a single logic coil in the target device.

The data field of the request frame consists of the relative address of the coil followed by the desired status for that coil. A hexadecimal status value of FF00 will activate the coil, while a status value of zero (H) will deactivate it. Any other status value is illegal.

If the controller is able to force the specified coil to the Requested state, the response frame will be identical to the request. Otherwise an exception response will be returned.

If the address 00 is used to indicate broadcast mode, all attached slaves will modify the specified coil address to the state required.

The following example illustrates a successful attempt to force coil 11 (decimal) OFF.

Request Message

Address	Function Code	Coil Offset		New Coil Status		CRC	
		Hi	Lo	Hi	Lo		
01	05	00	0A	00	00	ED	C8

Response Frame

Address	Function Code	Coil Offset		New Coil Status		CRC	
		Hi	Lo	Hi	Lo		
01	05	00	0A	00	00	ED	C8

Figure 10.16
Example of forcing a single coil

Preset single register (function code 06)

This function enables the host to alter the contents of a single holding register in the target device.

The data field of the request frame consists of the relative address of the holding register followed by the new value to be written to that register (most-significant byte first).

If the controller is able to write the requested new value to the specified register, the response frame will be identical to the request. Otherwise, an exception response will be returned.

The following example illustrates a successful attempt to change the contents of holding register 40003 to 3072 (0C00 Hex).

When slave address is set to 00 (broadcast mode), all slaves will load the specified register with the value specified.

Request Message

Address	Function Code	Register Offset		Register Value		CRC
		Hi	Lo	Hi	Lo	
01	06	00	02	0C	00	2D 0A

Response Frame

Address	Function Code	Register Offset		Register Value		CRC
		Hi	Lo	Hi	Lo	
01	06	00	02	0C	00	2D 0A

Figure 10.17
Example of presetting a single register

Read exception status (function code 07)

This is a short message requesting the status of eight digital points within the slave device.

This will provide the status of eight predefined digital points in the slave. For example this could be items such as the status of the battery, whether memory protect has been enabled or the status of the remote input/output racks connected to the system.

Request Message

Address	Function Code	CRC
11	07	.. ..

Response Frame

Address	Function Code	Coil Station	CRC
11	07	02	.. ..

Figure 10.18
Read exception status query message

Loopback test (function code 08)

The objective of this function code is to test the operation of the communications system without affecting the memory tables of the slave device. It is also possible to implement additional diagnostic features in a slave device (should this be considered necessary) such as number of CRC errors, number of exception reports etc.

The most common implementation will only be considered in this section; namely a simple return of the query message.

Request Frame

Address	Function Code	Data Diagnostic Code		Data		CRC
		Hi	Lo	Hi	Lo	
11	08	00	00	A5	37	.. ..

Response Frame

Address	Function Code	Data Diagnostic Code		Data		CRC
		Hi	Lo	Hi	Lo	
11	08	00	00	A5	37	.. ..

Figure 10.19
Loopback test message

Force multiple coils or digital outputs (function code 0F)

This forces a contiguous (or adjacent) group of coils to an ON or OFF state. The following example sets 10 coils starting at address 01 Hex (at slave address 01) to the ON state. If slave address 00 is used in the request frame broadcast mode will be implemented resulting in all slaves changing their coils at the defined addresses.

Request Frame

Address	Function Code	Address		Byte Count	Data Coil Status		CRC
		Hi	Lo		Hi	Lo	
01	0F	00	01	0F	FF	03	.. ..

Response Frame

Address	Function Code	Address		Number of Coils		CRC
		Hi	Lo	Hi	Lo	
01	0F	00	01	00	0A	.. ..

Figure 10.20
Example of forcing multiple coils

Force multiple registers (function code 10)

This is similar to the preset single register and the forcing of multiple coils. In the example below, a slave address 01 has 2 registers changed commencing at address 4011.

Request Frame

Address	Function Code	Address		Quantity		Byte Count	First Register		Second Register		CRC
		Hi	Lo	Hi	Lo		Hi	Lo	Hi	Lo	
01	10	00	0A	00	02	04	00	0A	01	02	.. ..

Response Frame

Address	Function Code	Address		Quality		CRC
		Hi	Lo	Hi	Lo	
01	10	00	0A	00	02	.. ..

Figure 10.21
Example of presetting multiple registers

Exception responses

Request frames containing parity or checksum errors are ignored – no response is sent by any device. If an otherwise valid request frame contains an illegal request (one not supported by the target slave unit), an exception response will be returned to the host.

The four fields of an exception response contain:

- The address of the responding controller
- The requested function number with its most-significant bit set to one
- An appropriate exception code
- The CRC-16 checksum

Table 10.3 lists the most important exception codes, which may be returned.

Code	Name	Description
01	Illegal Function	Requested function is not supported.
02	Illegal Data Address	Requested data address is not supported
03	Illegal Data Value	Specified data value is not supported
04	Failure in Associated Device	Slave PLC has failed to respond to a message
05	Acknowledge	Slave PLC is processing the command
06	Busy, rejected message	Slave PLC is busy

Table 10.3
Abbreviated list of exception codes returned

An example of an illegal request and the corresponding exception response is shown below. The request in this example is to READ COIL STATUS of points 514 to 521 (eight coils beginning an offset 0201H). These points are not supported in this PLC, so an exception report is generated indicating code 02, illegal address.

Request Message

Address	Function Code	Starting Point	Number of Points	CRC
01	01	02 01	00 08	6D B4

Exception Response Message

Address	Function Code	Exception Code	CRC
01	81	02	C1 91

Figure 10.22
Example of an illegal request

10.5   Allen Bradley Data Highway (Plus) protocol

Overview of Allen Bradley protocol

There are two main protocol standards used in Allen Bradley data communications:

The Data Highway protocol

This is a local area network (LAN) that allows peer-to-peer communications up to 64 nodes. It uses a half duplex (polled) protocol and rotation of link mastership. It operates at 57.6 kBaud.

The Data Highway Plus protocol

This is similar to the Data Highway Network although designed for fewer PCs and operates at a data rate of 57.6 kbaud. This has peer-to-peer communications with a token passing scheme to rotate link mastership among the nodes connected to that link.

Note that both protocol standards implement peer-to-peer communications through a modified token passing system called the floating master. This is a fairly efficient mechanism as each node has an opportunity to become a master at which time it can immediately transmit without checking with each mode for the requisite permission to commence transmission.

The Allen Bradley Data Highway Plus uses the three layers of the OSI layer model:

- Hardware (a physical layer)
- Data link layer protocol
- Application layer protocol

Physical layer (hardware layer)

This is based on twin axial cable with three conductors essentially in line with the RS-485 specifications.

Full duplex data link layer

Note that the asynchronous link can use either a full duplex (unpolled) protocol or a master slave communication through a half duplex (unpolled) protocol. Although both types of protocols are available the tendency today is to use the full duplex protocol as this explains the high performance nature of the link. Hence this protocol will be examined in more detail in the following sections.

Full duplex protocol is character orientated. It uses the ASCII control characters listed in the following table, extended to eight bits by adding a zero for bit number seven (i.e. the eighth bit).

The following ASCII characters are used:

Abbreviation	Hex Value
STX	02
ETX	03
ENQ	05
ACK	06
DLE	10
NAK	11

Table 10.4
ASCII characters used

Full duplex protocol combines these characters into control and data symbols. The following table lists the symbols used for full duplex implementation.

Symbol	Type	Description
DLE STX	control symbol	Sender symbol that indicates the start of a message.
DLE ETX BCC/CRC	control symbol	Sender signal that terminates a message.
DLE ACK	control symbol	Response symbol which signals that a message has been successfully received.
DLE NAK	control symbol	Response symbol which signals that a message was not received successfully.
DLE ENQ	control symbol	Sender symbol that requests retransmission of a response symbol from the receiver.
APP DATA	data symbol	Single character data values between 00-OF and 11-FF. Includes data from application layer including user programs and common application routines.
DLE DLE	data symbol	Symbol that represents the data value 10 Hex.

Table 10.5
Symbols used for full duplex mode

Format of a message

Note that response symbols transmitted within a message packet are referred to as embedded responses.

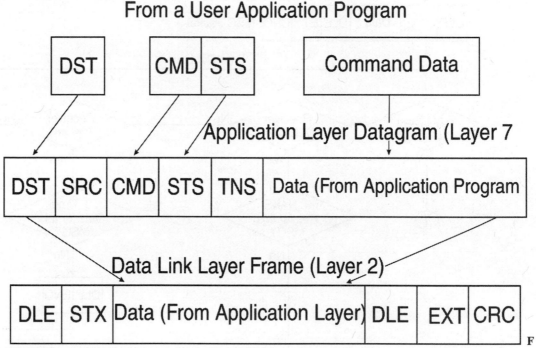

igure 10.23
Protocol structure

The CRC-16 calculation is done using the value of the application layer data bytes and the ETX byte. The CRC-16 result consists of two bytes. Refer to section 8.4 for more information on the cyclic redundancy check mechanism.

Note that to transmit the data value of 10H, the sequence of data symbols DLE DLE must be used. Only one of these DLE bytes and no embedded responses are included in the CRC value.

Message limitations

- Minimum size of a valid message is six bytes.
- Duplicate message detection algorithm – receiver compares the second, third, fifth and sixth bytes of a message with the same bytes in the previous message.

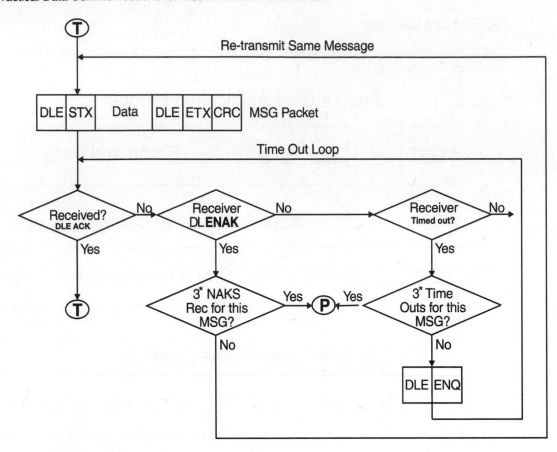

Figure 10.24
Software logic for transmitter

P	=	Recovery procedure
T	=	Ready to transmit next message
*	=	Default values used by module

Depending on the highway traffic and saturation level, there may be a wait for a reply from the remote node before transmitting the next message.

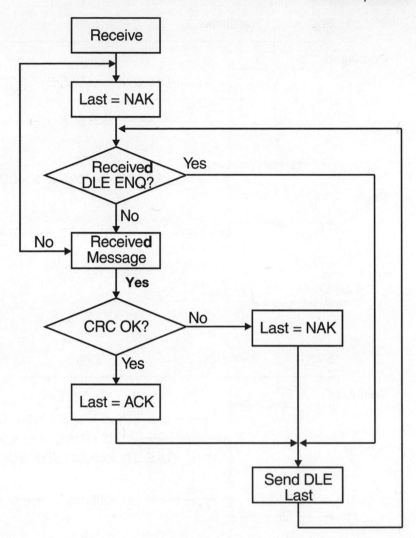

Figure 10.25
Software logic for receivers

Software logic for receivers

The following diagrams show typical events that occur in the communications process.

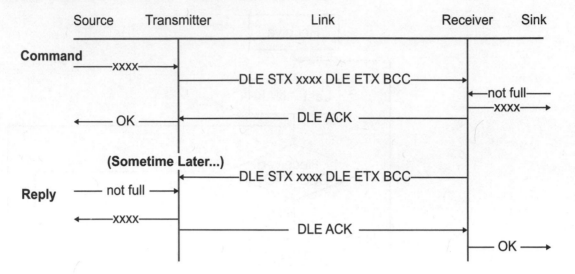

Figure 10.26
Normal message transfer

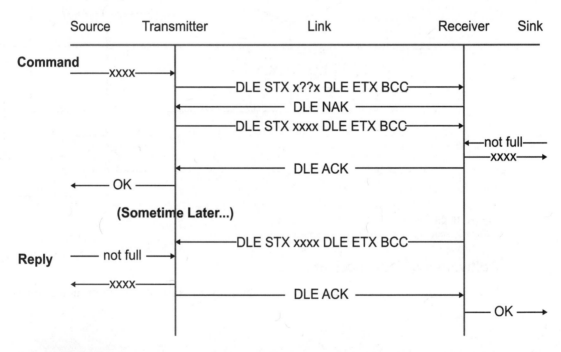

Figure 10.27
Message transfer with NAK

There are two types of application programs:

- Command initiators
- Command executors

Command initiators specify which command function to execute at a particular remote node.

The command executor must also issue a reply message for each command it receives. If the executor cannot execute a command it must send the appropriate error code.

The reply message may contain an error. The command initiator must check for this condition and depending on the type of error, retransmit the message or notify the user.

If the command executor reply is lost due to noise, the command initiator should maintain a timer for each outstanding command message. If the timer expires the command initiator should take appropriate action (notify the user or retransmit to executor).

If the application layer software cannot deliver a command message, it should generate a reply message with the appropriate error code and send the reply to the initiator. If it cannot deliver a reply message, the application layer should destroy the reply without notification to the command executor.

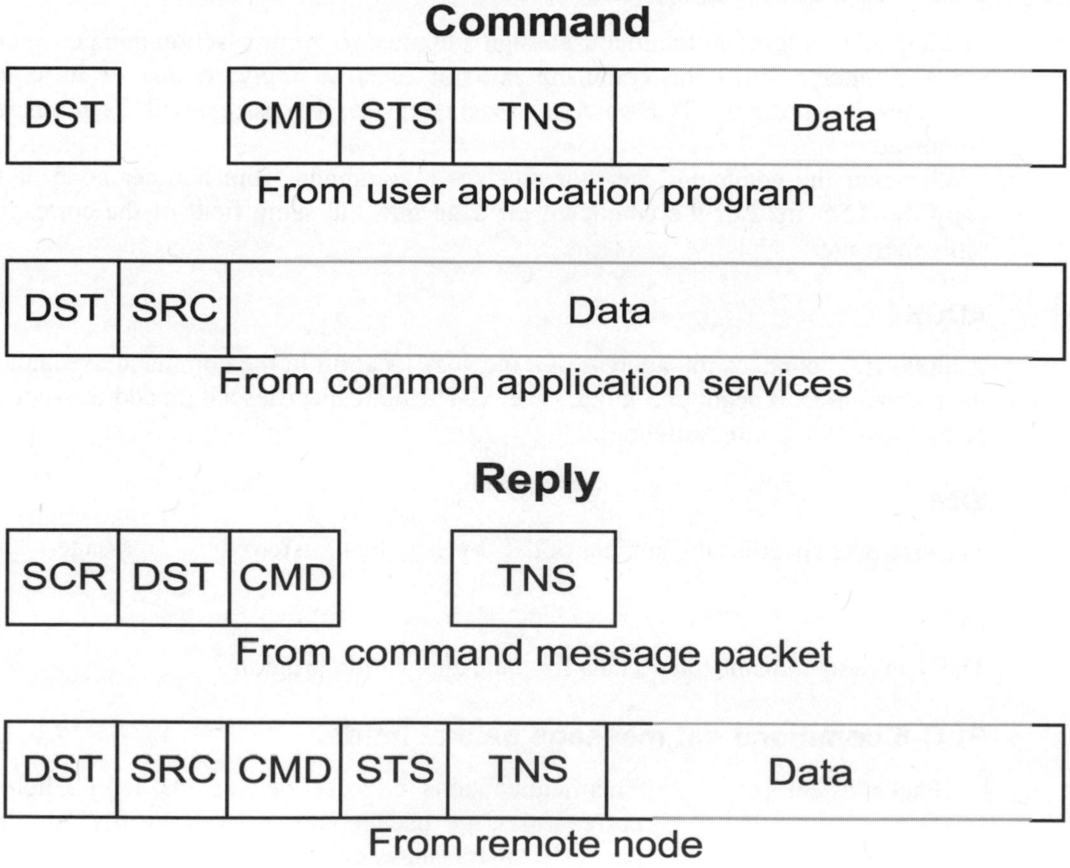

Command

From user application program

From common application services

Reply

From command message packet

From remote node

Figure 10.28
Basic command set message packet fields

Note that not all command messages have FNC, ADR, SIZE or DATA bytes. Not all reply messages have DATA or EXT STS bytes.

Explanation of Bytes

DST   Destination byte. This contains the ultimate destination of the node.

SRC   Source node of the message. Set this to zero as the KF2 interface module will set the byte to its own node number.

CMD   Command byte.

FNC   Function byte.

These together define the activity to be performed by the command message at the destination node. Note that bit five of the command byte shall always be zero (normal priority).

STS and EXT SYS – status and extended status bytes

In command messages the STS byte is set to zero. In reply messages the STS byte may contain a status code. If the four high bits of the STS byte are ones, there is extended status information in an EXT STS byte.

TNS – transaction bytes (two bytes)

The application level software must assign a unique 16 bit transaction number (generated via a counter). When the command initiator receives reply to one of its command messages, it can use the TNS value to associate the reply message with its corresponding command.

Whenever the command executor receives a command from another node, it should copy the TNS field of the command message into the same field of the corresponding reply message.

ADDR

Address field contains the address of a memory location in the command executor where the command is to begin executing. The ADDR field specifies a byte address (not a word address as in PLC programming).

Size

The size byte specifies the number of data bytes to be transferred by a message.

Data

The data field contains binary data from the application programs.

PLC-5 command set message packet fields

1 Packet offset: This field contains the offset between the DATA field of the current message packet and the DATA field of the first packet in the transmission.

2 Total trans: This field contains the total number of PLC-5 data elements transferred in all message packets initiated by a command.

Basic command set

The asynchronous link message packet formats to be used are delivered below:

In the lists below, privileged commands are initiated by computer and executed by PLCs. Non-privileged commands are initiated by a PLC or a computer. The CMD values listed are for non-priority command message packets.

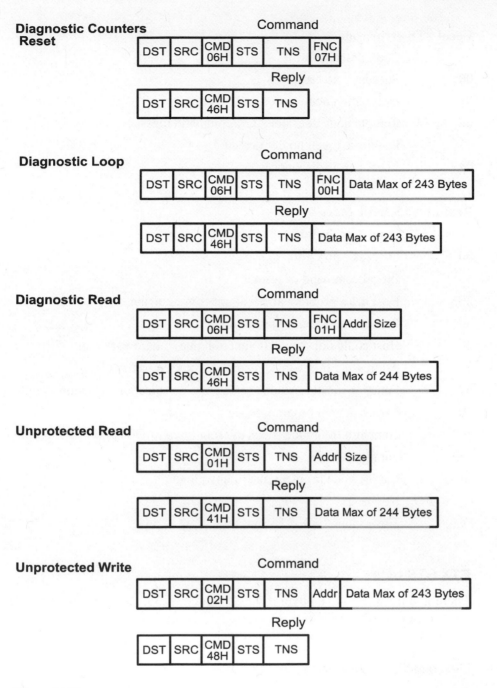

Figure 10.29
Basic command set for PLC-5

Synchronous link status code (STS, EXT STS)

The TS bytes provide information about the execution or failure of the corresponding command that was transmitted from the computer. If the reply returns a code of 00, the command was executed at the remote node. All other codes are divided into two types:

- Local error – local node is unable to transmit a message to the remote node.
- Remote error – remote node is unable to execute the command.

Local STS error code

	Code description
00	Success – no error
01	Destination node out of buffer space
02	Remote node does not ACK command message
03	Duplicate token holder detected
04	Local port is disconnected

Remote STS error codes

	Code description
00	Success – no errors
10	Illegal command or format
20	Host has a problem and will not communicate
30	Remote node is missing, disconnected
40	Host could not complete function due to hardware fault
50	Addressing problem or memory protect rungs
60	Function disallowed due to command protection selection
70	Processor is in program mode
80	Compatibility mode file is missing or communication zone problem
AO	Not used
BO	Remote node problem due to download
CO to EO	Not used
FO	Error in the ETX STS byte

ETX STS byte

There is only an EXT STS byte if the STS code is FO. If the command code is 00 to 08, there is not an EXT STS byte. Commands used in this implementation are in this range; hence the EXT STS byte is not being used.

Diagnostic counter for each module

Diagnostic counters are bytes of information stored in RAM in each Data Highway and Data Highway Plus module. When using the diagnostic read command a dummy value should be used for the address. The reply contains the entire counter block.

11

HART protocol

The highway addressable remote transducer (HART) protocol is one of a number of smart instrumentation protocols designed for collecting data from instruments, sensors, and actuators by digital communication techniques.

Objectives

When you have completed studying this chapter you will be able to:

- Describe the origin and benefits of the HART protocol
- Describe the three OSI layers of the HART protocol

11.1 Introduction to HART and smart instrumentation

Smart (or intelligent) instrumentation protocols are designed for applications where actual data is collected from instruments, sensors, and actuators by digital communication techniques. These components are linked directly to programmable logic controllers (PLCs) and computers.

The HART (highway addressable remote transducer) protocol is a typical smart instrumentation Fieldbus that can operate in a hybrid 4–20 mA digital fashion.

HART is, by no means, the only protocol in this sphere. There are hundreds of smart implementations produced by various manufacturers – for example Honeywell, which compete with HART. This chapter deals specifically with HART. For information about the other Fieldbus protocols refer to Chapter 12.

At a basic level, most smart instruments provide core functions such as:

- Control of range/zero/span adjustments
- Diagnostics to verify functionality
- Memory to store configuration and status information (such as tag numbers etc.)

Accessing these functions allows major gains in the speed and efficiency of the installation and maintenance process. For example, the time consuming 4–20 mA loop check phase can be achieved in minutes, and the device can be readied for use in the process by zeroing and adjustment for any other controllable aspects such as the damping value.

11.2   Highway addressable remote transducer (HART)

This protocol was originally developed by Rosemount and is regarded as an open standard, available to all manufacturers. Its main advantage is that it enables an instrumentation engineer to keep the existing 4–20 mA instrumentation cabling and to use simultaneously the same wires to carry digital information superimposed on the analog signal. This enables most companies to capitalize on their existing investment in 4–20 mA instrumentation cabling and associated systems; and to add the further capability of HART without incurring major costs.

HART is a hybrid analog and digital protocol, as opposed to most Fieldbus systems, which are purely digital.

The HART protocol uses the frequency shift keying (FSK) technique based on the Bell 202 communications standard. Two individual frequencies of 1200 and 2200 Hz, representing digits 1 and 0 respectively, are used. The average value of the sine wave (at the 1200 and 2200 Hz frequencies), which is superimposed on the 4–20 mA signal, is zero. Hence, the 4–20 mA analog information is not affected.

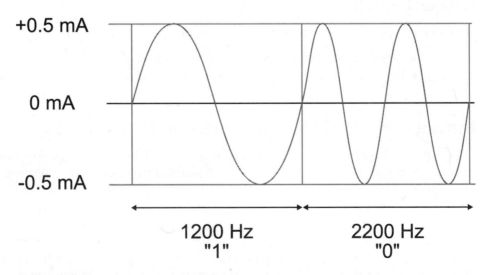

Figure 11.1
Frequency allocation of HART protocol

The HART protocol can be used in three ways:

- In conjunction with the 4–20 mA current signal in point-to-point mode
- In conjunction with other field devices in multidrop mode
- In point-to-point mode with only one field device broadcasting in burst mode

Traditional point-to-point loops use zero for the smart device polling address. Setting the smart device polling address to a number greater than zero creates a multidrop loop.

The smart device then sets its analog output to a constant 4 mA and communicates only digitally.

The HART protocol has two formats for digital transmission of data:

- Poll/response mode
- Burst (or broadcast) mode

In the poll/response mode the master polls each of the smart devices on the highway and requests the relevant information. In burst mode the field device continuously transmits process data without the need for the host to send request messages. Although this mode is fairly fast (up to 3.7 times/second) it cannot be used in multidrop networks.

The protocol is implemented with the OSI model (see Chapter 9) using layers 1, 2 and 7. The actual implementation is covered in this chapter.

11.3 Physical layer

The physical layer of the HART Protocol is based on two methods of communication.

- Analog 4–20 mA
- Digital frequency shift keying (FSK)

Analog 4 to 20 mA communications

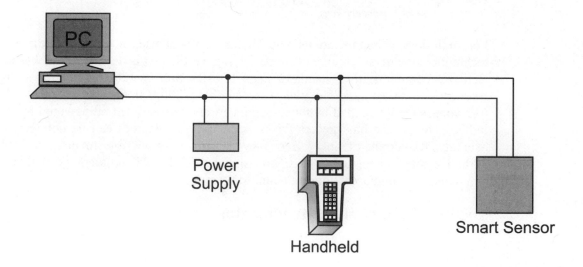

Figure 11.2
HART point-to-point communications

The basic communication of the HART protocol is the 4–20 mA current system. This analog system is used by the sensor to transmit an analog value to the HART PLC or HART card in a PC. In a 4–20 mA the sensor outputs a current value somewhere between 4 and 20 mA that represents the analog value of the sensor. For example, a water tank that is half full – say 3400 kilolitres – would put out 12 mA. The receiver would interpret this 12 mA as 3400 kilolitres. This communication is always point-to-point, i.e. from one device to one other. It is not possible to do multidrop communication using this method alone. If two or more devices put some current on the line at the same time, the resulting current value would not be valid for either device.

Digital multidrop communications

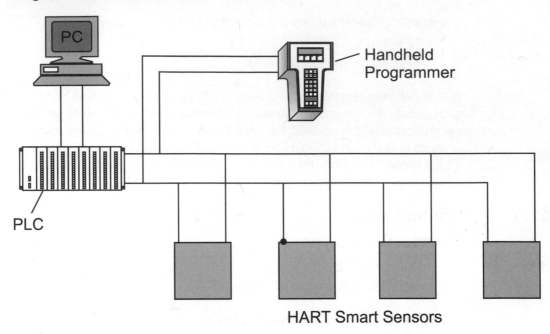

PC

Handheld
Programmer

PLC

HART Smart Sensors

Figure 11.3
HART multi-point communications

For multidrop communications, the HART protocol uses a digital/analog modulation technique known as frequency shift keying (FSK). This technique is based on the Bell 202 communication standard. Data transfer rate is 1200 baud with a digital '0' frequency (2200 Hz) and a digital '1' frequency (1200 Hz). Category 5 shielded, twisted pair wire is recommended by most manufacturers. Devices can be powered by the bus or individually. If the bus powers the devices, only 15 devices can be connected. As the average DC current of an ac frequency is zero, it is possible to place a 1200 Hz or 2200 Hz tone on top of a 4–20 mA signal. The HART protocol does this to allow simultaneous communications on a multidrop system.

The HART handheld communicator

Figure 11.4
HART handheld controller

The HART system includes a handheld control device. This device can be a second master on the system. It is used to read, write, range and calibrate devices on the bus. It can be taken into the field and used for temporary communications. The battery operated handheld has a display and key input for specific commands.

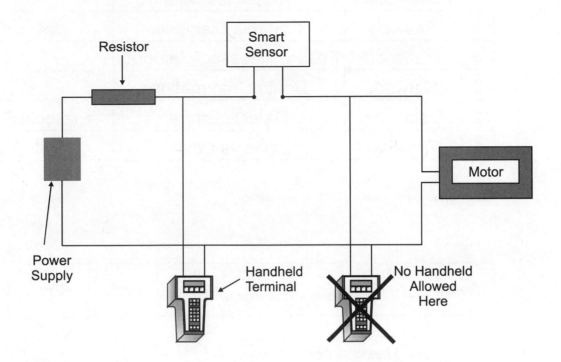

Figure 11.5
HART handheld connection method

The HART field controller in Figure 11.5 is wired in series with the field device (valve positioner or other actuator). In some cases, a bypass capacitor may be required across the terminals of the valve positioner to keep the positioner's series impedance below the 100 Ω level required by HART specifications. Communications with the field controller requires the communicating device (handheld terminal or PC) to be connected across a loop impedance of at least 230 Ω. Communications is not possible across the terminals of the valve positioner because of its low impedance (100 Ω). Instead, the communicating device must be connected across the transmitter or the current sense resistor. (Taken from the HART applications guide by the HART Communications Foundation 1999 *www.hartcomm.org*.)

11.4   Data link layer

The data link frame format is shown in Figure 11.7.

	Layer	Description	HART ™
7	Application	Serves up formatted data	Hart Commands
6	Pesentation	Translates Data	
5	Session	Controls Dialogue	
4	Transport	Ensures Message Integrity	
3	Network	Routes Information	
2	Data Link	Detects Errors	Protocol Rules
1	Physical	Connects Device	Bell 202

Figure 11.6
HART protocol implementation of OSI layer model

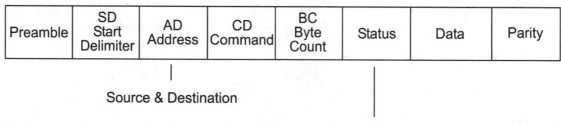

Figure 11.7
HART data link frame format

Two-dimensional error checking, including both vertical and longitudinal parity checks, is implemented in each frame. Each character or frame of information has the following parameters:

- 1 start bit
- 8 data bits
- 1 odd parity bit
- 1 stop bit

11.5   Application layer

The application layer allows the host device to obtain and interpret field device data. There are three classes of commands:

- Universal commands
- Common practice commands
- Device specific commands

Examples of these commands are listed below.

Universal commands

- Read manufacturer and device type
- Read primary variable (PV) and units
- Read current output and per cent of range
- Read up to 4 predefined dynamic variables
- Read or write 8-character tag, 16-character descriptor, date
- Read or write 32 character message
- Read device range, units and damping time constant
- Read or write final assembly number
- Write polling address

Common practice commands

- Read selection of up to 4 dynamic variables
- Write damping time constant
- Write device range
- Calibrate (set zero, set span)
- Set fixed output current
- Perform self-test
- Perform master reset
- Trim PV zero
- Write PV units
- Trim dac zero and gain
- Write transfer function (square root/linear)
- Write sensor serial number
- Read or write dynamic variable assignments

Instrument specific commands

- Read or write low flow cut-off value
- Start, stop or clear totalizer
- Read or write density calibration factor
- Choose PV (mass flow or density)
- Read or write materials or construction information
- Trim sensor calibration

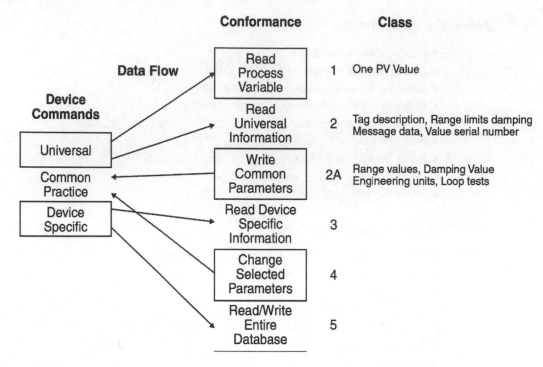

Figure 11.8
HART application layer implementation

Summary of HART benefits

- Simultaneous analog and digital communications
- Allows other analog devices on the highway
- Allows multiple masters to control same smart instrument
- Multiple smart devices on the same highway
- Long distance communications over telephone lines
- Two alternative transmission modes
- Flexible messaging structure for new features
- Up to 256 process variables in any smart field device

11.6 Typical specification for a Rosemount transmitter

Communication specifications

Method of communication:	Frequency shift keying (FSK). Conforms to Bell 202 modem standard with respect to baud rate and binary '1' and binary '0' frequencies.
Baud rate:	1200 bps
Binary '0' frequency:	2200 Hz
Binary '1' frequency:	1200 Hz
Data byte structure:	1 start bit
	8 data bits
	1 odd parity bit
	1 stop bit

Digital process variable rate:	poll/response mode:	2.0 per second
	burst mode:	3.7 per second
No. of multidropped devices:	loop powered:	15 max.
	individually powered:	no limit
Multi-variable specification:	max. 256 process variables per smart device	
Communication masters:	max. 2	

Hardware recommendations

Minimum cable size:	24 AWG, (0.51 mm diameter)
Cable type:	single pair shielded
	or multiple pair with overall shield
Single twisted pair length:	3048 meters max. (3335 yards)
Multiple twisted pair length:	1524 meters max. (1667 yards)

The following formula can be used to determine the maximum cable length:

$$L = \left[\frac{65 \times 10^6}{RC} \right] - \left[\frac{Cf + 10000}{C} \right]$$

Where:
L	=	max. length (meters)
R	=	total resistance (Ω), inclusive of barriers
C	=	cable capacitance (pF/m)
C_f	=	max. shunt capacitance of smart field devices (pF)

Worked example

Assume that a Model 3051C smart pressure transmitter, for a Rosemount System 3 control system, is to be installed using a shielded twisted pair. Calculate the maximum cable length permitted for reliable operation.

R = 250 ohms
C = 164 pF/m
C_f = 5000 pF

$$L = \left[\frac{65 \times 10^6}{250 \times 164} \right] - \left[\frac{5000 + 10000}{164} \right]$$

$L = 1494\ meters$

12

Open industrial Fieldbus and DeviceNet systems

Fieldbus and DeviceNet are communications standards that enable communications between smart or intelligent instruments and a master device such as a PLC. This chapter examines the different Fieldbus systems on the market.

Objectives

When you have completed studying this chapter you will be able to:

- Describe the origin and benefits of Fieldbus and DeviceNet systems
- List and describe the three network classes of Fieldbus and DeviceNet systems
- Describe the characteristics of the following standards (including OSI layers)
  - Actuator sensor interface (AS-i)
  - Seriplex
  - CANbus and DeviceNet
  - Interbus-S
  - Profibus
  - Factory information bus (FIP) and WorldFIP
  - Foundation Fieldbus

12.1   Introduction

There are currently several hundreds of analog and digital standards available for communication between data acquisition and control devices. These field devices communicate using both open and proprietary standards. Traditionally suppliers have produced and sold complete systems that included hardware, software and proprietary protocols. These closed systems made it difficult if not impossible to connect devices of different manufactures. The introduction of open, non-proprietary protocol standards has seen the beginnings of truly open and interoperable systems.

For simplicity the word Fieldbus will be used to refer to both Fieldbus and DeviceNet systems in this chapter. DeviceNet generally refers to the on/off and simpler digital devices whilst Fieldbus tends to encompass instrumentation systems, which need to transfer 16-bit data as a minimum.

A universal open protocol standard is thought by some to be the most desirable conclusion to the problem of multiple Fieldbus systems. The benefits to end users being that all devices would talk using the same protocol and therefore the user could buy any product and plug it in to any system without interfacing problems.

This chapter includes:

- A history of Fieldbus systems
- Classes of Fieldbuses
- The OSI model and Fieldbus systems
- Interoperability
- Examples of various Fieldbus protocols

Before examining the different protocols, it would be helpful to ask why there is considerable effort, time and money being invested in searching for a 'perfect' digital communication network. Why are there several approaches and not just one unified effort? Aren't there enough standards and what is wrong with the one's we have? To answer these questions we need to look at the evolution of digital technology and in particular digital communication technology.

12.2   Overview

Looking at these technologies from a historical perspective, it becomes clear that they are relatively new and, more importantly, still evolving. As technology progresses, more complicated and smaller systems are developed. These new applications and systems reveal shortcomings in the existing technology. This requires the technology to be modified or improved to meet these new demands.

The current approach to cabling a typical control system is shown in Figure 12.1. The concept of Fieldbus is illustrated in Figure 12.2. The figure shows how the instruments are connected with a communication cable. There are numerous benefits not only with regard to minimization of the cables but in greatly increased levels of data available to the operator of the instrument.

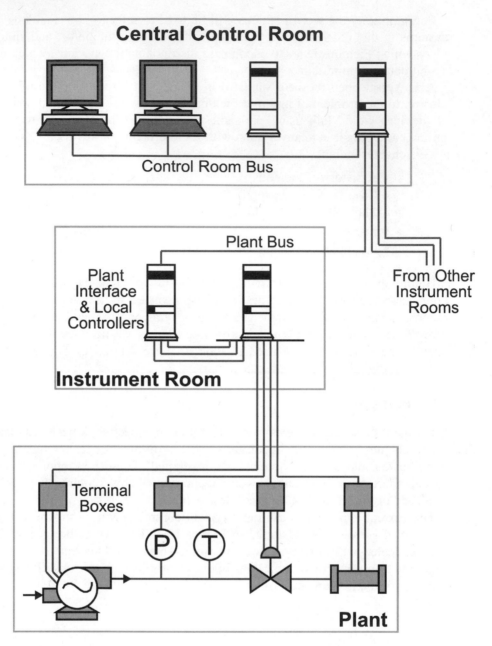

Figure 12.1
Current approach to cabling of a typical control system

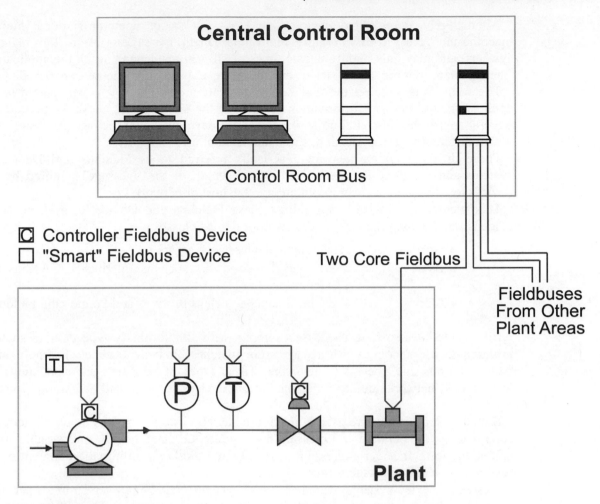

Figure 12.2
Fieldbus approach to cabling of a typical control system

Benefits of the modern Fieldbus

There are real benefits to be gained from the emerging networks, including:

- Greatly reduced wiring costs
- Reduced installation and startup time
- Improved on-line monitoring and diagnostics
- Easy change-out and expansion of devices
- Improved local intelligence in the devices
- Improved interoperability between manufacturers

Classes of Fieldbus networks

It would seem at first that a single Fieldbus system would be beneficial to all users, but this is not the case. Very simple field devices such as proximity switches, limit switches, and basic actuators only require a few bits of digital information to communicate an 'off' or 'on' state. These are usually associated with real time control applications where update times of a few milliseconds are required. The associated electronics necessary to communicate with these systems can be simple, compact and inexpensive enough to be integrated in the device itself.

Alternatively, complicated devices like PLCs, DCSs, or operator stations (human-machine-interfaces, HMIs), require multi-byte length messages (up to 256 in some systems) and may only require update times of 10–100 ms depending on the application. These systems require larger packets due to a large amount of data to be transferred.

The solution is to select the digital communication network that is best suited to the application, and integrate information up through the higher speed networks as required. Several approaches in digital networks have been developed over the last few years, each with a different target application, speed and technology.

These different approaches are generically referred to as Fieldbus and DeviceNet systems and are typically categorized by the length of the 'message' required by the devices to adequately convey information to the host or network.

This method of categorization allows these Fieldbus and DeviceNet systems to be placed in one of the following three network oriented classes:

- Bit: Sensor level devices such as AS-i
- Byte: Device level instruments such as Interbus-S, CANbus and DeviceNet
- Message: Field level devices such as Profibus and Foundation Fieldbus

Bit oriented systems are used, for example, with simple binary type devices such as proximity sensors, contact closures (pressure switches, float switches, etc.), simple push-button stations and pneumatic actuators. These types of networks are also known as 'sensor bus' networks due to the nature of the devices (sensors and actuators) typically used.

Byte oriented systems are used in much broader applications such as motor starters, bar code readers, temperature and pressure transmitters, chromatographs and variable speed drives due to their larger addressing capability and the larger information content of the several byte length message format.

Message oriented systems, which are those systems containing over 16 bytes per message, are used in interconnecting more intelligent systems such as PCs, PLCs, operator terminals and engineering workstations where uploading and downloading system or device configurations is required, or in linking the above mentioned networks together.

The OSI model and Fieldbus systems

The ISO/OSI is an internationally accepted communications reference model and as such has been universally accepted by all Fieldbus systems committees as a starting point in the design process.

As outlined, the OSI model allocates specific tasks and defines the interface for each layer. The model used in an industrial system is a simplified version with only three layers: application, data link and physical (see Figure 12.3). In addition to the three OSI model levels, a user layer is required in Fieldbus systems, to incorporate the function blocks. This is discussed later.

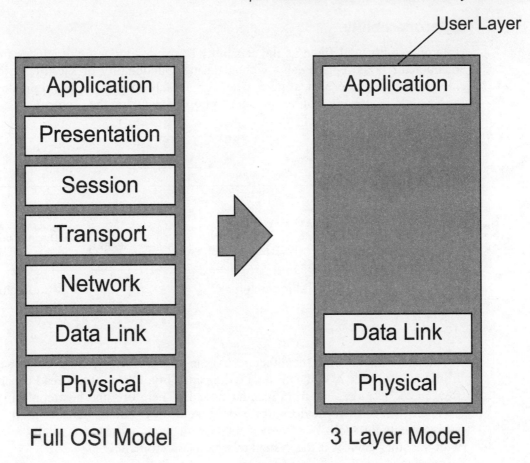

Figure 12.3
OSI and simplified OSI models

The functions of each layer are:

Physical layer

This layer defines the voltages and physical connections. Data received from the data link layer is encoded as electrical information on the actual wire. Similarly, electrical signals received from the wire are passed as binary data to the data link layer.

Data link layer

This defines the protocol and error detection part of the protocol, where the messages sent on the wire are encoded and messages received from the wire are decoded.

Application layer

This layer defines the content messages and the services required supporting them.

Network and transports layers have been omitted by almost every producer of Fieldbus protocols. This means that without a Network layer the protocol cannot 'internetwork' as can be done with the TCP/IP protocol. Therefore most industrial Fieldbus protocols are not directly able to communicate over multiple interconnected networks as with Ethernet and TCP/IP.

Interoperability

Interoperability is defined as the capability of using similar field devices from different manufacturers as replacements without losing functionality or sacrificing the degree of integration with the host system. The user is able to choose the right devices for an application independent of the supplier, control system and the protocol.

Refer to Figure 12.4.

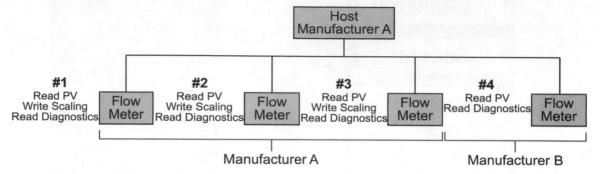

Figure 12.4
A Non-interoperable system

The host system, from manufacturer A, can access flow meters at addresses 1, 2 and 3 from manufacturer A with full read/write capability, but only has read capability for the flow meter at address 4, from manufacturer B. Therefore, the host control system treats each of these field devices differently and they could not be used as effective replacements for each other. Only if the flow meter at address 4 is totally interchangeable with the other devices is the system considered interoperable.

Interoperability is valuable because:

- It allows the end user to select different manufacturers' devices in an interchangeable manner with no discernible differences in the use of each device.
- The concept allows the easy integration of new field devices into control strategies, as they become available.

Importantly, a communication hierarchy, such as the OSI model, cannot address the issue of interoperability. Standardization of the physical, data link and application layers will ensure information can be exchanged among devices on a Fieldbus network. It is the User layer that actually specifies the type of data or information and how it is to be used. Hence, specification of the user layer is vital to ensure complete performance of a Fieldbus system (although it is not part of the OSI communications model).

Protocol review

The following sections include a short review of selected open Fieldbus standards. These include:

- Actuator sensor interface (AS-i)
- Seriplex
- CANbus, DeviceNet and SDS
- Interbus-S
- Profibus
- Foundation Fieldbus

12.3   Actuator sensor interface (AS-i)

The AS-i is a master/slave, open system network developed by eleven manufacturers. These manufacturers created the AS-i Association to develop an open Fieldbus specification. Some of the more widely known members of the AS-i association include Pepperl-Fuchs, Allen-Bradley, Banner Engineering, Datalogic Products, Siemens, Telemecanique, Turck, Omron, Eaton, and Festo. The number of AS-i Association members continues to grow. The AS-i Association also certifies that products under development for the network meet the AS-i specifications. This will assure compatibility between products from different vendors.

AS-i is a bit oriented communication link, designed to connect binary sensors and actuators. Most of these devices do not require multiple bytes to adequately convey the necessary information about the device status, so the AS-i communication interface is designed for bit-oriented messages to increase message efficiency for these types of devices.

The AS-i interface is an interface for binary sensors and actuators, designed to interface binary sensors and actuators to microprocessor based controllers using bit length 'messages'. It was not developed to connect intelligent controllers together as this would be far beyond the limited capability of short bit length message streams.

Modular components form the central design of AS-i. Connection to the network is made with unique connecting modules requiring minimal, or in some cases no tools, and provide for rapid, positive device attachment to the AS-i flat cable. Provision is made in the communications system to make 'live' connections, permitting the removal or addition of nodes with minimum network interruption.

Connection to higher-level networks is made possible through plug-in PC, PLC cards or serial interface converter modules.

The following sections examine these features of the AS-i network in more detail.

The physical layer

AS-i uses a two-wire untwisted, unshielded cable, which serves as both communication link and power supply for up to thirty-one slaves. A single master module controls communication over the AS-i network, which can be connected in various configurations such as bus, ring, or tree (see Figure 12.5). The AS-i flat cable has a unique cross-section that permits only properly polarized connections when making field connections to the modules (see Figure 12.6). Other types of cable may be used for the AS-i network providing they meet the AS-i cable specification. A special shielded cable is also available for high noise environments.

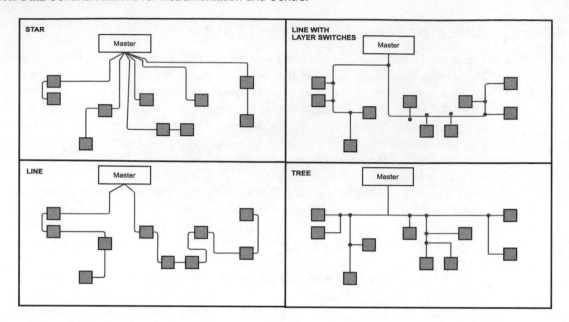

Figure 12.5
AS-i topographical examples

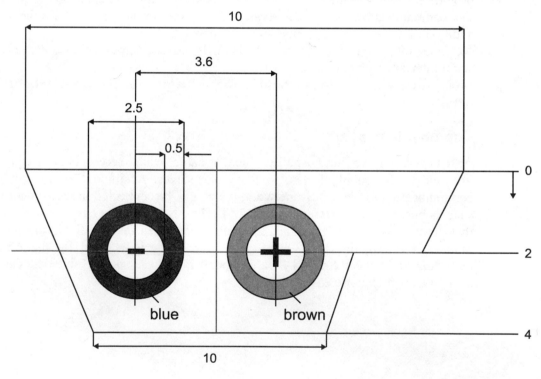

Figure 12.6
AS-i cable cross-section

Each slave is permitted to draw a maximum of 65 mA from the 30 V dc power supply. If devices require more than this, separate supplies must be provided for each such device. With a total 31 slaves drawing 65 mA, a total limit of 2 A has been established to prevent excessive voltage drop over the 100 m permitted network length. A 16 AWG cable is specified to ensure this condition.

The slave (or field) modules are available in four configurations:

- Input modules for 2- and 3-wire dc sensors or contact closure
- Output modules for actuators
- Input and output (I/O) modules for dual purpose applications
- Field connection modules for direct connection to AS-i compatible devices

The I/O modules are capable of accepting up to 4 I/O per slave, and a total of 124 I/O for the network.

A unique design allows the field modules to be connected directly into the bus while maintaining network integrity (see Figure 12.7). The field module is composed of an upper and lower section; secured together once the cable is inserted. Specially designed contact points pierce the self-sealing cable, providing bus access to the I/O points and/or continuation of the network. True to the modular design concept, two types of lower sections and three types of upper sections are available to permit 'mix-and-match' combinations to accommodate various connection schemes and device types (see Figure 12.8). Plug connectors are utilized to interface the I/O devices to the slave (or with the correct choice of modular section screw terminals) and the entire module is sealed from the environment with special seals where the cable enters the module. The seals conveniently store away within the module when not in use.

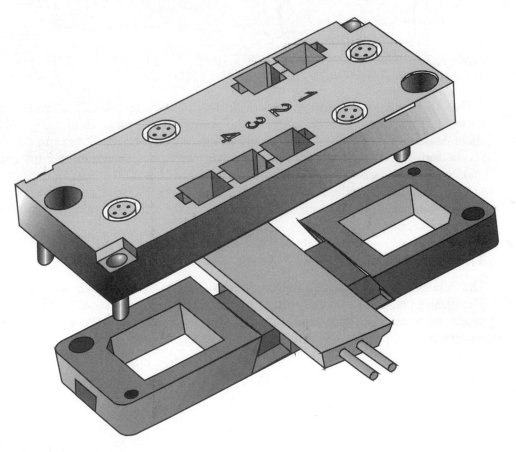

Figure 12.7
AS-i cable to device connections (1)

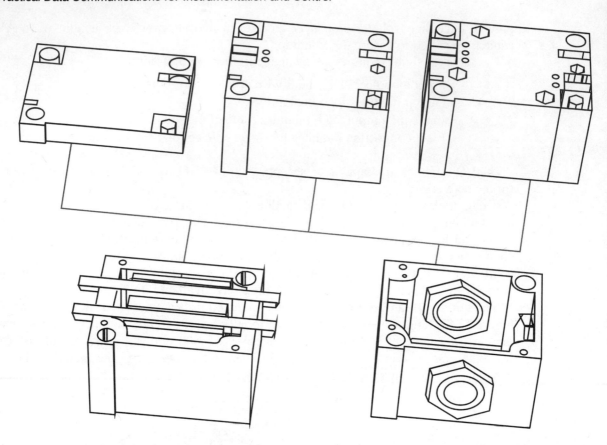

Figure 12.8
AS-i cable to device connections (2)

The AS-i network is capable of a transfer rate of 167 kbps. Using an access procedure known as 'master-slave access with cyclic polling', the master continually polls all the slave devices during a given cycle to ensure rapid update times. For example, with all 31 slaves and 124 I/O points connected, the AS-i network can ensure a 5 ms cycle time, making the AS-i network one of the fastest available.

A modulation technique called 'alternating pulse modulation' provides this high transfer rate capability as well as high data integrity. This technique is described in the following section.

The data link layer

The data link layer of the AS-i network consists of a master call-up and slave response. The master call-up is exactly fourteen bits in length while the slave response is 7 bits. A pause between each transmission is used for synchronization, error detection and correction. Refer to Figure 12.9 for example call-up and answer frames.

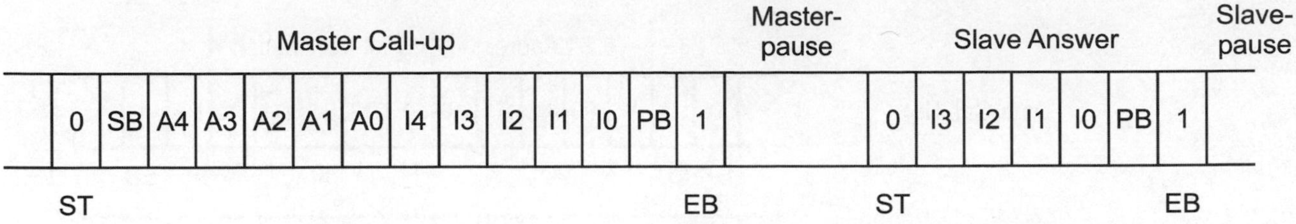

		Master Call-up												Master-pause		Slave Answer						Slave-pause
0	SB	A4	A3	A2	A1	A0	I4	I3	I2	I1	I0	PB	1			0	I3	I2	I1	I0	PB	1

ST   EB   ST   EB

ST	Startbit:	marks the beginning of a master call-up =0: valid start bit =1: not allowed
SB	Control Bit:	marks the data/parameter/address call-ups respectively the command call-up =0: data/parameter/address call-up =1: command call-up
A0...A4	Address:	address of the called slave (5 bit)
I0....I4	Information	the 5 information bits contain the information per each call-up type, which is communicated to the ASI slave. Details are described for each current report.

Figure 12.9
AS-i packet format

Various code combinations are possible in the information portion of the call-up frame. It is these code combinations that are used to read and write information to the slave devices. Examples of some of the master call-ups are listed in Figure 12.10. A detailed explanation of these call-ups is available from the AS-i association literature and is only included here to illustrate the basic means of information transfer on the AS-i network.

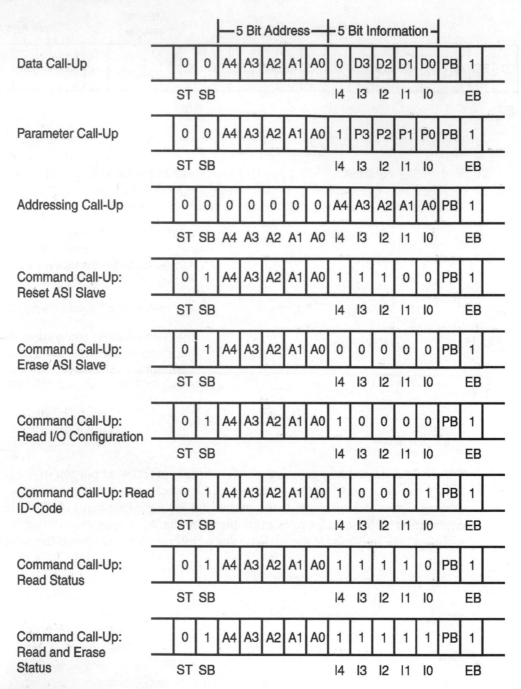

Figure 12.10
AS-i packet format continued

The modulation technique used by AS-i is known as 'alternating pulse modulation' (APM). As the information frame is of a limited size, providing conventional error checking was not possible and therefore the AS-i developers chose a different technique to insure high-level data integrity.

Referring to Figure 12.11, the coding of the information is similar to Manchester II coding, but utilizing a 'sin squared' waveform for each pulse. This wave-shape has several unique electrical properties, which reduce the bandwidth required of the transmission medium (permitting faster transfer rates), and reduce the end of line reflections, common in networks using square wave pulse techniques. Also, each bit has an associated pulse during the second half of the bit period. This property is used as a bit level of error checking by all AS-i devices. The similarity to Manchester II coding is due to this technique having been used for many years to pass synchronizing information to a receiver along with the actual data.

In addition, AS-i developers also established an internal set of regulations for the APM coded signal, which is used to further enhance data integrity. For example, the start bit or first bit in the AS-i telegram must be a negative impulse and the stop bit a positive impulse. Two subsequent impulses must be of opposite polarity and the pause length between two consecutive impulses should be 3 ms. Even parity and a prescribed frame length are also incorporated at the frame level. So the 'odd' looking wave form, combined with the rules of the frame formatting, the set of regulations of the APM coded signal, and parity checking, work together to provide timing information and a high level of data integrity for the AS-i network.

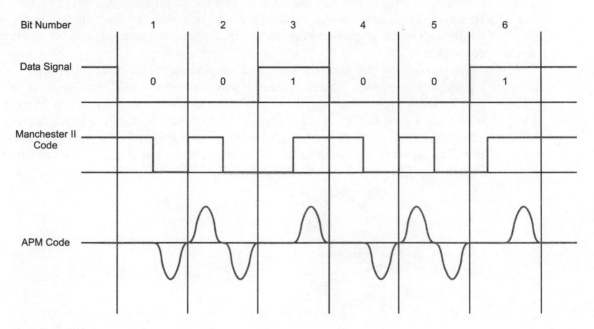

Figure 12.11
Alternating pulse modulation

Operating characteristics

AS-i node addresses are stored in non-volatile memory and can be assigned either by the master or one of the addressing or service units. Should a node fail, AS-i has the ability to automatically reassign the replaced node's address and in some cases reprogram the node itself allowing rapid response and repair times.

Since AS-i was designed as an interface between lower level devices, connection to higher-level systems enables the capability to transfer data and diagnostic information. Plug-in PC cards and PLC cards are currently available. The PLC cards allow direct connection with various Siemens PLCs. Serial communication converters are also available to enable AS-i connection to conventional RS-232, 422, and 485 communication links. Direct connection to a Profibus field network is also possible with the Profibus coupler, enabling several AS-i networks access to a high-level digital network.

Handheld and PC-based configuration tools are available which allow initial start-up programming and also serve as diagnostic tools after the network is commissioned. With these devices on-line monitoring is possible to aid in determining the health of the network and locating possible error sources.

12.4   Seriplex

Automated Process Control, Inc. developed the Seriplex control bus in 1987 specifically for industrial control applications. The Seriplex Technology Organization Inc. was formed to provide information concerning Seriplex, distribute development tools and the Seriplex Application Specific Integrated Circuit (AS-iC) chip, as well as technical assistance for Seriplex developers. Like other sensor (bit level) networks Seriplex was designed to interface lower level I/O devices over a dedicated cabling system, while providing the capability to connect to a host controller or higher-level digital networks. However, for simple control functions, a unique feature of the Seriplex network allows configuration in a peer-to-peer mode that does not require a host or supervisory controller.

Seriplex allows the implementation of simple control schemes without the need for a supervisory processor. This is done through the use of intelligent modules providing a link between inputs and outputs similar to logic gates, i.e. outputs can be programmed based on the status of certain inputs. If more complicated control is required, or supervisory functions desired, Seriplex may be connected to a host processor through interface adapters. Various PLC and PC plug-in cards are available for this interface (see Figure 12.12).

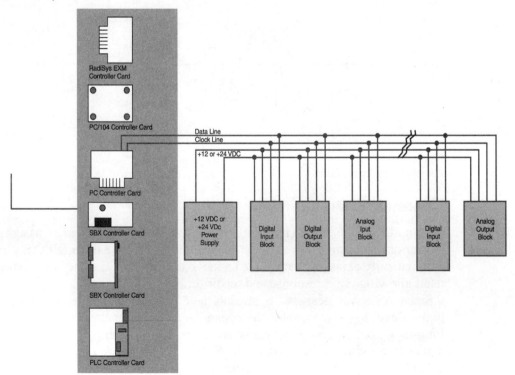

Figure 12.12
Seriplex system example

Various physical topologies are possible for connecting the modular components of the Seriplex network via a five-conductor cable, which provides power, data communications – and clocking signals. Over 7,000 binary I/O points or 480 analog channels (240 In, 240 Out) or various combinations can be supported by Seriplex over this cabling system. The basic configuration without multiplexing can support 255 digital I/O, 32 analog I/O or some combination thereof.

The following sections describe the Seriplex network in more detail.

The physical layer

The Seriplex cabling system consists of a single four-conductor cable with two AWG #22 shielded wires for data and clock signals and two AWG #16 wires for power and common. A shield drain wire is also provided for shield grounding. Clock rates from 16 to 100 kHz are selectable with newer versions capable of up to 200 kHz. Capacitance values of cable dramatically affect all communication systems and low capacitance cable designs are available from several manufacturers to maximize data transfer rates. Rates of up to 100 kHz over 500 feet are possible with Seriplex using low capacitance (16 pF/ft) cabling. However, 20 pF/ft cables would limit this distance at 100 kHz to 350 ft. The importance of low capacitance cabling cannot be over emphasized in any system.

12 V dc is provided by the cable to power the I/O devices in the first generation systems. Second generation systems operate on either 12 or 24 V dc, with the level selected by the user for the particular system used. Field connections are made through Seriplex modules located near the field devices.

Individual I/O addresses are programmed in the module to allow the network access to each point. A total of 255 usable addresses is available to the modules. Digital inputs and outputs use one address each. Each 8-bit analog module uses eight addresses (for one analog input or output). Multiplexing methods are employed to increase the total digital I/O count to 7,706 or analog I/O count to 480 or a combination of these.

Data and clock signals are transferred over the network in the form of 0 to +12 V digital pulses.

The data link layer

Two different methods of operation are possible with Seriplex, depending on the mode of operation. Both modes of operation use the unique access control method described in Mode 2 below.

In Mode 1, or peer-to-peer mode, modules can be logically inter-linked without the need for a host controller. In this case logical functions are implemented directly between modules. A separate clock module is required in this mode as there is no host to provide crucial clock line information. Module outputs can be logically programmed to function, based on the status of other modules' inputs. With this capability simple logic functions can be performed without the need of a host controller.

Mode 2 operation requires the host controller to provide timing clock signals. The receiver in each module counts the clock pulses. When the pulse count of the clock line equals the receiver's address, access is granted on the data link line for the receiver to read from and, in turn, write to, the host controller.

This access control method is unique in that a continuous 'train' of clock and data pulses cycle through the system. Access on the data line for individual addresses (bit status) is granted for a time period within the data stream based on the time slot of the address (see Figure 12.13). This 'continuous polling' starts with a synchronizing signal 8 clock cycles long and serves as notification that the 'polling' is about to begin. At the beginning of the cycle the data line is 'empty'. As the pulse count equals each module's address count, the modules 'dump' their bit values on the data line so that at the end of the cycle all information is available to the host. The frame size can be adjusted in length from 16 to 256 bits, in multiples of 16, to accommodate different size systems. Correct sizing of the system and resultant frame size can provide extremely fast update times for smaller networks.

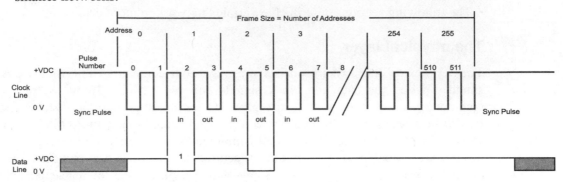

Figure 12.13a
Seriplex mode 1

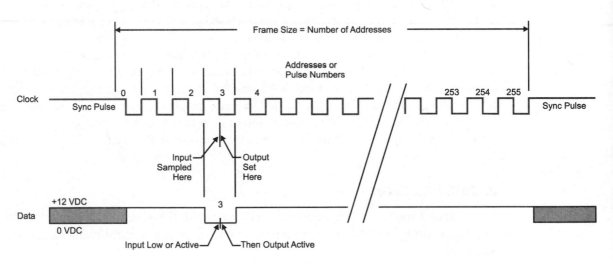

Figure 12.13b
Seriplex mode 2

Data echoing provides error detection on the bit level. The receiver echoes messages (which are typically one bit per address) to confirm correct data receipt. This is not automatic and is implemented in the application layer of the software by the applications programmer.

Operating characteristics

Seriplex is a bit oriented network system intended to link lower level devices both physically and, in Mode 1 operation, logically. These features are incorporated in the AS-I-C chip located in all Seriplex devices. Handheld programming devices are available to enable Seriplex device configuration.

Interface devices to higher-level field networks through host controllers or special Gateways are also available.

12.5   CANbus, DeviceNet and SDS systems

CANbus

The CAN network was developed in the automotive industry in response to the rapidly growing use of electronic-control systems in automobiles. As demands for fuel efficiency and safety increased, more and more electronic devices became part of the system. The need for multiple devices to pass information between them rapidly became a necessity. A type of serial data bus system was developed by Bosch to meet these demands. It was called the controller area network or CAN. CAN is formally specified in: 1) BOSCH CAN specification – Version 2.0, Part A, and 2) ISO 11898: 1993 – road vehicles – Interchange of digital information – controller area network (CAN) for high-speed communication. CAN has since been rapidly adapted to industrial applications.

CAN is a bus type network system which does not use a bus master or token passing schemes to access the bus. Instead it uses a unique access control method called 'non-destructive bit-wise arbitration'. This type of access control uses the station identifier bit pattern itself to gain access to the bus as shown in Figure 12.14. The priority of the station is determined by the addressing assignments during configuration of the network and allows the station with the highest priority preferred access. Unlike token passing or master-slave type arbitration schemes CAN is not deterministic, but defers to the station with highest priority making the lower priority stations wait for access.

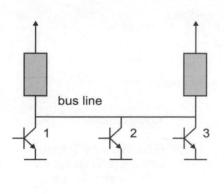
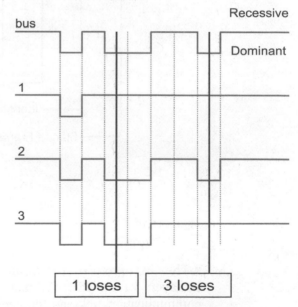

Figure 12.14
Bit arbitration example

Figure 12.14 shows the bit arbitration of a CAN type system. The devices 1, 2, and 3 try to transmit at the same time. Ground or a '0' is dominant. The results can be seen on the top waveform. Because device 1 puts out a '1' and it is dominated by the '0' from 2 and 3, it loses and stops transmitting. Then device 3 puts out a '1' and is dominated by 2. Therefore, 2 continues to transmit while 1 and 3 wait until the line is clear.

The CAN station that wins the arbitration continues to transmit its message frame uninterrupted with no corruption from the other stations arbitration attempts. This allows higher efficiency of data transfer over the network. A typical CAN message frame is shown in Figure 12.15. Notice the data field can be of variable length – up to 8 bytes. This makes CAN suitable for use with more sophisticated devices that may require several bytes to adequately convey their information content. CRC error checking and specific frame length requirements, as well as individual message acknowledgments, are used to ensure data integrity over the bus.

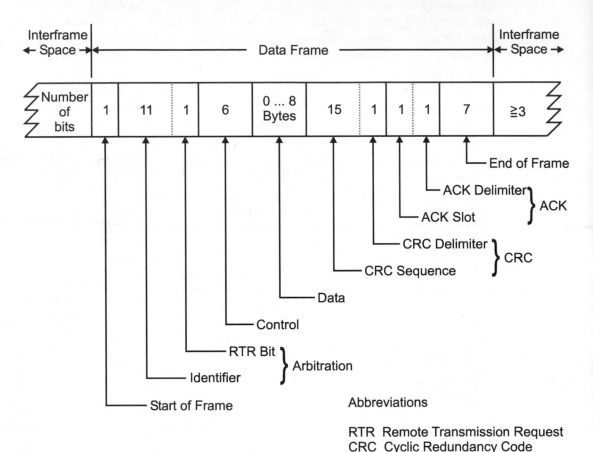

Figure 12.15
The canbus packet

The CAN protocol specifications cover only the physical (layer 1) and data link (layer 2) layers of the ISO/OSI model. Specifics concerning the physical medium for the communication link and the application layer (layer 7) are left to the designers of the systems as described below.

DeviceNet

DeviceNet, developed by Allen-Bradley, is a low-level device oriented network based on the CAN network. It is designed to interconnect lower level devices (sensors and actuators) with higher-level devices (controllers). The variable, multi-byte format of the

CAN message frame is well suited to this task as more information can be communicated per message than with the bit type systems.

The Open DeviceNet Vendor Association Inc. (ODVA) has been formed to issue DeviceNet specifications, ensure compliance with the specifications and offer technical assistance for manufacturers wishing to implement DeviceNet. Over 125 firms have either joined formally or have signed intent to become members. The DeviceNet specification is an open specification and available through the ODVA.

DeviceNet can support up to 64 nodes, supporting as many as 2048 total devices. A single, four-conductor cable provides both power and data communications. Various devices are available to interconnect I/O devices, and the network trunk-line cable allowing customized configurations.

As DeviceNet was designed to interface lower level devices with higher-level controllers, a unique adaptation of the basic CAN protocol was developed. This is similar to the familiar poll/response or master/slave technique, but still utilizes the speed benefits of the original CAN.

Figure 12.16 shows the DeviceNet profile in its relationship to the ISO/OSI Model. It is important to note that only Layers 1 and 2 are covered by the CAN protocol specification while the remaining layers were developed for the DeviceNet network.

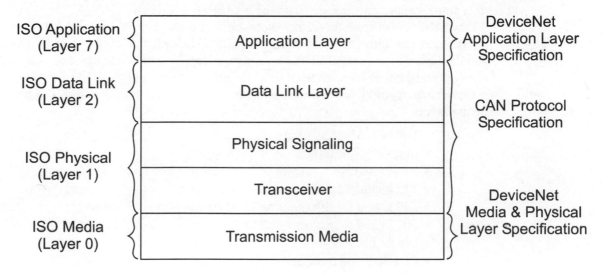

Figure 12.16
DeviceNet and the OSI model

The following sections examine these features of the DeviceNet network and protocol in more detail.

The physical layer

The DeviceNet cabling system consists of a single four-conductor cable in a bus topology providing both power and data communication. Data is transmitted over the #18 twisted pair. Power is provided over the #15 twisted pair. Both pairs have a foil shield and an overall braid with #18 drain wire. Terminating resistors are required on both ends of the trunk line. 24 V dc power is provided on the power bus and can support up to 3 Amp on the DeviceNet thin cable or 8 amp on the DeviceNet thick cable. The total length of trunk line allowed depends on which type of cable is used, the number of devices supported and the data rate. The voltage at each device should be at least 11 V dc or higher.

Data rates of 125, 250, and 500 kbps are possible with the corresponding network configuration shown in Figure 12.17. Various connectors can be used to connect devices to the network such as screw terminals, screw terminals or sealed screw tight connectors.

Data Rate	Trunk Distance	Drop Length	
		Maximum	Cumulative
125 k baud	500 meters (1600 ft.)		156 meters (512 ft.)
250 k baud	200 meters (600 ft.)	3 meters (10 ft.)	78 meters (256 ft.)
500 k baud	100 meters (300 ft.)		39 meters (128 ft.)

Figure 12.17
DeviceNet lengths and baud rates

To allow non-destructive arbitration during simultaneous transmission from two or more nodes the BOSCH CAN specification defines the two possible logic levels as 'dominant' and 'recessive'. During arbitration the dominant value will win access to the bus. For DeviceNet, the dominant level is represented by a logical '0' and the recessive level by a logical '1'. The electrical voltage levels representing these logic levels are taken from the ISO 11898 standard. CAN utilizes a balanced transmission system with data signals appearing as the difference between CAN_H and CAN_L.

DeviceNet specifications require isolation to prevent ground loops. As the circuitries in all devices are ultimately referenced to the V-bus signal, connection of the network should be earth grounded at the bus power supply only. All devices attached to the network must either be referenced to V− or otherwise ground isolated.

DeviceNet requires the following features to be incorporated within the physical and media layers:

- Use of CAN technology
- Support of both thick and thin drop line
- Ability to operate at a minimum of three data rates
- 125 kbaud for distances up to 500 meters (max.) (1640 feet)
- 250 kbaud for distances up to 200 meters (max.) (656 feet)
- 500 kbaud for distances up to 100 meters (max.) (328 feet)
- Linear bus topology
- Low loss and low delay cable
- Shielded twisted pair cable, containing both power and signal pairs
- Small size and low cost
- Support of up to 64 nodes
- Support drop lines of up to 6 meters (20 feet) in length
- Node removal without disturbing or interrupting network operation
- Simultaneous support of isolated and non-isolated physical layers

DeviceNet uses two types of pre-made cables, thick cable and thin cable. The thick cable is a large gray cable used as long trunk runs between devices. The thin cable is a small and usually short yellow cable that connects thick cables to devices. The thick and thin cables are connected together by a large black 'T' junction. All cables and connections have threaded ring connectors.

The data link layer

The data link layer is specified in the CAN protocol specification (see Figure 12.16). The format of the data link layer (frame format) is fixed by this specification. However, the method used to encode the identifier and data fields in the CAN message packet is left to the application layer developer as described in the following section. The method of communication is based on the producer/consumer approach where one station (*the producer) places data on the bus at regular intervals and this is then read by the consumer station on the network.

The application layer

The CAN specification does not dictate how information within the CAN message frame fields are to be interpreted – this was left up to the developers of the specific application software. In the case of DeviceNet a unique method was developed to allow for two types of messages to exist.

Through the use of special identifier codes (bit patterns), master is differentiated from slave. Also, sections of this field tell the slaves how to respond to the master's message. For example, slaves can be requested to respond with information simultaneously in which case the CAN bus arbitration scheme assures the timeliest consecutive response from all slaves in decreasing order of priority. Or, slaves can be polled individually, all through the selection of different identifier field codes. This technique allows the system implementers more flexibility when establishing node priorities and device addresses.

System operation

Several devices are available to allow connection of DeviceNet to higher-level devices. For example, Allen-Bradley has developed PLC plug in cards to function as DeviceNet scanners. These devices support a master/slave configuration communicating with slave devices through either the strobe or poll methods. Two separate DeviceNet channels (or networks) can be supported. These modules also perform limited diagnostics on the network, and chassis communication link report this information to higher-level controllers. An interface is also available, which allows a PC to act as another node on the Network.

With the DeviceNet flex I/O adapter up to 128 non-DeviceNet compatible devices can communicate to other DeviceNet I/O and PLC controllers. Other types of DeviceNet compatible products are also being marketed which are connected directly to the network with a minimum of configuration effort.

Smart distributed system (SDS)

The smart distributed system (SDS) was developed by Honeywell and is a low-level device oriented network based on the CAN network. It is designed to interconnect lower level devices (sensors and actuators) with higher-level devices (controllers). The variable, multi-byte format of the CAN message frame is well suited to this task since more intelligence can be communicated per message than with the bit type systems.

The SDS 'partners' program has been formed, and in cooperation with Honeywell issues the SDS specifications, ensures compliance and offers technical assistance for manufacturers wishing to implement SDS. The SDS specification is an open specification and available through Honeywell and the SDS 'partners' program.

The SDS network can connect up 126 devices on a single bus. Each group of 16 I/O is interfaced to a higher-level device (PLC, for example) through the interface terminal strip

(ITS). The ITS provides the physical interface between the network bus and individual I/O points on the PLC I/O cards. Plug-in cards are also available to interface the bus directly to the PC. This choice between interfaces gives the designer a method for integrating the SDS with an existing PLC system.

The SDS utilizes the CAN network to allow devices to report information only when there is a need, e.g. a change of state of an input to the controller. This approach reduces traffic on the network by minimizing polling inquires from the controller to the slave devices.

As with other CAN based systems the SDS network uses the OSI Layers 1 and 2 (physical and data link layers) of the CAN protocol and develops the SDS application layer (OSI Layer 7) for its specific target application area, integrating lower level devices with higher level controllers.

The following sections examine these features of the smart distributed system (SDS) network and protocol in more detail.

The physical layer

The SDS cabling system consists of a single, four conductor, shielded, cable in a bus topology providing both power and data communication. Both data and power pairs are twisted and an overall shield is provided for noise protection. Terminating resistors are required on both ends of the bus. 12 to 24 V dc power is provided on the power bus to support field devices. The total length of trunk line allowed depends on which type of cable is used, the number of devices supported and the data rate.

Various data rates are possible with the corresponding network configuration restrictions. Several connector types can be used to connect devices to the network such as screw terminals or sealed screw tight connectors.

The data link layer

The data link layer is specified in the CAN protocol specification (see Figure 12.16). The format of the data link layer (frame format) is fixed by this specification. However, the method used to encode the identifier and data fields in the CAN message packet is left to the application layer developer.

The application layer

The CAN specification does not dictate how information within the CAN message frame fields is to be interpreted; this was left up to the developers of the specific application software. In the case of SDS, various codes within the identifier frame allow for communication between slave devices and controllers.

Through the use of special identifier codes (bit patterns) unique addresses are established for each device. Source and destination addresses of messages are distinguished by the setting (1 or 0) of the most significant CAN identifier bit, called the SDS Direction bit. A 0 designates the address, what follows is the destination; a 1 designates it as the source of the message. The application protocol allows any device, which needs to read the message (or 'consumer') access to this information as it appears on the network. There can be more than one consumer of a given message.

Conversely, when a device senses a change of state it can put that information on the network as soon as it can gain access to the bus consistent with the CAN arbitration procedure. This device is known as a 'producer' in the CAN protocol.

Through these unique CAN identifier field code patterns, SDS provides the functions and unique capabilities of this flexible and fast device level networking system.

System operation

Several features of SDS are implemented at the system level to speed startup and monitor the 'health' of the devices and network. One of these is the Autobaud. Through this special function of the bus manager, (a designated device controller on the network, usually the host controller) a unique message packet is sent immediately after initial bus power-up. This allows all the other devices to monitor the time length of the frame and determine the baud rate setting of the controller. Each device can then adjust its baud rate accordingly to ensure all devices operate at the same data rate.

Continuous monitoring for 'missing' devices and defective devices is implemented by periodic polling. If a device fails to report within a specified period of time the host will flag the device missing warning. Polling will continue until the device reappears.

Another monitoring feature is the periodic device self-test control. Periodically, instead of device polling, the host will substitute the self-test command and examine the device's diagnostic registers in the reply for device errors.

Several devices developed through the SDS 'partners' program have enabled direct SDS connection to various higher level devices such as PLCs, PCs, VMEbus systems, starters and pilot devices. Interfaces to new devices are certainly possible in the future for SDS as this network continues to find new applications in the industry.

12.6   Interbus-S

The Interbus-S is an open device level network that allows connection of up to 4096 digital I/O points over a distance of up to 400 m. Through a unique frame transfer protocol these points can be updated in as little as 14 ms, faster speeds are possible with a lower I/O count. It is a timed ring topology with subsystem drops (tree structure) allowing connections of up to 256 stations. Data rate is 500 kbps.

The variable length frame format allows message frames of up to 512 bytes enabling communication between intelligent I/O devices. Integration to the higher-level Fieldbus networks is also within the capability of this network.

The Interbus-S Club, founded in 1993 was established to maintain and advance the Interbus-S network standard. The organization provides Interbus-S specifications to potential developers and assists with technical information.

The following sections examine these features of the Interbus-S (IBS) network and protocol.

The physical layer

The Interbus-S cabling system specification allows for either twisted pair copper or fiber optic cable connected to each station in a ring topology. Communication is serial and frame transmission is accomplished through a unique register shifting procedure developed specifically for the Interbus-S network. Two types of communication buses are used as part of the same network – local bus and remote bus. Each bus type carries the same signals but at different electrical levels. Local bus operates at TTL voltage levels and is designed for short distances typically within a control enclosure. Remote bus utilizes RS-485 voltage levels and is designed to communicate over much longer distances – up to 1300 ft (400 m). Both buses operate at 500 kbps transfer speed and a special module; the BK module is required to translate the two signal levels (Figure 12.18).

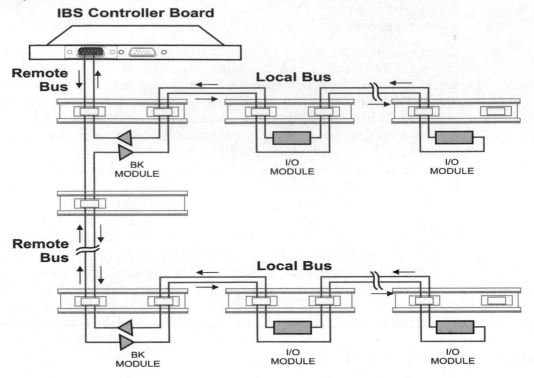

Figure 12.18
Interbus-S layout example

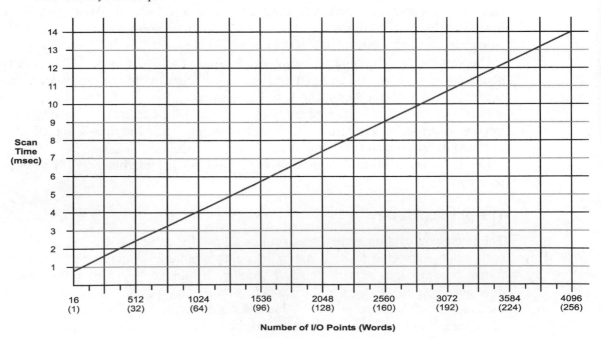

Figure 12.19
Interbus-S scan cycles and I/O points

Communication is performed through scan cycles. Each scan cycle shifts messages through each station in increasing order through the network. Data is read from the message and written to the message during each cycle making for very fast response times (see Figure 12.19).

The data link layer

The data link layer protocol provides for full duplex transmission. The complete message frame is clocked through the network on each cycle. No arbitration or contention for access to the network is required since each station has access during each cycle. All input and output data are updated and transferred during each scan cycle. CRC error checking is performed between each network connection allowing identification of the error source. Both digital and analog data are supported as well as client/server messaging.

The network does individual station addressing automatically during initialization of the network, eliminating the need to manually assign these during system startup. This is accomplished through an identification (ID) cycle that tells the controller the type and physical order of location of the stations on the network (see Figure 12.20).

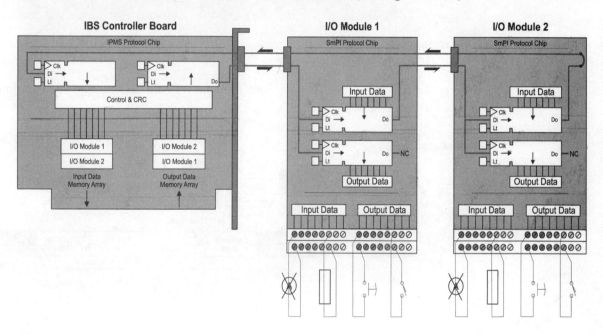

Figure 12.20
Interbus block diagram example

The application layer

Interbus-S supports network and module diagnostics and monitoring through a special 'telegram' message sent out after each byte is shifted. All stations monitor this message simultaneously. This unique function is accomplished by a 'telegram control' switch in each station (see Figure 12.21) which automatically activates after each 8-bit shift allowing not only simultaneous reception of the message by all stations, but synchronization information as well.

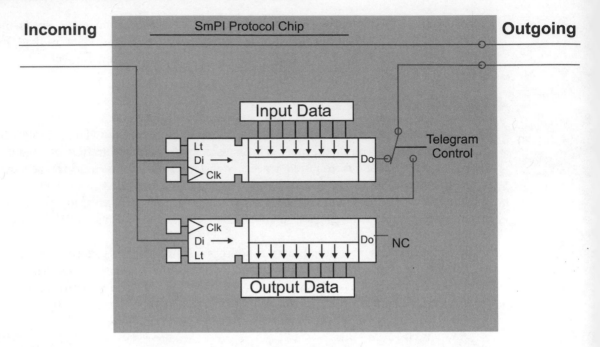

Figure 12.21
Interbus-S register flow diagram

Figure 12.22 shows the ID and scan transmission frames noting the location of the various field parameters.

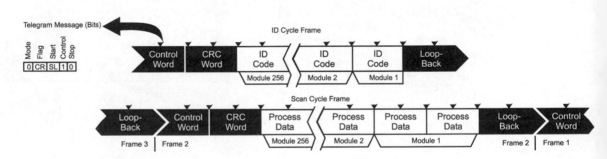

Figure 12.22
Interbus-S packet configuration

12.7 Profibus

Introduction to Profibus

Profibus is an open standard Fieldbus defined by the German DIN 19245 Parts 1 and 2. It is based on a token bus/floating master system. There are three different types of Profibus – FMS, DP and PA. Fieldbus message specification (FMS) is used for general data acquisition systems. DP is used when fast communications are needed. And PA is used in areas when intrinsically safe devices and intrinsically safe communications are needed. Figure 12.23 outlines the structure of the various versions of Profibus.

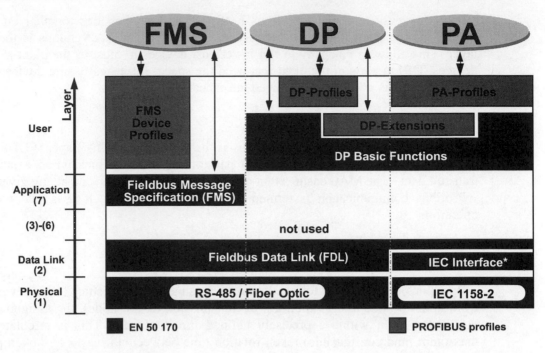

Figure 12.23
Profibus protocol architecture

The physical layer

The physical layer specifies the type of Profibus transmission medium. The RS-485 voltage standard is defined for the FMS and DP versions of Profibus. The IEC 1158-2 standard is used in the PA version. For FMS and DP a maximum number of 255 stations are possible.

- FMS (RS-485):   187.5 kbps   General use
- DP (RS-485):   500 kbps /1.5 Mbps/12 Mbps   Fast devices
- PA (IEC 1158-2):   31.25 kbps   Intrinsically safe

Basic properties of the RS-485 voltage standard for Profibus

Topology:	Linear bus, terminated at both ends
Medium:	Twisted pair shielded cable
Wire size:	18 AWG (0.8 mm)
Attenuation:	3 dB/km at 39 kHz
Number of stations:	32 stations without repeaters extendible to 127
Bus length:	max. 1200 meters (3940 feet) extendible to 4800 meters (7900 feet) at slow rates
Speed:	1200 to 12 Mbps
Connector:	Phoenix type screw or 9-pin D-sub connector

IEC 1158-2 is a standardized current standard used in special areas of a factory or plant that require intrinsically safe devices. IEC 1158-2 works by modulating a Manchester encoded bipolar NRZ ±10 mA signal on top of a 9 to 32 dc voltage. This 10 mA creates a ±1 volt signal that is read by each of the devices on the bus.

It is very easy to connect Profibus FMS, DP and PA versions together on the same system, as the main difference between the FMS, DP and PA versions is the physical layer. This allows a company to run lower cost devices in most of the plant (FMS), fast devices (DP) in parts of the plant that need the speed. Intrinsically safe devices (PA) are used in the areas of the plant that need intrinsically safe devices.

The data link layer

The data link layer is defined by Profibus as the Fieldbus data link Layer (FDL).

The medium access control (MAC) part of the FDL defines when a station may transmit data. The MAC ensures that only one station transmits data at any given time.

Profibus communication is termed hybrid medium access. It uses two methods of operation:

- Token passing
- Master/slave.

By means of a software token, the token passing method ensures (which when passed from node to node assigns the right of transmission to that node), the assignment of the bus access right within a precisely defined time interval. This is circulated with a maximum (and configurable) token rotation time between all masters. Token passing is especially useful for communication between complex automation masters who require equal rights on the bus. The token is passed in a defined sequence (in order of increasing addresses).

The master/slave method allows the master that currently has the token to communicate with the associated slave devices. The master can then read from or write data to the slave devices.

A typical configuration is shown in Figure 12.24.

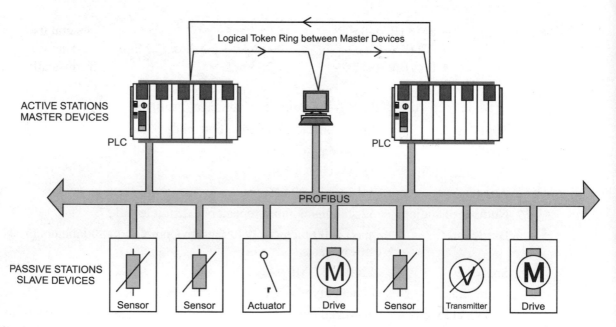

Figure 12.24
Typical architecture of a Profibus system

During the Bus system startup phase, the task of the active station's MAC is to detect logical assignment and establish the token ring. The MAC also handles adding or deleting stations (which have become inactive), deleting multiple nodes with the same address and multiple or lost tokens.

The application layer

This consists of two sections:

- Fieldbus message specification (FMS)
- Lower layer interface (LLI)

The application layer is defined in DIN 19245 part 2.

The Profibus communication model

The part of the application process in a field device that is readable for communication is called the virtual field device (VFD). The VFD contains the communication objects that may be manipulated by the services of the application layers. The objects of a real device that are readable for the communication (variables, programs, data domains) are called communication objects.

All communication objects of a Profibus station are entered into its local object dictionary (source OD). There are two types:

- Static communication objects
- Dynamic communication objects

Static communication objects are defined in the static object dictionary. They may be predefined by the manufacturer of the device, or defined during the configuration of the bus system. Static communication objects are used mainly for communication in the field area. Profibus recognizes the following static communication objects:

- Simple variable
- Array – sequence of simple variables of the same type
- Record – sequence of simple variables, not necessarily of the same type
- Domain – data range
- Event

Dynamic communication objects are entered into the dynamic part of the OD (list of variable lists of program invocations). They may be predefined or defined, deleted or changed by the application services in the operational phase.

Profibus supports the following dynamic communication objects:

- Program invocation
- Variable list (sequence of simple variables, arrays or records).

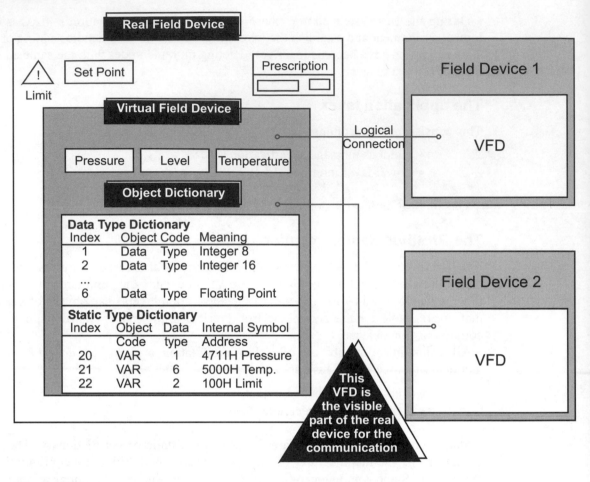

Figure 12.25
Virtual field device (VFFigD) with object dictionary (OD)

There are two methods for accessing the variables:

- Addressing by name (using a symbolic name)
- Physical addressing (to access a physical location in memory)

Profibus defines logical addressing (by symbolic name) as the preferred method as this increases the speed of access.

Application services

From the point of view of an application process, the communication system is a service provider offering various application services – the FMS services. The FMS describes the communication objects, the application services, and the resulting models from the viewpoint of the communication partner. There are two types of services:

- Confirmed services: These are only permitted on connection-oriented communication relationships.
- Unconfirmed services: These are used on connectionless communication relationships such as broadcast and multicast.

Note:
Refer to 'Connectionless-oriented' and 'Connection-oriented' under Lower layer interface following for an explanation of these terms.

Service primitives in the Profibus standard describe the execution of the services. The services can be divided into the following groups:

- Context management services allow establishment and release of logical connections
- Variable access services permit access to simple variables, records, arrays and variable lists
- Domain management services enable the transmission of contiguous memory areas
- Program invocation management services allow the control of program execution
- Event management services make the transmission of alarm messages possible
- VFD support services permit device identification and status report
- OD management services permit object dictionaries to be read and written

Lower layer interface (LLI)

The LLI conducts the data flow control and connection monitoring as well as the mapping of the FMS services onto the layer 2 with consideration of the various types of devices.

The user communicates with other application processes over the logic channels, the communication relationships. For the execution of the FMS and FMA7 services the LLI provides various types of communication relationships.

There are two types of communication relationships:

- **Connection-oriented relationships**
  This requires a connection establishment phase (or initiate service) before the connection can be used for data transmission. When the connection is no longer required, it may be released with the abort service (or connection release phase). The connection attribute distinguishes between defined connections where the communication partner is fixed at a configuration time and may not be changed and an open connection where the communication partner is dynamically defined in the connection establishment phase.

- **Connectionless-oriented relationships**
  Cyclic data transfer means exactly one variable is permanently read or written over a connection. A typical application for cyclic data transfer is the periodic update of the remove inputs and outputs of a PLC. Acyclic data transfer means an application sporadically accesses various communication objects over a connection.

Communication relationship list (CRL)

The CRL contains the description of all communication relationships of a device independent of the time of their usage.

Network management

In addition to the application services and FMS models, Profibus includes specifications for network management (Fieldbus management layer 7, FMA7).

FMA7 functions are defined in three groups:

- **Context management**
  This allows the establishment and release of management connections.

- **Configuration management**
  This allows the CRL to be loaded and read, access to variable, statistic counters and parameters of the layers 1 and 2, identification of communication components of the stations and registration of stations.

- **Fault management**
  This allows the indication of faults and events and the reset of stations.

Profibus profiles

For the various application fields it is necessary to adopt the functionality actually needed for the real world. A profile includes application specific definitions of the meanings of the communication functions, as well as the interpretation of status and error indications.

Profiles for the following application fields are available:

- Building automation
- Drive control
- Sensors and actuators
- Programmable logic controllers
- Textile machines

These enable different manufacturers, which use the same profile, to have full interoperability with the different devices on a common interconnecting Profibus.

Gateway

Gateways are required to link other protocols to the Profibus system. Some gateways are easy to implement, such as one from Profibus to the higher-level MAP. This is due to both standards adhering to the OSI model and the good relationship of Profibus to the definitions of the MAP layer 7 functions (using MMS).

Implementation features

No special hardware components are required to implement the Profibus protocol, providing the microprocessor has a UART serial interface. The implementation can range from a simple slave device, which has one microprocessor (such as an Intel 8051). This executes both the protocol and the application task; to a complex master device, which has communication, functions (and protocol) implemented on one processor. The application tasks are performed on a separate processor. For time critical applications, it is possible to implement the functions of layers 1 and 2 using a special hardware circuit (e.g. AS-iCs or a Motorola 68302).

12.8   Factory information bus (FIP)

The FIP is the result of work carried out by companies located primarily in France, Italy and Belgium. US companies such as Honeywell are involved with French manufacturers in developing the World FIP standard (see next section).

The FIP standard aims for very high transmission rates and strictly defined scanning intervals.

Bus access method

The broadcasting approach is used, with a central unit (called the bus arbitrator) co-coordinating the transmissions. This means that it is not necessary to give each device a unique address. A variable (processed by the transmitter only) is transmitted on the bus by one transmitter and is read by any number of receivers situated on the same bus.

The bus arbitrator has three operating cycles:

- **Cyclic traffic**
  The bus arbitrator names a set of variables using a table command.

- **Aperiodic traffic**
  The bus arbitrator calls on request variable from every device.

- **Message service**
  The arbitrator gives the right to transmit to a device, which requested it during the previous cyclic traffic period.

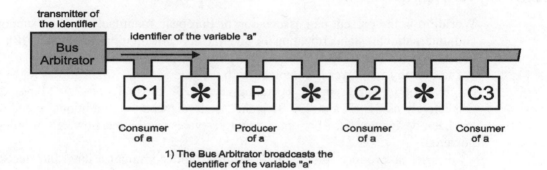

1) The Bus Arbitrator broadcasts the identifier of the variable "a"

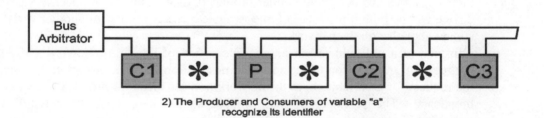

2) The Producer and Consumers of variable "a" recognize its identifier

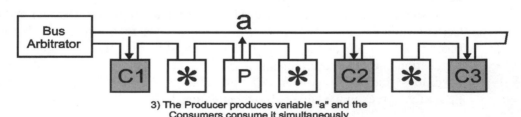

3) The Producer produces variable "a" and the Consumers consume it simultaneously

＊ Other network stations not involved in the production or consumption of variable "a"

Figure 12.26
Operation of FIP

The physical layer

The FIP standard allows twisted pairs, optical fiber or coaxial cable in a bus topology to a maximum distance of 2 km. Speeds can vary between 31.25 and 1000 kbps. A maximum number of 256 devices is allowed on the bus network.

The data link layer

The data link layer is non-proprietary FIP.

Installations

There are a number of installations in France and Italy that are being used to evaluate the FIP standard.

The FIP standard has evolved into the WorldFIP standard as discussed in the next section.

12.9   WorldFip

WorldFip is the present day association of European manufacturers that support the use and international standardization of the factory information protocol (FIP). FIP is a communications protocol developed and presently in use in Europe.

The FIP physical layer is compliant with the IEC S50.02, which allows for twisted pair, or fiber optic cable media operating at 31.25 kbps, 1, or 2.5 Mbps, designated respectively as S1, S2 and S3. S2 is the standard speed. An additional speed of 5 Mbps has been designated for fiber optic media. Devices can be bus powered or independently powered.

FIP uses a producer-distributor-consumer type communications and access control model for transferring time critical information throughout the network. Devices and their variables are designated either producers or consumers of specific variables. One device can be both a producer of one variable but a consumer of another variable located somewhere else on the network.

Instead of a poll and response type integration of the entire network and then routing the required information to the specified destination, the FIP bus arbitrator simply places the request for a variable on to the network in a broadcast fashion. All devices 'hear' the broadcast. The producer of that variable then places it on the network again in a broadcast form. It is then available to all consumers of that particular variable – see Figure 12.26. This procedure allows rapid access of all variables in a timely and determined manner while ensuring no collisions and therefore a very efficient use of the network capabilities.

This requires a configuration and scheduling table within the arbitrator and the devices. Certain variables may need to be polled more often than others and this is taken into account in the scheduling table. In fact, the table can be configured for the specific application and time requirements of the process, making FIP very adaptable to changing conditions and new applications. The table is defined during initial network configuration. An example table is shown in Figure 12.27 for reference.

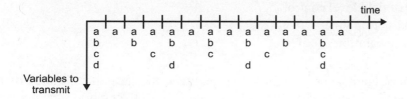

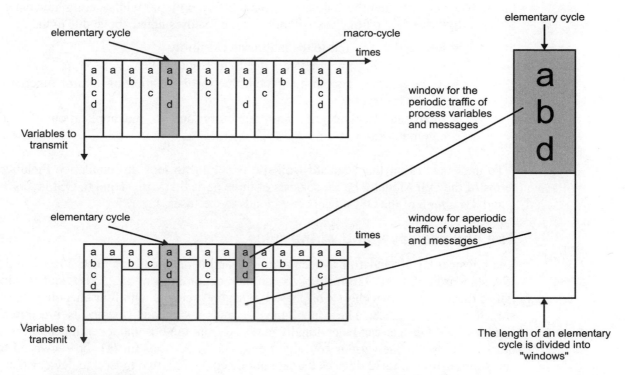

Figure 12.27 FIP
Configuration and scheduling table

FIP uses the Manchester coding to transfer data and synchronizing information. The unique frame start and stop sequences are used to help the receivers distinguish clearly the start and end of the data frames from random noise that may occur on the network. The unusual pattern is clearly different than almost any randomly occurring noise pattern.

12.10   Foundation Fieldbus

Introduction to Foundation Fieldbus

Prior to 1994, two organizations were independently trying to develop a communication solution standard for the process control industry. These organizations were WorldFip (North America) and Interoperable Systems Project (ISP). In September 1994 these two organizations merged to become the Foundation Fieldbus (FF).

Foundation Fieldbus takes full advantage of the emerging 'smart' field devices and modern digital communications technology allowing end user benefits such as:

- Reduced wiring
- Communications of multiple process variables from a single instrument
- Advanced diagnostics
- Interoperability between devices of different manufacturers

- Enhanced field level control
- Reduced startup time
- Simpler integration

The concept behind Foundation Fieldbus is to preserve the desirable features of the present 4–20 mA standard (such as a standardized interface to the communications link, bus power derived from the link and intrinsic safety options) while taking advantage of the new digital technologies. This will provide the features noted above due to the:

- Reduced wiring due to the multidrop capability
- Flexibility of supplier choices due to interoperability
- Reduced control room equipment due to distribution of control functions to the device level
- Increased data integrity and reliability due to the application of digital communications.

To understand how this standard works, it is helpful to look at Foundation Fieldbus in terms of the OSI Model. The FF consists of three parts that correspond to OSI layers 1, 2, 7 and 8. Layer 8 of the OSI model corresponds to the 'user' layer.

The physical layer and wiring rules

The physical layer standard has been approved and is detailed in the IEC 1158-2 and the ISA standard S50.02-1992. It supports communication rates of 31.25 kbps and 10 Mbps. All of these use the Manchester bi-phase L encoding scheme with four encoding states as shown in Figure 12.28. The use of the N+ and N− encoding states is illustrated in Figure 12.29. Devices can be optionally powered from the bus under certain conditions as detailed below for the various configurations. The 31.25 kbps (or H1, or low-speed bus) can support from 2 to 32 devices that are not bus powered, two to twelve devices that are bus powered or two to six devices that are bus powered in an intrinsically safe area. Repeaters are allowed and will increase the length and number of devices that can be put on the bus. The H2 or high speed bus options are not currently being implemented, but have been superseded by the high speed Ethernet (HSE) standard. This is discussed later in this section.

The low speed bus was intended to utilize existing plant wiring and is referred to as Type B wiring (shielded twisted pair) and with #22 AWG can be used for segments up to 1200 m (3936 feet). The higher speeds require higher grade cabling and are referred to as Type A. For Type A cable (shielded twisted pair) for H1 #18 AWG can be used up to 1900 meters (6232 feet). Two additional types of cabling are specified and are referred to as Type C (multi-pair twisted without shield) and Type D (called multi-core, no shield). Type C using #26 AWG cable is limited to 400 meters (1312 feet) per segment and Type D with #16 AWG is restricted to segments less than 200 meters (660 feet).

- Type A #18 AWG   1900 m (6232 feet)
- Type B #22 AWG   1200 m (3936 feet)
- Type C #26 AWG   400 m (1312 feet)
- Type D #16 AWG multi-core   200 m (660 feet)

The Foundation Fieldbus wiring method is floating balanced and equipped with a termination resistor combination connected across each end of the transmission line. Neither of the wires should ever be connected to ground. The terminator consists of a

100 Ω quarter watt resistor and a capacitor sized to pass 31.25 kHz. As an option one of the terminators can be center tapped and grounded to prevent voltage build-up on the Fieldbus. Power supplies must be impedance matched for FF. Off the shelf power supplies must be conditioned. If a 'normal power supply' is placed across the line it will load down the line due to its low impedance. This will cause the transmitters to stop transmitting.

Fast response times for the bus are one of the FF goals. For example, at 31.25 kbps on the H1 bus response times as low as 32 microseconds are expected (this will vary based on the loading of the system but will average between 32 ms and 2.2 ms with an average approximately 1 ms).

Spurs can be connected to the 'home run'. The length of the spurs depends on the type of wire used and the number of spurs connected. The maximum length is the total length of the spurs and the home run.

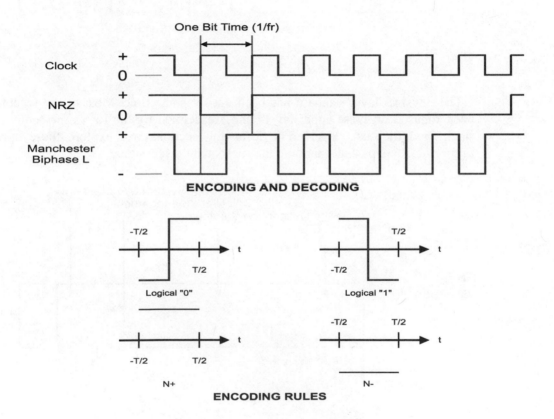

Symbols		Encoding
1	(ONE)	Hi-Lo transition (mid-bit)
0	(ZERO)	Lo-Hi transition (mid-bit)
N+	(NON-DATA PLUS)	Hi (No transition)
N-	(NON-DATA MINUS)	Lo (No transition)

ENCODING RULES

Figure 12.28
Foundation Fieldbus physical layer

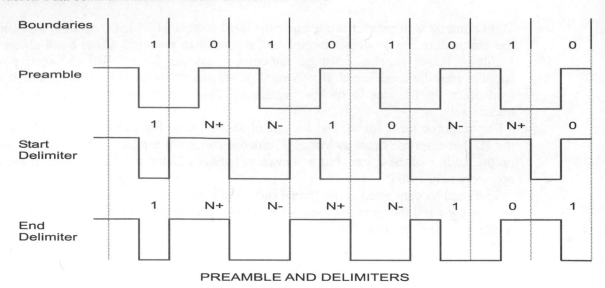

PREAMBLE AND DELIMITERS

Figure 12.29
Use of N+ and N– encoding states

The physical layer standard has been out for some time. Most of the recent work has been focused on these upper layers and are defined by the FF as the 'communications stack' and the 'user layer'. The following sections will explore these upper layers. Figure 12.30 helps understanding the subsequent discussions.

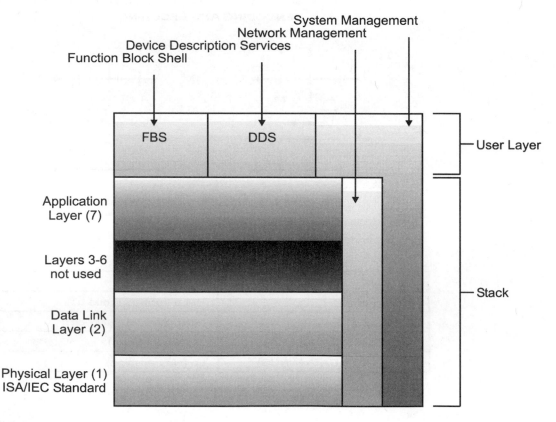

Figure 12.30
The OSI model of the FF protocol stack

The data link layer

The communications stack as defined by the FF corresponds to OSI layers two and seven, the data link and applications layers. The DLL (data link layer) controls access to the bus through a centralized bus scheduler called the link active scheduler (LAS). The DLL packet format is shown in Figure 12.31.

GENERAL PACKET LAYOUT

Preamble	Start Delimiter	Overhead + User Data	Frame Check Sequence	End Delimiter
1 Octet	1 Octet	1 to 256 Octets	2 Octets	1 Octet

Figure 12.31
Data link layer packet format

The link active scheduler (LAS) controls access to the bus by granting permission to each device according to predefined 'schedules'. No device may access the bus without LAS permission. There are two types of schedules implemented: cyclic (scheduled) and acyclic (unscheduled). It may seem odd that one could have an unscheduled 'schedule', but these terms actually refer to messages that have a periodic or non-periodic routine, or 'schedule'.

The cyclic messages are used for information (process and control variables) that require regular, periodic updating between devices on the bus. The technique used for information transfer on the bus is known as the publisher–subscriber method. Based on the user predefined (programmed) schedule the LAS grant permission for each device in turn access to the bus. Once the device receives permission to access the bus it 'publishes' its available information. All other devices can then listen to the 'published' information and read it into memory (subscribe) if it requires it for its own use. Devices not requiring specific data simply ignore the 'published' information.

The acyclic messages are used for special cases that may not occur on a regular basis. These may be alarm acknowledgment or special commands such as retrieving diagnostic information from a specific device on the bus. The LAS detects time slots available between cyclic messages and uses these to send the acyclic messages.

The application layer

The application layer in the FF specification is divided into two sub-layers – the Foundation Fieldbus access sublayer (FAS) and the Foundation Fieldbus messaging specification (FMS).

The capability to pre-program the 'schedule' in the LAS provides a powerful configuration tool for the end user since the time of rotation between devices can be established and critical devices can be 'scheduled' more frequently to provide a form of prioritization of specific I/O points. This is the responsibility and capability of the FAS. Programming the schedule via the FAS allows the option of implementing (actually, simulating) various 'services' between the LAS and the devices on the bus.

Three such 'services' are readily apparent such as:

- Client/server: with a dedicated client (the LAS) and several servers (the bus devices)
- Publisher/subscriber: as described above, *and*
- Event distribution: with devices reporting only in response to a 'trigger' event, or by exception, or other predefined criteria.

These variations, of course, depend on the actual application and one scheme would not necessarily be 'right' for all applications, but the flexibility of the Foundation Fieldbus is easily understood from this example.

The second sub-layer, the Foundation Fieldbus messaging specification (FMS), contains an 'object dictionary' which is a type of database that allows access to Foundation Fieldbus data by tag name or an index number. The object dictionary contains complete listings of all data types, data type descriptions, and communication objects used by the application. The services allow the object dictionary (application database) to be accessed and manipulated. Information can be read from or written to the object dictionary allowing manipulation of the application and the services provided.

The user layer

The FF specifies an eighth layer called the user layer that resides 'above' the application layer of the OSI model; this layer is usually referred to as Layer 8. In the Foundation Fieldbus this layer is responsible for three main tasks – network management, system management and function block/device description services. Figure 12.32 illustrates how all the layer's information packets are passed to the physical layer.

The network management service provides access to the other layers for performance monitoring and managing communications between the layers and between remote objects (objects on the bus). The system management takes care of device address assignment, application clock synchronization, and function block scheduling. This is essentially the time coordination between devices and the software, and ensures correct time stamping of events throughout the bus.

MESSAGE ENCODING/DECODING EXAMPLE

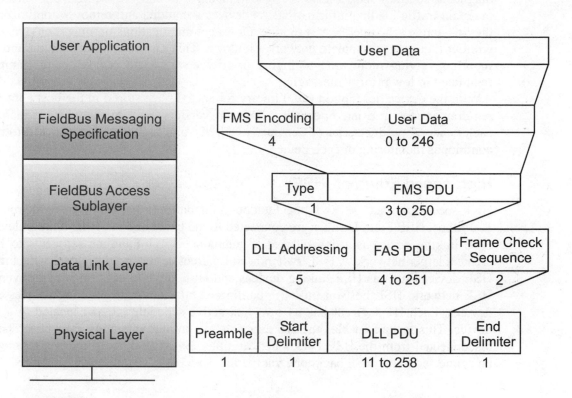

Figure 12.32
The passage of information packets to the physical layer

Function blocks and device description services provide pre-programmed 'blocks', which can be used by the end user to eliminate redundant and time-consuming configuration. The block concept allows selection of generic functions, algorithms, and even generic devices from a library of objects during system configuration and programming. This process can dramatically reduce configuration time since large 'blocks' are already configured and simply need to be selected. The goal is to provide an open system that supports interoperability and a device description language (DDL), which will enable multiple vendors, and devices to be described as 'blocks' or 'symbols'. The user would select generic devices then refine this selection by selecting a DDL object to specify a specific vendor's product. Entering a control loop 'block' with the appropriate parameters would nearly complete the initial configuration for the loop. Advanced control functions and mathematics 'blocks' are also available for more advanced control applications.

Error detection and diagnostics

FF has been developed as a purely digital communications bus for the process industry and incorporates error detection and diagnostic information. It uses multiple vendors' components and has extensive diagnostics across the stack from the physical link up through the network and system management layers by design.

The signal method used by the physical layer timing and synchronization is monitored constantly as part of the communications. Repeated messages and the reason for the repeat can be logged and displayed for interpretation.

In the upper layer, network and system management is an integral feature of the diagnostic routines. This allows the system manager to analyze the network 'on-line' and maintain traffic loading information. As devices are added and removed, optimization of the link active scheduler (LAS) routine allows communications optimization dynamically without requiring a complete network shutdown. This ensures optimal timing and device reporting, giving more time to higher priority devices and removing, or minimizing, redundant or low priority messaging.

With the device description (DD) library for each device stored in the host controller (a requirement for true interoperability between vendors) all the diagnostic capability of each vendors' produces can be accurately reported and logged and/or alarmed to provide continuous monitoring of each device.

High-speed Ethernet (HSE)

High-speed Ethernet (HSE) is the Fieldbus Foundation's backbone network running at 100 Mbps. HSE Field Devices are connected to the backbone via HSE linking devices. A HSE linking device is a device used to interconnect H1 Fieldbus segments to HSE to create a larger network. A HSE switch is an Ethernet device used to interconnect multiple HSE devices such as HSE linking devices and HSE field devices to form an even larger HSE network. HSE hosts are used to configure and monitor the linking devices and H1 devices. Each H1 segment has its own link active scheduler (LAS) located in a linking device. This feature enables the H1 segments to continue operating even if the hosts are disconnected from the HSE backbone. Multiple H1 (31.25 kbps) Fieldbus segments can be connected to the HSE backbone via linking devices.

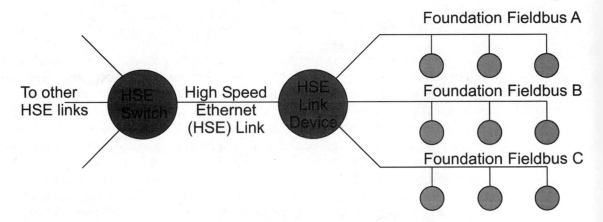

Figure 12.33
High-speed Ethernet and Foundation Fieldbus

13

Local area networks (LANs)

A network is a system for interconnecting various devices, usually in such a way that all users have access to common resources (such as printers) and can communicate with each other. In the 1970s, networks were developed to link terminals to a mainframe computer, laying the foundations for computer networks in general. This chapter is concerned with local area networks (LANs), with a special focus on networks generally used in industrial data communications.

Objectives

When you have completed studying this chapter you will be able to:

- Explain the difference between circuit switched and packet switched networks
- Describe the different network topologies:
    - Star
    - Ring
    - Bus
- Discuss the physical and protocol issues arising when taking different approaches to networking
- Describe baseband and broadband transmission techniques
- Describe the Ethernet standard:
    - Topology
    - Collision avoidance
    - Protocol operation
    - Hardware requirements
    - Performance
- Describe the token ring standard
- Describe the token bus standard
- Describe internetwork connections
- Network operating systems and architectures

13.1 Overview

There are three broad classes of network, although the distinction between them is blurred and they tend to overlap:

- Local area networks (LANs)
  LANs are usually confined to one building or group of buildings within a radius of a few hundred meters. All devices on a LAN are connected to a common transmission medium such as coaxial cable. Transmission speeds are typically up to hundreds of Mbps.

- Metropolitan area networks (MANs)
  A MAN covers a city or metropolitan area, and may have several LANs connected to it. Transmission speeds are generally up to hundreds of Mbps and almost always use optical fiber cable.

- Wide area networks (WANs)
  WANs may cover thousands of kilometers and involve several different transmission media (such as optical fiber, satellite links, microwave and coaxial cable). Transmission speeds vary greatly. An example WAN is the public telecommunications system, which now has 200 Mbps optical fiber links between capital cities and major centers. However, many WAN circuits (such as twisted pair telephone lines) may be limited to a few thousand bits per second.

Refer to Internetwork connections, Chapter 13.17, for a brief discussion about the use of WANs to link LANs.

13.2 Circuit and packet switching

The two basic types of networks are 'circuit' switched and 'packet' switched. In a circuit switched network, a connection is established between the two ends and maintained for the duration of the message exchange (an example is the public telephone system). The advantage is a guarantee of continuity, while the disadvantage is cost. The circuit is tied up even when no one is talking or the transmission rate may be slow.

A packet switched network does not establish a direct connection. Instead, the message is broken up into a series of packets or frames, sometimes known as protocol data units (PDUs). These are transmitted one at a time, each carrying the destination address. Depending on the network conditions, they may take different routes to the destination, and may arrive out of order. It is the job of the protocol software to reassemble the packets in the right order. Packet switching is cheaper as it makes better use of the resources; the physical communications links carry packets from multiple sources concurrently.

Packet-switched network can offer either connectionless or connection-oriented communications, depending on the protocols used. Datagram services are provided in connectionless transmissions. Because connectionless transmissions cannot guarantee delivery of messages, receipt verification is the responsibility of a higher-level protocol. Contrary to this, a virtual circuit is a temporary connection between two points. It appears as a dedicated line to the uses, but actually uses packet switching to accomplish transmission. The virtual circuit is maintained as long as the connection exists.

This book concentrates on the packet switching networks, as they are used almost exclusively in data communications networks.

13.3   Network topologies

The way in which nodes are interconnected is known as the network topology. The three most common topologies are:

- Star
- Ring or loop
- Bus (or multidrop)

Star topology

In the star configuration, there is a central node or hub and all the outlying nodes communicate back to it on separate communication links. A typical example of a star network is a timesharing mainframe computer system, where the central node is the computer itself and the outlying nodes are the user terminals. With the star network, each outlying node is connected to the central node via its own cable. Figure 13.1 shows the topology of a star network.

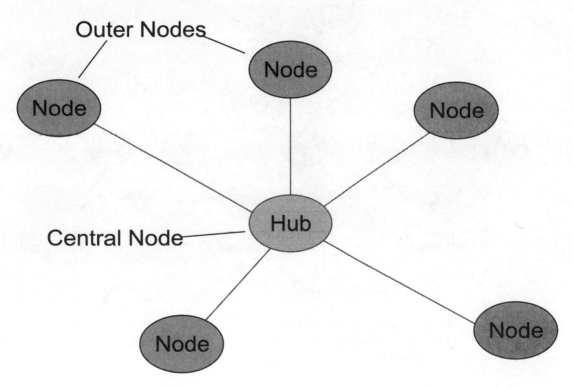

Figure 13.1
Example of a star network topology

The central hub must have the capacity to simultaneously send and receive messages. Contention problems are overcome by using buffers in the computer at the hub of the star network.

The major disadvantages of the star network are:

- If the central hub is disabled, the entire system is inoperable.
- Depending on the physical layout, the costs of cabling a star network tend to be higher than some of the alternatives.

Ring topology

Nodes in a ring network are connected node to node, ultimately forming a loop. The data flow is often arranged to be unidirectional, with each node passing data on to the next node and so on. It is essential that each node, when receiving, is capable of removing data from the ring so that it does not circle through the network indefinitely.

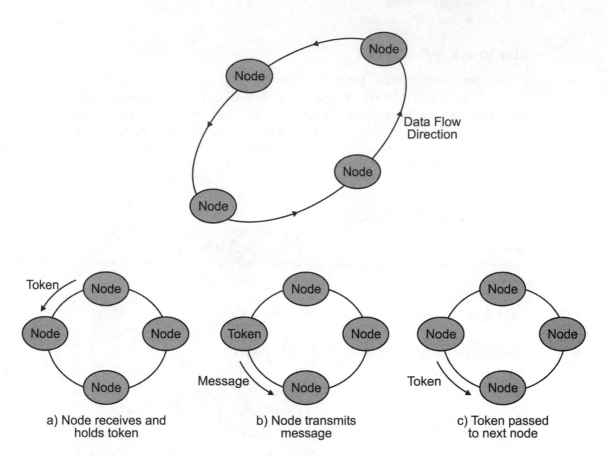

Figure 13.2
Example of ring network topology

The basic ring network is unable to function if a node is disabled or the ring broken. Consequently, if modifications or additions are to be made to the network, a complete system shutdown is necessary. Figure 13.2 shows the topology of a ring network.

Bus topology

The bus topology consists of a communication path with nodes connected to it like leaves off a branch. The nodes are not physically inserted into the bus, as is the case with ring topology, but are 'teed-off' the bus.

Bus networks can be bi-directional or unidirectional. Figure 13.3 shows the topology of some typical bus networks.

Data on the bus can be 'seen' by all the nodes, but only the destination node or nodes will copy the information.

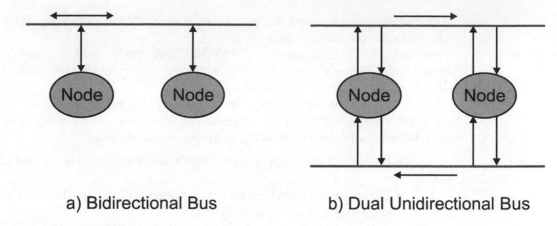

a) Bidirectional Bus   b) Dual Unidirectional Bus

Figure 13.3
Examples of bus network topology

In the event of a disabled node, either due to a failure or an access control malfunction (i.e. the destination node not taking its information from the bus), the network can continue to operate as before, without the malfunctioning node.

As any of the devices on the network can send data at any time, there is a potential 'traffic control' problem. A number of protocols have been developed to regulate the access of devices to the bus network to avoid data collisions. Ethernet is one of the best known and most successful of the LANs using the bus topology. It is described in more detail in the following sections.

13.4   Media access control mechanisms

The three main methods of controlling access to the medium are:

- Master-slave (or poll-response) mode
- Token passing
- CSMA/CD

Master-slave (or poll-response) mode

This is a common method of gathering data from an instrument connected to a bus. When one master node is connected to several slave nodes, the master-slave method works as follows:

- The master node sends a message addressed to the first slave, in a sequence of nodes, requesting data (or writing data to the slave).
- All the slave nodes on the bus read the message, but if the destination address of the message does not match their own address, they discard the message.
- The correctly addressed slave node reads the message and checks for any errors (e.g. frame size and CRC or BCC errors).
- If the addressed slave node does not immediately respond, the master node typically attempts to transmit three read requests (in total) before retrying to transmit a message to the next slave node in the sequence.
- The master node then cycles through all the nodes on the bus requesting data (or writing data) to each of them. A complete cycle is referred to as a polling-cycle.

The master node may need to retry a failed node every few polling-cycles to see if it has recovered and now wants to transfer data.

Depending on the configuration, the physical link can be used in a half-duplex or full-duplex mode. For example, using RS-232 would allow full-duplex mode for only two nodes, (one at each end of the link).

The advantage of this approach is that the software is easy to setup as the master node is in total control and will never receive unsolicited messages. The polling-cycle approach is very efficient if any one, or all, of the following situations exist:

- Each slave has a constant predictable amount of data to be read and written to it.
- There is no need to transfer data between slave nodes (but only from the slave node back to the master node).
- A slave node will never need to jump ahead of the polling queue to transfer urgent data back to the master node.
- There is a very low probability of failure of a slave or master node.

The disadvantage of this approach is that when a slave node has an unpredictable amount of data and may want to get onto the bus at certain times to transfer critical data to the master or any other node.

The polling-cycle approach is commonly used for instruments connected back to a central node (such as a PLC or computer). The physical topology used is typically a bus type network.

Token passing

Token passing is a common methodology for industrial control systems requiring a guaranteed transfer of data between nodes that are network peers.

There are two types of messages:

- Token messages (which transfer control of the network from one node to the next)
- Application messages (which transfer actual data from one node to the other)

The approach would work as follows for a group of nodes connected on a bus network:

- A node receives the token message from a neighbor node, indicating that it now has control.
- The token will remain at this node for a specified maximum length of time or until it has transferred it messages, whichever is the shorter period of time.
- This node now transmits messages to other nodes with which it wants to communicate.
- The token is forwarded to the next node and the process repeats itself.

The use of a token ensures that each node on the network will be allowed to transmit within a given time slot. This is called deterministic operation.

Tokens can be used on any network topology. Examples are:

- Arcnet (star)
- Modbus (bus)
- IBM token ring (ring)

This is very common method where transfer of data (such as alarm conditions) is vital and worst-case access time to the medium must be guaranteed.

Carrier sense multiple access/collision detection (CSMA/CD)

CSMA/CD is the simplest method of passing data, on a bus, between nodes that want to communicate in a peer-to-peer fashion. It is becoming increasingly popular in industrial systems because of its ease of implementation and low cost.

The system works as follows for nodes connected to a bus:

- A node that wants to transmit first listens for any bus activity. If there is no activity detected the node will transmit a message.
- As the node transmits a message it compares the message being sent with that being present on the bus. If it detects a mismatch it will immediately stop transmitting as this signifies an error on the system (either due to noise or to another node or nodes transmitting at the same time).
- If there is a collision, the affected node will stop transmitting and wait for a random period of time before trying to transmit again.
- The node backs off for a random time. This reduces the risk of a collision when the node tries to retransmit.

The advantages of CSMA/CD are its simplicity and speed, when the bus is lightly loaded.

The disadvantage with CSMA/CD is that all activity on the bus is unpredictable (and the greater the loading of the bus the more unpredictable it becomes). In fact, during extreme high-activity periods the bus may almost cease to function and no data transferred.

CSMA/CD uses a bus topology and operates over a half-duplex connection. The most common example of CSMA/CD is Ethernet (or the IEEE 802.3). Canbus (or DeviceNet and SDS) use a similar version called CSMA/CA carrier sense multiple access/collision avoidance, as discussed in section 12.5. The difference here is that collisions are not destructive and the highest priority node continues transmitting and all others stop.

13.5 Transmission techniques

Two main methods used for the transmission of information over a LAN are baseband and broadband.

Baseband

This is also known as time division multiplexing (TDM). Only one device is allowed to transmit at any one time and can use the entire bandwidth of the system. No carrier is used, so the signal (e.g. the output from a UART) is directly applied to the medium.

As one communicating device takes over the medium during transmission, there has to be a time limitation on individual device access so that other devices can take their turns.

Broadband

Broadband is also known as frequency division multiplexing (FDM). The system bandwidth is divided into channels that do not overlap, meaning that many pairs of devices can communicate simultaneously and they usually retain their channel until the message transfer is complete. As only a part of the system bandwidth is available, data transfer rates for individual communications are less than for TDM using the same physical setup.

Data is transmitted by injecting a carrier (sine) wave on to the medium and modulating the carrier with the data – be that by frequency modulation, amplitude modulation, or phase modulation.

Coaxial cable and optical fiber are preferred for FDM because the bandwidth of a twisted pair is generally not sufficient to make the technique worthwhile. However, special new twisted pair types such as TwistLAN offer up to 16 Mbps, making them also suitable for FDM use.

13.6   Summary of LAN standards

The most important standard for LAN interfaces and protocols is IEEE 802, a series administered by the IEEE 802 LAN Standards Technologies Committee. The standard has several sections, each with its own co-coordinating committee. Some standards have been superseded by ISO standards as shown in brackets in the descriptions below.

- IEEE 802.1 (ISO 8802.1)
  Details how the other 802 standards relate to one another and to the ISO/OSI reference model. As with the ISO/OSI model, the IEEE 802.1 specification describes interface layers or communication interfaces between different hierarchical levels of devices and activities.

- IEEE 802.2 (ISO 8802.2)
  Standard has divided the ISO/OSI data link layer into two sublayers and defines the functions of the logic link control (LLC) sublayer and the media access control (MAC) sublayer. An IEEE 802.2 interface defines services that fall into the following categories:

Interpreting message packets

Generating appropriate responses to errors and acknowledgment of message packets
Both Ethernet and IBM's token ring are compatible with this section.

- IEEE 802.3 (ISO 8802.3)
  This standard defines the carrier sense multiple accessing with collision detection (CSMA/CD) protocol, which is the protocol used by Ethernet and is described in the section on Ethernet.

- IEEE 802.4 (ISO 8802.4)
  This standard defines the token passing bus access method. Physically this looks like a bus network with the operation like that of a token ring network. Nodes on the network see themselves as being arranged in a logical loop, with each node assigned an address. Collision problems that occurred in CSMA/CD are solved as only one token can exist on the network at one time, and only one node may own the token. Token holding times exist so that no node may own the period for prolonged periods of time. Token bus is superior to CSMA/CD for networks with heavy loads, because each node has a regular time to communicate.

- IEEE 802.5 (ISO 8802.5)
  This standard defines a token ring access method, similar to that used by IBM for their token ring standard.

13.7 Ethernet

Ethernet was developed by Xerox in the early 1970s and standardized by Xerox, Digital Equipment and Intel in 1978. It uses CSMA/CD as a medium access control method.

The relevant standards are:

- Ethernet V2 (Bluebook)
- IEEE 802.3 (ISO 8802.3)

It should be noted that Ethernet is a trade name for a proprietary LAN system Ethernet Version 2, commonly called 'Blue Book Ethernet', and although it is now virtually a standard in itself, it is not the only LAN of that type. The IEEE '802.3 LAN' has a slight difference in its frame format, but is commonly known as 'Ethernet'. This convention is used here.

To ensure every node address is unique, the IEEE assigns blocks of addresses to manufacturers, who are responsible for ensuring that no address is duplicated. The 48-bit address is incorporated into the firmware. There should therefore never be any address conflicts on a LAN.

Below are listed some of the hardware variations of Ethernet, which are explained more fully later in 'Ethernet hardware requirements' in sections 13.10 and 13.13

- Standard (or thick) Ethernet (10Base5) uses 10 Mbps baseband operation on coaxial cable with a maximum segment length of 500 m.
- Thin Ethernet (10Base2) uses 10 Mbps baseband operation on coaxial cable with a maximum segment length of 185 m.
- 10BaseT uses unshielded twisted pair cables and operates at 10 Mbps with the use of a wiring hub onto which each node connects.
- 100BaseT is similar to 10BaseT but operates at 100 Mbps.
- 1000BaseT (gigabit Ethernet) is similar to Fast Ethernet, but operates at 1000 Mbps.
- 1BaseT is similar to 10BaseT but limited to a 1 Mbps data rate. Obsolete.
- Broadband Ethernet (10Broad36) uses FDM with maximum segment length of 3600 meters (11,800 feet). Obsolete.
- 10BaseF is a 10 Mbps baseband system operating on optic fibers.

Ethernet topology

Standard and thin Ethernet use a bus topology, in which each node attaches to the communications cable as shown in Figure 13.4. In this arrangement the removal of a node does not affect operation of the LAN; it simply means that the node will then not reply to frames addressed to it.

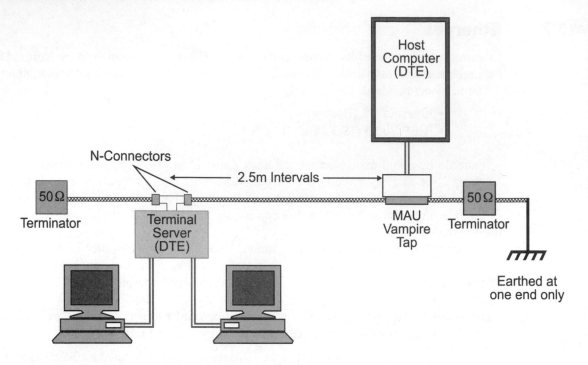

Figure 13.4
Standard Ethernet (10Base5) bus topology

10BaseT (Ethernet) uses a star configuration in which each node is connected via two twisted pairs to a wiring hub as shown in Figure 13.5. A hub may be connected to some other type of Ethernet; operation is the same as for the bus configuration.

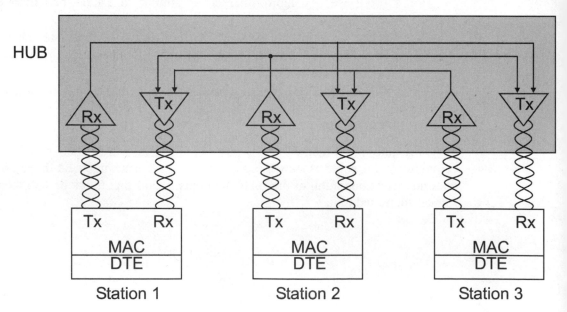

Figure 13.5
10BaseT star topology

13.8   Medium access control

As every node has access to the LAN at all times, it is possible that two nodes may transmit simultaneously, thus disrupting communications. Ethernet uses the carrier sense multiple access with collision detection (CSMA/CD) technique to detect and minimize collisions between frames.

A node always waits until there is no carrier present on the line (indicating that no other node is transmitting) before sending a frame. It then monitors the cable as it transmits, and any difference between the signals it is sending and those it is receiving indicates that a collision has occurred (that is, another node has started transmission).

If a collision is detected, the node enforces it by sending a random bit pattern (the jam sequence), then waits for a short random interval before trying again. As each of the two nodes is waiting for a different interval, the chances are that a collision will not occur the next time.

As a node may have to wait for signal propagation to reach the other end of the bus before a collision occurs, there is a minimum length for messages to ensure collision detection in all circumstances. The hardware is required to pad short messages to reach this minimum length.

Figure 13.6 illustrates the sequence of events when a collision occurs. Note that node 1 must wait for twice as long as it takes for the signal to propagate to node 2, and must be transmitting all this time in order to be able to detect a collision.

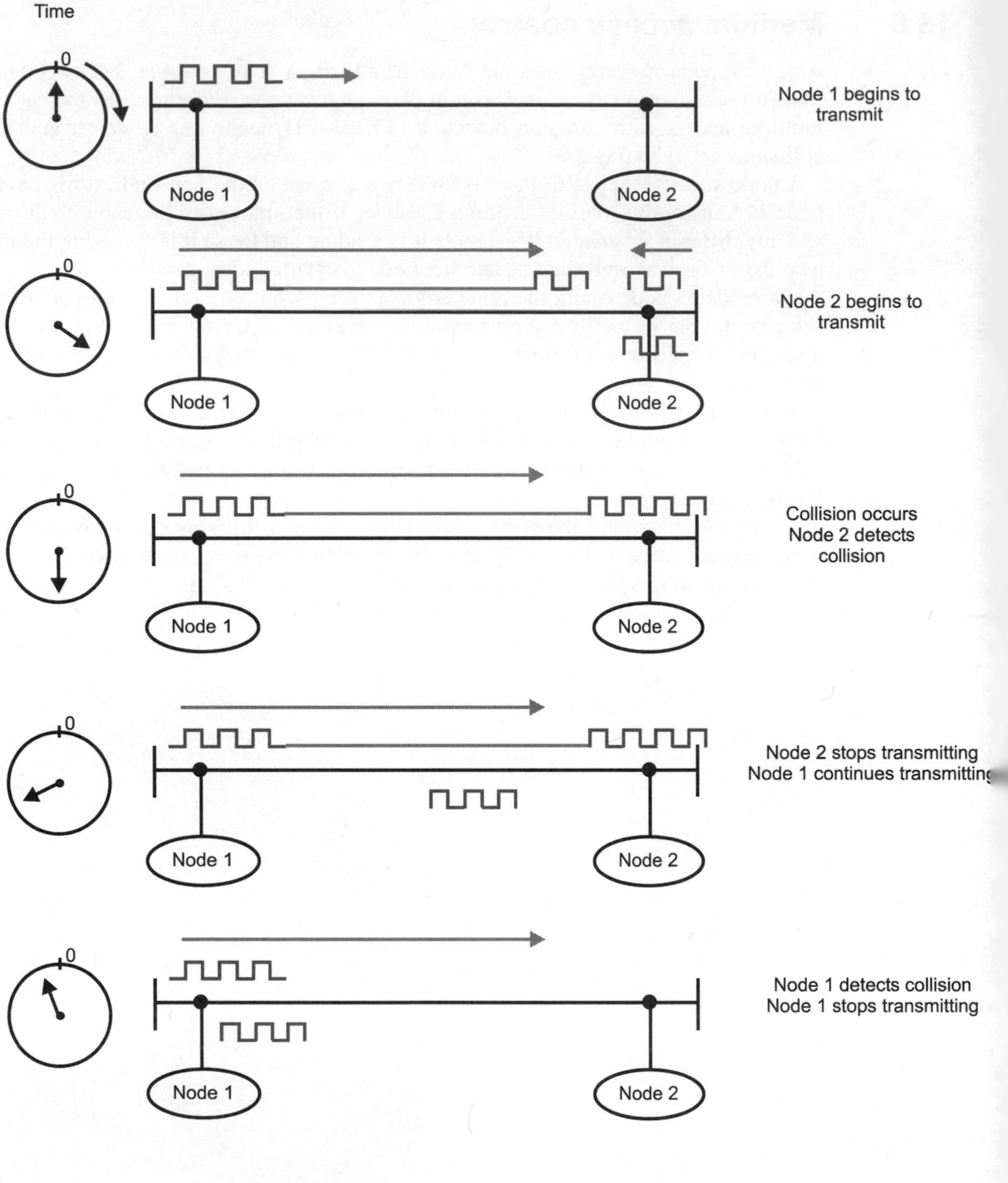

Time

Figure 13.6
CSMA/CD collision

13.9   Ethernet protocol operation

Data frame

All data transfer is in the form of a packet or frame. It consists of an envelope containing
control information (such as synchronization bytes and addresses) and the actual message

data. Each node examines the destination address and reads the data if the frame is directed at that node.

An Ethernet data frame has the structure shown in Figure 13.7.

Preamble	Start Delimiter	Destination Address	Source Address	Length	Data	CRC
7 Bytes	1 Byte	2 or 6 Bytes	2 or 6 Bytes	2 Bytes	46 - 1500 Bytes	4 Bytes

Figure 13.7
Format of an IEEE 802.3 frame

Preamble

This field comprises seven bytes, each with the binary value 10101010. Its purpose is to allow all receiving MAC units to synchronize with the frame.

Start of frame delimiter (SFD)

The SFD indicates the start of a frame and has the binary value 10101011.

Destination address

This may be either 16 bits or 48 bits, depending on how the system is configured. In practice it is almost always 48 bits. It must be the same length for every node. The node with this address will read the data.

Source address

This may be either 16 bits or 48 bits, depending on how the system is configured. In practice it is almost always 48 bits. This is the address of the node that sent the data.

Length indicator

The two-byte length indicator specifies how many bytes are in the data field.

Data

This is the actual message data and can be from 46 to 1500 bytes in length. The minimum value is determined by the need for collision detection and the maximum value limits the access time for any one node to 1, 2 milliseconds. If the actual data is less than 46 bytes it must be 'padded' up to 46 bytes.

Frame check sequence

This is a 32-bit cyclic redundancy check value used for error detection.

Transmission sequence

       1 The MAC unit applies the frame envelope to the data.
       2 The MAC unit monitors the line until no carrier is sensed.
       3 The node sends the frame as a bit stream, via the transceiver unit.
       4 The transceiver monitors the line in order to detect a collision.

If there is a collision, the transmitter sends a jam sequence, terminates transmission and tries again after a short random interval.

Reception sequence

1 The MAC unit detects an incoming signal from the transceiver.
2 The carrier sense signal is switched on to inhibit any transmission by the MAC unit.
3 The MAC unit uses the incoming preamble bytes to achieve synchronization.
4 The destination address is checked, and reading continues if it matches this node.
5 The data is read, and the length checked against the length indicator. The FCS is calculated and checked against the FCS field value. If both are OK, the message is passed on.

Data encoding and transmission

Clock signals are encoded into the transmitted bit stream and extracted by each receiver for synchronizing purposes. That is, the system is 'self-clocking'. The encoding and decoding process is illustrated in Figure 13.8. The original data stream varies between two levels and does not return to zero symbols. This is known as a non-return to zero (NRZ) signal. This data is encoded for transmission using Manchester (or phase) encoding where a binary 1 is encoded as a 'low to high' transition in the middle of each bit and a binary 0 as a 'high to low' transition. There is always a transition in the middle of each bit and this is used to extract the received clock signal. This received clock signal defines when the receiver samples the encoded data, which is either high (for binary 1) or low (for binary 0), which ensures a correct reproduction of the original data.

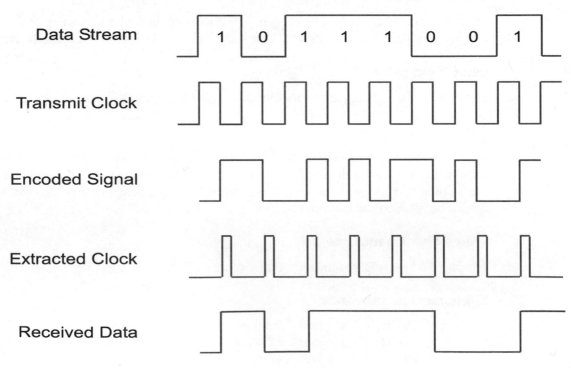

Figure 13.8
Manchester encoding and decoding

13.10 Ethernet hardware requirements

The hardware requirements will depend on the variety of Ethernet being used; these are described below.

Standard (Thick) Ethernet

The standard 10Base5 Ethernet requires a 10.28 mm outside diameter 50 ohm cable (RG-8) that can carry a clamp-on tap forming a T junction. A transceiver media attachment unit (MAU) is joined to the tap. A transceiver drop cable or attachment unit interface (AUI) connects the MAU to the media access control (MAC) unit. In the case of an instrumentation system, the MAC unit is built into the controller card and includes the protocol control firmware. This makes up the communication subsystem, acting as a DTE.

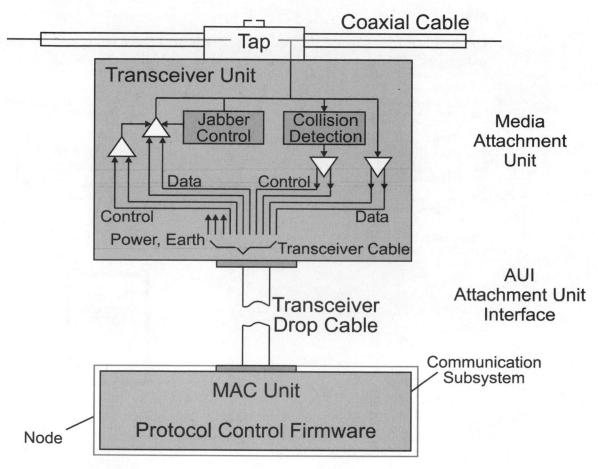

Figure 13.9
Standard Ethernet hardware

Within the transceiver, the jabber control section disconnects the node from the bus if it detects excessive transmission activity. This prevents a faulty node from disrupting all network communications. The collision detection section implements the CSMS/CD method described previously.

MAU

The MAU can be attached to the cable in one of two ways:

- A 'vampire' tap. In older systems it was necessary to predrill a hole for connection to the center conductor of the coaxial cable. Modern systems use a simple insulation displacement method whereby 'spikes' are driven into the screen and center conductor by means of a small wrench or screwdriver.
- An N-connector MAU has two female N-connectors. The cable has to be cut and fitted with two female N connectors. This approach is preferable in a dirty environment such as a factory.

An MAU can have one, two or four ports. The standard requires a minimum interval of 2.5 meters (8.2 feet) between MAUs. A 500 meter (1640 feet) segment (the maximum) can have up to 100 MAUs connected.

AUI

The AUI is a 15-conductor shielded cable, consisting of five individually shielded pairs and may be up to 50 meters (164 feet) long. Figure 13.10 illustrates an AUI.

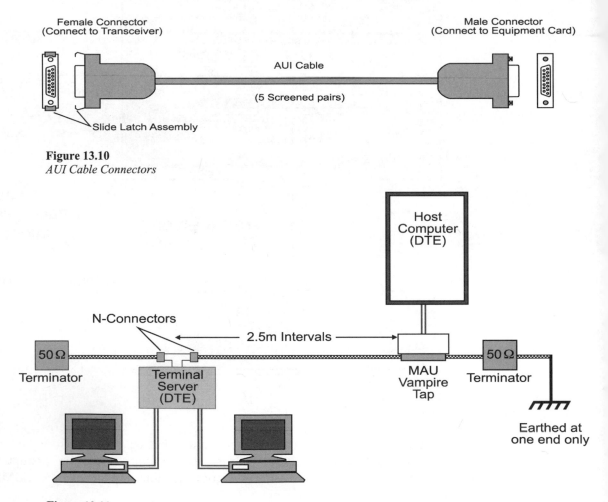

Figure 13.10
AUI Cable Connectors

Figure 13.11
Standard Ethernet cabling example

It is recommended that all cabling within a system be taken from the same reel to avoid reflections due to impedance mismatches. Splicing is achieved via male N-connectors on the cable ends, joined by a female-female connector barrel.

Thin Ethernet

A thin Ethernet system uses 50 Ω RG-58 A/U or C/U cable and can have a maximum segment length of 185 m (607 feet). It is designed for lower installation costs.

The setup is the same as for standard Ethernet, except that the minimum spacing between MAUs is 0.5 m (1.7 feet) and up to 30 MAUs are allowed per segment. A segment cable should not be spliced, but joined with barrel connectors.

At a controller, the MAU and AUI are typically integrated into the card as shown in Figure 13.12.

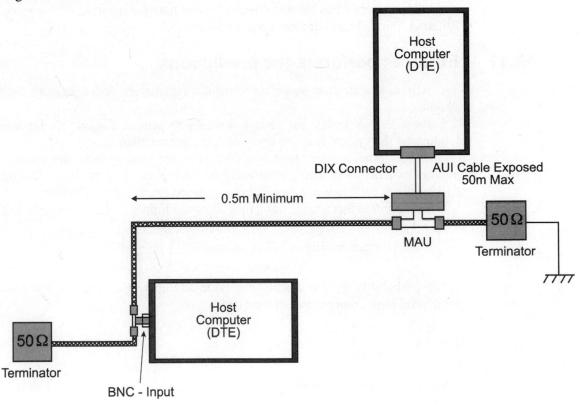

Figure 13.12
Thin Ethernet cabling example

10Base5 components can be used on 10Base7, but the MAU cannot use a vampire tap or N-connector. The cable is different therefore a BNC T-piece is used.

10BaseT

A 10BaseT system provides lower installation costs again and uses twisted pair cables. Each node is connected to a central hub by two pairs (one for receiving and one for transmitting). The hub can be up to 100 m (330 feet) away. Connection to the interface cards is by modular RJ-45 8-pin plugs as shown in Figure 13.13.

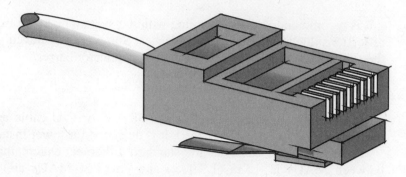

Figure 13.13
RJ-45 connector for twisted pairs

A hub provides a bus between the nodes and may incorporate AUI connectors for thick Ethernet, thin Ethernet or fiber optic transceivers.

13.11   Ethernet performance predictions

We will now calculate some performance parameters for a heavily loaded Ethernet system.

Assume that N nodes are always queuing to send a packet. N represents the total loading on the system and it is assumed to be greater than 1.

On an Ethernet system there is a contention interval or time slot during which nodes attempt to transmit before guaranteeing to have control of the bus. During this time of 512 bit times collisions are possible due to propagation delays on the network.

If it is assumed that a node has a probability of transmitting in any time slot of

$\dfrac{1}{N}$, then the probability of being delayed is: $\left[1 - \dfrac{1}{N} \right]$

The probability P, that exactly one node attempts to transmit in the time slot, and is successful in gaining control of the bus is:

$$P = \left[1 - \left\{ \frac{1}{N} \right\} \right]^{N-1}$$

So, the probability of waiting no time before gaining control of the bus is P.

The probability of waiting only one time slot is $P(1-P)$ and the probability of waiting N time slots is $P\left[(1-P)^N \right]$.

This is a geometric progression that has a mean of $\dfrac{(1-P)}{P}$

Therefore, the mean number of time slots a node must wait before getting access to the bus is:

$$S = \frac{(1-P)}{P}$$

For two stations queuing continuously $P = \left\{ 1 - \dfrac{1}{2} \right\}^{2-1} = 0.5$

and $\quad S = \dfrac{1-P}{P} = \dfrac{0.5}{0.5} = 1$

The efficiency E is given by the percentage of time the network is transmitting useful data. Each packet has a maximum of 192 overhead bits, consisting of:

7 bytes preamble
1 byte   SFD
6 bytes destination address
6 bytes source address
2 bytes length
4 bytes CRC

The maximum packet size is 1526 bytes or 12208 bits while a minimum sized packet is 72 bytes or 576 bits. Between each frame is a gap of 96 bit times or 9, 6 microseconds at 10 Mbps.

For a maximum sized frame:

$$E = \frac{Useful\ data}{Useful\ data + waiting + interframegap}$$

$$E = \frac{12\ 208}{12\ 208 + 512 + 96} = \frac{12\ 208}{12816} = 95\%$$

For a minimum length frame: $\quad E = \dfrac{656}{656 + 512 + 96} = 52\%$

This reduction in efficiency is caused by the increasing proportion of time for the fixed overheads and interframe gap.

13.12 Reducing collisions

The main cause of collisions is the signal propagation time between nodes. Effectively, the delay in a node picking up the fact that there has been a collision can be up to twice the propagation delay between the two nodes. This round trip time is often referred to as the collision window. This time is usually fairly short – in the order of a few microseconds. With the maximum configuration for Ethernet consisting of five 500 meters (1640 feet) cables, four bit repeaters, ten transceivers and ten 50 meters (164 feet) transceiver cables, the round trip propagation through all the cables, transceivers and repeaters can be as much as 50 microseconds. This is equivalent to about 500 bit times at 10 Mbps. Note that the minimum length of an Ethernet message is 64 bytes, or 512 bits, which represents 51.2 ms. Hence, collisions can always be detected within one packet.

The main reasons for collision rates on an Ethernet network are the:

- Number of packets per second
- Signal propagation delay between transmitting nodes
- Number of nodes initiating packets
- Bandwidth utilization

A few suggestions on reducing collisions in an Ethernet network are:

- Keep all cables as short as possible
- Keep all high activity sources and their destinations as close as possible. Possibly isolate these nodes from the main network backbone with bridges/routers and switches to reduce backbone traffic
- Use buffered repeaters rather than bit repeaters
- Check for unnecessary broadcast packets which are aimed at non existent nodes
- Remember that the monitoring equipment to check out network traffic can contribute to the traffic (and the collision rate)

13.13 Fast Ethernet

Fast Ethernet systems operate at 100 Mbps on different forms of physical media and they retain the existing Ethernet MAC layer.

IEEE 802.3u standard defines:

- 100BaseTX, which uses two pairs of category 5 UTP or STP and is the most commonly used standard.
- 100BaseFX, which uses two pairs of multimode (or single node) fiber.
- 100BaseT4, which uses four pairs of category 3, 4 or 5 UTP. This is no longer used.

The IEEE also has a standard 802.3y that defines 100BaseT2, which was to use two pairs of category 3, 4 or 5 UTP. This system has not been developed commercially.

The original 802.3 MAC layer was defined independently of the various physical layers it supports. The MAC layer defines the CSMA/CD access mechanism, and defines most parameters in terms of bit-time intervals, which are independent of speed. With 100 Mbps systems, the interframe gap and the time to transmit the frame are reduced to one tenth that of the 10 Mbps system.

The 100BaseTX systems are star wired to a hub in the same configuration as the 10Base T systems. Modern copper based systems use 10/100 nodes and hubs. Here the node and the hub exchange capability information and adjust to the appropriate speed (i.e. 10 Mbps or 100 Mbps).

The system is designed to operate over 100 meters (328 feet) of Cat 5 UTP between node and hub. The maximum size network in which collisions can be detected is 250 meters (820 feet), being one tenth the size of the maximum 10BaseT network. This effectively means that networks greater than 200 meters (656 feet) need to be logically connected by store and forward devices such as bridges, routers or switches. This is discussed in section 13.17.

Fast Ethernet is becoming cheaper and is now widely used for industrial applications. Gigabit Ethernet has been commercially available since 1998.

13.14 Token ring

The token ring LAN was developed by IBM in the 1980s, and involves the passing of a 'token' a special frame as the method of collision avoidance.

The relevant standards are:

- IBM token ring
- IEEE 802.5 (token ring)

Topology

This is a ring network in which the transmission medium forms a closed loop. Data is transmitted in one direction as shown in Figure 13.14.

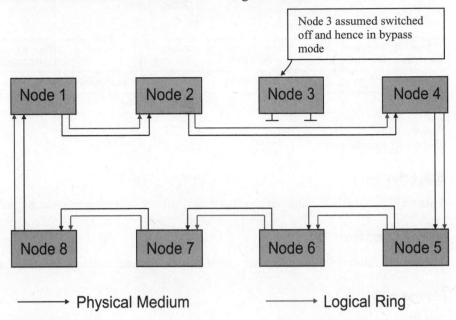

Figure 13.14
Token ring topology

Note that failure at any node will put the network out of action unless special precautions are taken, making a plain token ring unsuitable for industrial applications.
The ring can be modified in the following ways to improve reliability:

- Each node can have a set of bypass relay switches so that ring continuity is maintained even when a node fails or is switched off
- Ring architecture can be duplicated to provide two communication paths

Media access control

A special empty frame called a token is passed from one node to another, and a node can transmit data only when it holds the token. After confirming transmission of a data frame, the node generates a new token and sends it to the next node. This means that collisions cannot occur.

Protocol operation

As mentioned above, a node cannot transmit a data frame unless it holds the token.
There is a time limit on how long a node can hold the token.

Data frame

A token ring data frame has the structure shown in Figure 13.15.

Start Delimiter	Access Control	End Delimiter
1 Byte	1 Byte	1 Byte

(a) Control Token

Start Delimiter	Access Control	Frame Control	Destination Address	Source Address	Information	Frame Check Sequence	End Delimiter	Frame Status
1 Byte	1 Byte	1 Byte	2 or 6 Bytes	2 or 6 Bytes	0 or more Bytes	4 Bytes	1 Byte	1 Byte

(b) Message Frame

Figure 13.15
Token Ring data frame

Start delimiter

This field indicates the start of the frame and is encoded with non-data symbols.

Access control

The field contains access and reservation sub-fields and controls access to the ring.

Frame control

This distinguishes normal data frames from protocol control (MAC) frames.

Destination address

This may be either 16 bits or 48 bits, depending on how the system is configured. It must be the same length for every node. The node with this address will read the data.

Source address

This may be either 16 bits or 48 bits, depending on how the system is configured. It must be the same length for every node.

Information

This is the actual message data and has no upper limit apart from that related to the maximum time the node is allowed to hold a token.

Frame check sequence

This is a 32-bit cyclic redundancy check value used for error detection. It is calculated over the entire frame apart from the Start and End Delimiters.

End delimiter

Signals the end of the frame.

Token format

A token has the format shown in Figure 13.15(a). Field descriptions are as for the data frame.

Transmission sequence

1 The MAC unit formats a frame ready for transmission.
2 When the node receives a token with the same priority as the frame, the MAC unit transmits the frame.
3 The MAC unit receives the frame back and checks that it has been read by the destination node.
4 The MAC unit forwards the token to the nearest downstream active neighbor, which repeats the procedure.

Reception sequence

1 The MAC unit receives a frame from the nearest active upstream neighbor.
2 If the destination address matches the node, the MAC unit copies the frame and sets an access control flag to indicate that it has read the information. It passes the frame on to the nearest downstream active neighbor.
3 If the destination address does not match the node, the MAC unit passes the frame on without taking any action.

Hardware requirements

The hardware layout for a token ring system is shown in Figure 13.16.

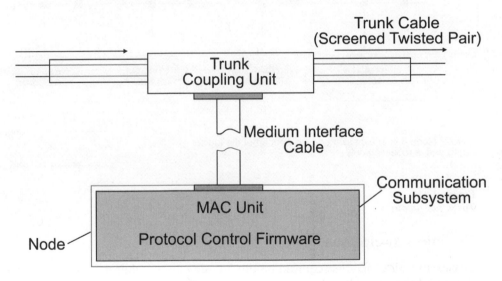

Figure 13.16
Token ring hardware

Although they use a logical ring structure, token rings are actually arranged in a hub (star) topology, with each node connected to a central hub or MAU (multistation access unit). MAUs can be interconnected by using special RI (ring in) and RO (ring out) ports on the MAUs.

13.15 Token bus

Token bus network provides guaranteed access for all nodes at regular intervals and can prioritize frame transmission. Operation is either broadband or a single-channel carrier band. Token bus architecture supports the following:

- Both carrier band (single channel) and broadband networks
- Operation over either 75 Ω coaxial or fiber optic cable
- Network speeds of 1, 5, 10 and 20 Mbps depending on the medium
- Four priority levels for regulating access to the network medium
- Four physical layer medium configurations; two carrier band (full bandwidth), one broadband and an optical configuration.

An open standard for token bus is IEEE 802.4. However, most token bus systems are proprietary standards such as MAP, Allen Bradley, Data Highway Plus, Modbus Plus or Honeywell TDC300.

Topology

The physical topology is that of a bus, but overlaying that is a logical ring topology (based on addresses) as illustrated in Figure 13.17.

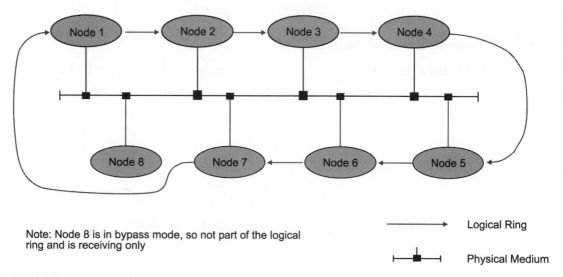

Note: Node 8 is in bypass mode, so not part of the logical ring and is receiving only

⟶ Logical Ring

├──■──┤ Physical Medium

Figure 13.17
Token bus topology

Collision avoidance

Possession of a token is required before a node can transmit a frame, in the same way as described for token ring.

13.16   Token bus protocol operations

A node waits until it receives the token, which is passed around the logical ring from node to node. Each node knows its predecessor and successor; and the physical position of nodes on the bus is of no consequence as long as all nodes read all messages. The token is passed from higher addresses to lower addresses.

At initialization, each node puts a message on the bus demanding to be the token holder. The node with the highest address is successful, and then starts the procedure described in 'addition of nodes' to setup the logical ring and start token rotation.

When a node has messages to transmit, it can keep the token up to a predefined time and may send as many frames as desired within that time limit. If the node finishes transmissions before the timeout is reached, it must immediately generate a new token and send it to its successor on the logical ring. This scheme is useful for industrial communications networks because the maximum time a node must wait before it can send a message is known.

Messages can be assigned one of four priority classes, and the higher priority messages are always transmitted first. This means that during heavy network traffic the lower priority messages may be delayed. Token rotation time increases, as each node tends to hold the token for the maximum period to reduce the message backlog.

Messages can be broadcast to all nodes or multicast to groups of nodes.

Data frame

A token bus data frame has the structure shown in Figure 13.18

Preamble	Start Delimiter	Frame Control	Destination Address	Source Address	Information	Frame Check Sequence	End Delimiter
0 to n Bytes	1 Byte	1 Byte	2 or 6 Bytes	2 or 6 Bytes	0 to 819 Bytes	4 Bytes	1 Byte

Figure 13.18
Token bus data frame

Preamble field

This is a field that allows the MAC unit of the receiver to synchronize with the frame.

Start delimiter

This field indicates the start of the frame and is encoded with non-data symbols.

Frame control

This distinguishes normal data frames from protocol control (MAC) frames.

Destination address

This may be either 16 bits or 48 bits, depending on how the system is configured. It must be the same length for every node. The node with this address will read the data. Individual node addresses have the least significant bit (LSB) set to 0; multicast (group) addresses have the LSB set to 1. A broadcast message has all bits set to 1.

Source address

This may be either 16 bits or 48 bits, depending on how the system is configured. It must be the same length for every node.

Data

This is the actual message data and can be up to 8191 bytes long.

Frame check sequence

This is a 32-bit cyclic redundancy check value used for error detection. It is calculated over the entire frame except the preamble and start and end delimiters.

End delimiter

Signals the end of the frame and is encoded identically to the start delimiter.

Addition of nodes

The protocol includes a mechanism for the automatic addition of nodes as they come on line. At regular intervals the current token holder sends an inquiry control frame containing its own address and that of the node it has listed as its successor. Three results are possible:

- There is no response within a defined period, indicating that there is no new node between the transmitting node and its successor. The node sends the token to its successor as usual.
- A new node sends a message giving its address as the new successor. The node that originated the inquiry updates its list to make the new node its successor and sends the token to it.
- Two or more nodes attempt to send a message indicating they are the new additions. An arbitration procedure then follows to work out the addresses and logical assignments. This will also resolve the problem of two nodes with incorrectly set identical addresses.

As addition of nodes can be time consuming, it should be carried out when network loading is not too heavy.

Removal of nodes

When a node is to leave the network, it waits for the token then sends an advisory message to its predecessor, giving its successor's address. The predecessor then sends future tokens to the leaving node's successor and the node can leave the network.

Node failure

When a node sends the token, it listens for a response from its successor and acts as follows:

- If there is no transmission within a defined period, the node sends the token again.
- If there is still no response, the successor is assumed to be out of action and the node sends a general inquiry message to determine the faulty node's successor. If the latter responds, the transmitting node removes the faulty node from its list and substitutes the successor; it then sends the token.
- If there is still no response, the node tries again. Should there still be no response, the node attempts to build a new logical ring by requesting all active nodes to respond.

Hardware requirements

The hardware components for a token bus node are shown in Figure 13.19.

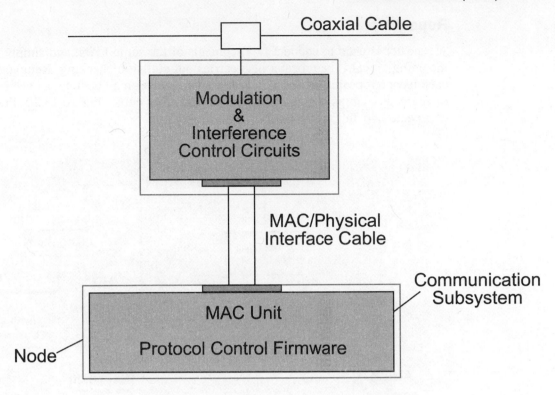

Figure 13.19
Token bus hardware

Token bus configurations use a co-axial broadband cable, which has 75 Ω impedance. The cable is quite often quad shielded with a foil/braid/foil/braid composition around the signal wire and the dielectric, as protection under even the worst operating conditions.

Cable segments are linked using 'F' connectors.

Some token bus configurations use optical fibre as the transmission medium. This arrangement uses a star configuration in which the center of the star may be a node (active star) or a coupler (passive star). In an active star, each node in the star sends its transmissions to the central node, which then broadcasts the transmission to all the other connected nodes.

In a passive star, the coupler, or signal re-director, is created by fusing the fibers coming from each of the nodes. This fusion creates paths between all nodes, so that any transmission from a node will automatically reach all the other nodes.

13.17 Internetwork connections

In many cases LANs need to have connections to other networks. For example, a company with offices in each capital city will have a LAN at each site and usually interconnections between them via a WAN. Messages can be sent to nodes on any LAN, and to the user there appears to be just a single large network. However, depending on the speed of the WAN links, operations between remote LANs may be considerably slower than those between nodes on the same LAN.

Various items of equipment are used for connections within and between networks, as described in the following sections.

Repeaters

A repeater is used to connect two segments of the same LAN, and simply retransmits an incoming signal. The repeater also carries out collision checking. Remote segments may each have a repeater, joined by a link; also a repeater may operate between different types of segment such as coaxial cable and optical fiber cable. Figure 13.20 illustrates the use of repeaters in an Ethernet LAN.

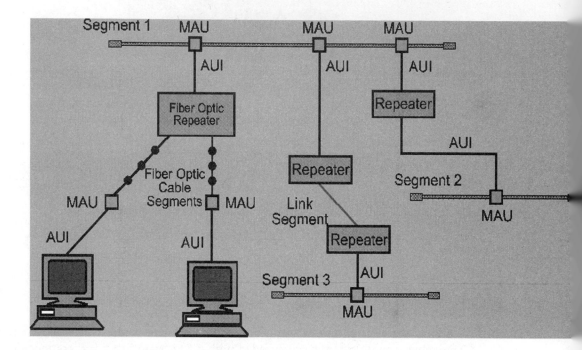

Figure 13.20
Use of repeaters

Bridges

A bridge connects two networks, or two segments of one network. It acts as a node as far as each side is concerned. The data link layer protocol has to be the same on each side, but the physical media can be different.

Bridges are more intelligent than repeaters and have software that ensures that noise and truncated packets are not passed across. The most commonly used bridges maintain address lists so that only packets addressed to the other side are retransmitted. In addition, they usually have 'learning' algorithms that allow them to build up and maintain complete address lists that can respond to changes in the network. Splitting a network with such bridges can produce a dramatic reduction in traffic density.

Another use for bridges is to extend a network. Two fully stocked networks (that is, with the maximum number of nodes and segments) can be joined by a bridge and made into effectively one large network. In fact it will be more efficient than an equivalent single network would be because the bridge ensures that only the necessary messages are repeated to one side or the other.

Switches

On a normal hub, all ports are interconnected and hence all users connected to that hub share the same available bandwidth. Any traffic on a given port will be 'seen' by all users connected to the hub.

A switching hub (or 'switch'), on the other hand, only forwards each packet to the relevant port, based on the hardware address information in the header. A switch therefore acts as a multi-port bridge.

The advantage of using a switch to interconnect several LANs is that each LAN retains its full bandwidth, whilst still allowing communication between all hosts.

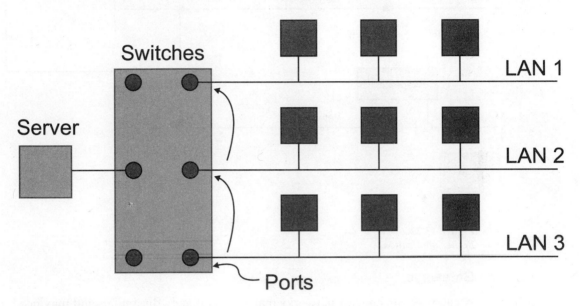

Figure 13.21
Switch applications

Routers

A router transfers data between networks that have the same network layer protocols (such as TCP/IP) but not necessarily the same physical or data link protocols. Routers maintain tables of addresses in the networks to which they are attached, and route each packet to the appropriate network depending on its destination address.

When a packet arrives at one side, the router translates the data link protocol if necessary, and then sends the packet on to the next node on the route. Figure 13.22 illustrates the use of routers.

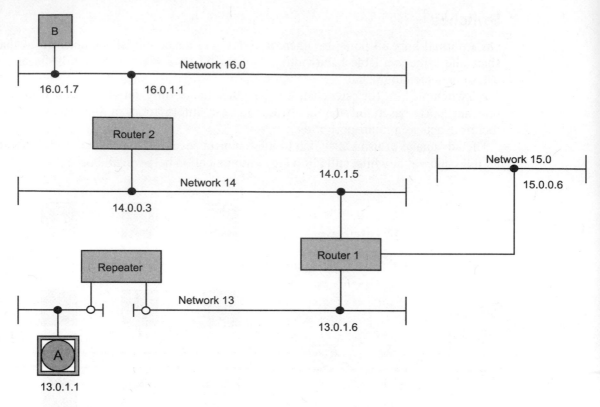

Figure 13.22
Router applications

Gateways

A gateway can connect networks that are completely dissimilar, and may need to translate all seven OSI protocol layers. Gateways therefore have the highest overhead and lowest performance of the internetworking devices.

13.18   Network operating systems

This section looks at the features of a network operating system (NOS) and examines how implementation relates to the OSI model. It also looks at the various systems' architectures and their associated protocols and, finally, briefly explores some of the commercially available network operating systems. Though no one network operating system is recommended, the features are compared from an academic point of view.

Chapter 9 demonstrated that data communication between any two systems could be described in the open systems interconnection reference model (OSI/RM) architecture as proposed by the International Standards Organization. This model lays down the framework on which any network can be based to ensure interoperability. It describes the requirements of the different logical entities, which interact with one another, but can be implemented independently. The associated protocols contain the details of the exact implementations.

When manufacturers or software providers implement those protocols, the result is an operating environment, which makes the underlying complexities of a particular system or its interconnections transparent to a user.

A network operating system is the software necessary to integrate the various components of a network into a single entity to which users have access. It manages the resources of a network, schedules the services and tries to ensure an error free session for every user. Simply, a NOS is a network resource manager.

In an ideal situation, no user need be aware of the connection details or the mechanism by which various things are co-ordinated. For example, to copy file1 to file2, having both copies to reside on the same physical medium, such as a disk, one could type:

copy file1 file2

However, if there are different physical disks or volumes present, the source and destinations might need to be designated in a way appropriate to the command interpreter being used. For example, in MSDOS, a file might be copied from one disk to another by means of a command such as:

copy c:myfile.dat a:yourfile.dat

It follows that if the source and destination can be represented uniquely; it should not matter to which machine each medium is physically connected. Any resource attached to any computer in a network can be made to appear as a common resource of the network and the way it is achieved is transparent to the user. This is what a NOS is typically aimed to provide. It presents an operating environment in which any valid user is able to utilize the resources of a network as if it is attached to a local node.

So, a NOS extends the functions of a conventional operating system. On local stand-alone computers, the operating environment is provided by the local operating system. It:

- Manages files
- Manages the computer's memory
- Manages the peripherals attached to it
- Manages programs as processes
- Does a few housekeeping functions

Additional functionality is provided (the more sophisticated operating systems become) depending on the complexity of the hardware platform.

In addition, a network operating system manages the communications between network components. This can be achieved in two ways. Either the local resources are managed by the local operating system and the communication management is provided through add-on modules, or, local as well as network resources are managed as an integrated environment.

As many manufacturers started providing network services before the OSI/RM was established, the approaches developed to cater for the networking needs of industry are not uniform or even compatible. However, most of the vendors provide implementation to most of the popular approaches, thereby allowing interoperability, though the trend is to adopt a uniform approach based on the OSI/RM.

13.19   Network architectures and protocols

Though there are many Novell network and Windows NT operating systems architectures in the market, the following three are the most popular.

- OSI/RM
- TCP/IP
- SNA

Transmission control protocol/Internet protocol (TCP/IP)

This is the result of a US Department of Defense initiated project to implement a global network, interconnecting various local area networks or individual computers. In this sense this is a demonstrated open system model. The architecture is based on a four-layer model. The layers are:

- Application layer
  This comprises the session, presentation and application layers of the OSI model.
- Services layer (host-to-host)
  This represents the transport layer of the OSI model.
- Internetwork layer
  This represent the network layer of the OSI model.
- Network interface layer
  It represents the physical and datalink layers of the OSI model.

Figure 13.22 shows how TCP/IP relates to the OSI/RM. It can be seen that although the requirement for communication is the same or similar in reality, they are classified differently in different architectures. Hence it is not possible to define the exact equivalence of layers in different architectures.

Due to the support of the US Department of Defense and later by the National Science Foundation, TCP/IP has become the *de facto* open systems standard. It is widely used on the Internet and is supported by many PLC manufacturers for industrial interfacing.

Systems network architecture (SNA)

SNA is a layered architecture similar to OSI/RM. SNA is part of an IBM corporate design philosophy, which laid the framework for the data communication development of its products. The scope of SNA is so broad that none of its products implements the entire architecture. Each product implements only those elements of SNA that pertain to the function of that particular product. As SNA is a layered architecture, integration of all these products to form a network is only a matter of the proper configuration. The layers are described below.

- End user layer
  This describes the end user requirement of communication, much like the application layer of OSI.
- Function management layer
  All the requirements of translation in terms of coding or file formats, together with their management, are described in this layer and are functionally equivalent to the presentation layer of OSI.
- Data flow control layer
  This takes care of the control aspects of creating an end-to-end connection/session.
- Transmission control
  This describes the end-to-end data transmission details such as reliability and integrity.
- Path control layer
  The actual sequencing of information packets and its routing on the network are described in this layer and is functionally equivalent to the transport and network layers put together of OSI.

- Data link layer
  This is essentially similar to that of OSI.
- Physical layer
  The description of the physical media details as in OSI.

SNA, when originally marketed, had only five layers and did not have end-user and physical layers. This was because all traditional mainframe operating systems provided the end user interface as part of the environment and the communication requirements were described without them. Also, since IBM had proprietary network standards, where the physical media details were described, there were no additional layers described. However, a seven layer integrated SNA model was proposed later and became an accepted standard. Other vendors had to reconcile with this and had to give SNA connectivity due to market pressures.

While OSI/RM is only a functional model, SNA is a functional as well as an implemented proprietary standard. It is dynamic and is growing in order to accommodate developments in the networking field. Advanced peer to peer communication (APPC) is the result of such a development and is primarily designed to provide enhanced capabilities for a distributed environment.

13.20 NOS products

NETBIOS and NETBEUI

NETBIOS is an interface, an upper-level protocol developed by IBM. It provides a standard interface to the lower networking layers and functionally covers the top three layers (session, presentation and application) in the OSI reference model.

NETBIOS can also serve as an API (Application Program Interface) for data exchange. It provides programmers with access to resources for establishing a connection between two machines or between two applications on the same machine.

NETBIOS provides four types of service

- Naming, for creating and checking group and individual names (hardware or symbolic names)
- Datagram support, for connectionless communication
- Session support, for connection-orientated communication
- General services (e.g. resetting adapter states, canceling application commands, etc.)

NETBEUI (network basic extended user interface) is a protocol developed originally for use on IBM token ring networks. NETBEUI can communicate with standard (IEEE 802.2 LLC) protocols at the lower layers.

NETBEUI can be used instead of TCP/IP or SPX/IPX in a protocol stack. It is faster than TCP/IP and SPX/IPX and is ideally suited for small LANs but it is unfortunately not routable.

SPX/IPX

Novell has a major share of the commercial LAN market today with its NetWare product ranges. Comparison of the NetWare components SPX and IPX to OSI/RM is given in Figure 13.23.

Examples

OSI Model			Application Software		TCP/IP Protocol Suite				Netware	

Comparison table / figure:

#	OSI Model		TCP/IP Protocol Suite	Netware
1	Application / Presentation	Application Software	Telnet / FTP / SMTC / SNMP	Netware Core Services Protocol (NCP)
2	Session			Netbios
3	Transport	Communications Protocol	TCP	SPX
4	Network		IP	IPX
5	Data Link	Network Driver	LLC / MAC	
6	Physical	Network Card	Physical	

Figure 13.23
Comparison of Netware to OSI/RM

Banyan vines

Unlike other vendors, Banyan decided to implement its network operating system on the popular UNIX kernel.

LAN operating systems from IBM

IBM supports several LAN technologies by its PC LAN program and the OS/2 LAN Server.

3Com's LAN manager

3Com based its NOS on the LAN manager developed by Microsoft.

LAN operating systems from Microsoft

Microsoft has introduced a few remarkable products in the network operating systems field such as Windows for Work Groups, Windows NT, Windows NT Advanced Server and Windows 95/98.

Summary

The function of a NOS is to provide an environment in which computer systems communicate irrespective of the different hardware or software available at the local nodes. The international standards movement that resulted in OSI/RM has influenced all the big players in the networking market and the market is leaning towards this common standard slowly. Hence the integrated/add-on approaches adopted by the main players in the LAN market today, namely Novell, Banyan, 3Com, Microsoft and IBM.

Appendix A

Numbering systems

A generalized number system

A number system is formed by allocating symbols to specific numerical values. Any group of symbols can be used with the total number of symbols for a number system called the **base** of the system.

The three most common bases are:

- Binary with two symbols (0 and 1) and hence a base of 2.
- Hexadecimal with sixteen symbols (0,1,2...9,A, B....F) and hence a base of 16.
- Decimal with ten symbols (0,1,2...9) and hence a base of 10.

When numbers with different bases are being used in the same descriptive text they sometimes have the subscript referring to the base being used, as in 3421.19_{10} for a decimal or base 10 number.

Numerical symbols have to be combined in a certain way to represent other combinations of numbers. The decimal numbering system has the structure laid out in Table A.1 for weighting each digit in the number 3421.19_{10} in a combination of numbers written together.

Exponential notation is used here, for example: 10^2 means 100 and 10^{-3} means 0.001.

Weight	10^4	10^3	10^2	10^1	10^0	.	10^{-1}	10^{-2}	10^{-3}	10^{-4}	10^{-5}
	0	3	4	2	1	.	1	9	0	0	0

Table A.1
Decimal weighting structure

The most significant digit (or MSD) in this number is 3. This refers to the left most digit that has the greatest weight (10^3 or 1000) assigned to it.

The least significant digit (or LSD) in this number is 9. This refers to the right most digit that has the least weight (10^{-2} or 0.01) assigned to it.

This represents the number calculated below:

$$...0\times10^4+3\times10^3+4\times10^2+2\times10^1+1\times10^0+1\times10^{-1}+9\times10^{-2}+0\times10^{-3}+...$$

Binary numbers

Binary numbers are commonly used with computers and data communications because they represent two states – either ON or OFF. For example, the EIA-232-C standard has two voltages assigned for indicating ON (say, –5 Volts,) or OFF (say, +5 Volts). Any other voltages outside a narrow band around these voltages are undefined.

The word **bit**, referred to often in the literature, is a contraction of the words **binary digit**.

The same principles for representing a binary number apply as in section 1 above. For example, the number 1011.1_2 means the following using Table A.2.

Weight	2^4	2^3	2^2	2^1	2^0	.	2^{-1}	2^{-2}	2^{-3}	2^{-4}	2^{-5}
	0	1	0	1	1	.	1	0	0	0	0

Table A.2
Binary weighting system

This translates into the following number:

$$.....0\times2^4+1\times2^3+0\times2^2+1\times2^1+1\times2^0+1\times2^{-1}+0\times2^{-2}+.....$$

The most significant bit (MSB) in the above number is the left most bit and is 1 with weighting of 2^3. The right most bit is the least significant bit (LSB) and is valued at 1 with a weighting of 2^{-1}.

Conversion between decimal and binary numbers

Table A.3 gives the conversion between decimal and binary numbers. Note that the binary equivalent of decimal 15 is written in binary form as 1111 (using 4 bits). This 4 bit binary grouping will have significance in hexadecimal arithmetic later. As expected binary 0 is equivalent to decimal 0.

Decimal number	Binary equivalent
0	0
1	1
2	10
3	11
4	100
5	101
6	110
7	111
8	1000
9	1001
10	1010 contd...

contd... 11	1011
12	1100
13	1101
14	1110
15	1111

Table A.3
Equivalent binary and decimal numbers

The procedure to convert from a binary number to a decimal number is straightforward. For example, to convert 1101.01_2 to decimal; use the weighting factors for each bit to make the conversion.

$$1101.01_2 = 1 \times (2^3) + 1 \times (2^2) + 0 \times (2^1) + 1 \times (2^0) + 0(2^{-1}) + 1 \times (2^{-2})$$

This is equivalent to:

$$1101.01_2 = 1 \times (8) + 1 \times (4) + 0 \times (2) + 1 \times (1) + 0 \times (\frac{1}{2}) + 1 \times (\frac{1}{4})$$

This then works out to:

$$1101.01_2 = 8 + 4 + 0 + 1 + 0.25$$

$$1101.01_2 = 13.25$$

The conversion process from a decimal number to a binary number is slightly more complex. The procedure here is to repeatedly divide the decimal number by 2 until the quotient (the result of the division) is equal to zero. Each of the remainders forms the individual bits of the binary number.

For example, to convert decimal number 43_{10} to binary form:

2	43 remainder 1 (LSB)
2	21 remainder 1
2	10 remainder 0
2	5 remainder 1
2	2 remainder 0
2	1 remainder 1 (MSB)
	0

Table A.4
Illustration of decimal to binary conversion

This translates a number 43_{10} to 101011_2.

Hexadecimal numbers

Most of the work done with computers and data communications systems is based on the Hexadecimal number system, with the base of 16 and uses the sequence of symbols:

0,1,2,3,4,5,6,7,8,9,A,B,C,D,E,F

Hence, the number of $FA9.02_{16}$ would be represented as below in Table A.5

Weight	16^4	16^3	16^2	16^1	16^0	.	16^{-1}	16^{-2}	16^{-3}	16^{-4}	16^{-5}
	0	0	F	A	9	.	0	2	0	0	0

Table A.5
Hexadecimal weighting structure

This translates into the following number:

$$.....0 \times 16^4 + 0 \times 16^3 + F \times 16^2 + A \times 16^1 + 9 \times 16^0 + 0 \times 16^{-1} + 2 \times 16^{-2} + ...$$

The most significant digit (MSD) in the above number is the left most symbol and is F with weighting of 16^2. The right most symbol is the least significant digit (LSD) and is valued at 2 with a weighting of 16^{-2}.

Conversion between binary and hexadecimal

The conversion between binary and hexadecimal is effected by modifying Table A.6 to Table A.6 below:

Decimal number	Hexadecimal equivalent	Binary equivalent
0	0	0000
1	1	0001
2	2	0010
3	3	0011
4	4	0100
5	5	0101
6	6	0110
7	7	0111
8	8	1000
9	9	1001
10	A	1010
11	B	1011
12	C	1100
13	D	1101
14	E	1110
15	F	1111

Table A.6
Relationship between decimal, binary and hexadecimal numbers

As can be seen from the table, the binary numbers are grouped in fours for the largest single digit hexadecimal character or symbol. A similar approach of grouping bits in fours is followed in expressing a binary number as a Hexadecimal number.

In converting the binary number 1000010011110111_2 to its hexadecimal equivalent the following procedure should be adopted. First, break up the binary number into groups of four commencing from the least significant bit. Then equate the equivalent Hex symbol to it (derived from Table A.6 above).

$$1000010011110111 \text{ becomes:}$$

1000	... 0100	... 1111	... 0111_2
8	... 4	... F	... 7_{16}

or $84F7_{16}$.

In order to convert a hexadecimal number back to binary the procedure used above must be reversed.

For example, in converting from C9A4 to binary this becomes:

C	... 9	... A	... 4_{16}
1100	... 1001	... 1010	... 0100_2

or 1100100110100100_2.

Binary arithmetic

Addition

Knowledge of binary addition is useful although it can be cumbersome. It is based on the following four combinations of adding binary numbers:

0	0	1	1
0	1	0	1
0	1	1	0 and carry 1

The carry 1 (or bit) is the only difficult part of the process. This addition of the individual bits of the number should be done sequentially from the LSB to the MSB (as in normal decimal arithmetic).

An example of addition is given below:

$$
\begin{array}{l}
1010001001_2 \\
\underline{0011101010_2} \\
1101110011_2
\end{array}
$$

Subtraction

The most commonly used method of binary subtraction is to use 2's complement. This means that instead of subtracting two binary numbers (with the attendant problems such as having 'carry out' bits); the addition process is applied.

For example, take two numbers and subtract the one from the other as follows:

12	which is equivalent to:	1100
-4	Subtrahend	-0100
8	Result	1000

The two's complement is found by first complementing all the bits in the subtrahend and then adding 1 to the least significant bit.

Complementing the number results in 0100 becoming: 1011.

Add 1 to the least significant bit gives a two's complement number of: 1100.

Add 1100_2 to 1100_2 as follows:
```
1100
1100
1000     carry 1
```
(This is the same result as above).

Exclusive-OR (XOR)

Exclusive-OR is a procedure very commonly used with binary numbers in the error detection sequences of data communications. The result of an XOR operation on any two binary digits is the same as the **addition** of two digits **without the carry bit**. Consequently, this operation is sometimes also called the Modulo-2 adder. The truth table for XOR is shown below:

Bit 1	Bit 2	XOR
0	0	0
0	1	1
1	0	1
1	1	0

Table A.7
Exclusive-OR truth table

Hardware/firmware and software

The hardware refers to the physical components of a device, such as a computer, sensor, controller or data communications system. These are the physical items that one can see.

The software refers to the programs that are written by a user to control the actions of a microprocessor or a computer. These may be written in one of many different programming languages and may be changed by the user from time to time.

The firmware refers to the 'microprograms', usually residing in a read-only memory (ROM) and which normally cannot be changed by the user. The firmware usually controls the sequencing of a microprocessor. Consequently, it is a combination of hardware and software.

A port is the place of access to a device or a network used for the input or output of digital data signals.

Appendix B

Cyclic redundancy check (CRC) program listing

```c
#include <malloc.h>
#include <stdlib.h>
#include <stdio.h>

#ifndef DAT                 // to avoid multiple definitions due to order
#define DAT                 // of #includes
struct dat      // this is a data structure used for passing between objects
    {
    int    addr,                        // address of device
           fcn,                         // number of function
           dcount,                      // number of elements in data
           crc;                         // crc check code
    char   *data;                       // pointer to data
    };
#endif

static unsigned CRC16=0xA001; // Polynomial used for CRC-16 checksum

union doub {                                    // union for CRC check
    unsigned i;                         // as an unsigned word
    char c[2];                          // as two characters
    struct bits{                        // as a bitfield
```

```c
            unsigned msb:1;              // most significant bit
            unsigned:14;
            unsigned lsb:1;              // least significant bit
            } b;
      };


void whatcrc(struct dat *d, int mode)    // calculate a CRC given a dat structure
      {
      char *msg;                   // buffer to message
      int i,j,len;         // counters and length of message
      union doub sck,byt;

      len=(mode?2:0)+2+(d->dcount); // calculate length (data is the only field
without fixed length)
      msg = (char *)malloc(len);   // allocate space for message buffer
      if (!msg)                    // didn't happen? Say so and quit.


            {
            printf("Sorry, but I couldn't allocate memory\n");
            exit(1);
            }


      // Load the msg buffer
      msg[0]=d->addr;              // load the addr field as byte 1
      msg[1]=d->fcn;               // load the fcn field as byte 2


      for (i=0;i<d->dcount;++i)
            msg[i+2]=d->data[i];


      if (mode)
            {
            msg[i+2]=(d->crc&0xFF00)>>8;
            msg[i+3]=d->crc&0x00FF;
            }


      // CRC check algorithm live!

      sck.i=0xFFFF;                       // set initial remainder
      for (i=0;i<len;++i)                 // for each byte in buffer
```

```
        {
        byt.c[0]=msg[i];      // put the current character at end of working byt
        byt.c[1]=0;                   // set start of byt to 0
        sck.i^=byt.i;                 // set sck to sck XOR byt (MOD-2 maths)
        for (j=0;j<8;++j)             // for each bit
              {
              if (sck.b.msb)
                    {
                    sck.i>>=1;              // shift the remainder right 1 bit
                                                (divide by 2)
                    sck.b.lsb=0;            // and set the MSB to 0
                    sck.i^=CRC16;           // set remainder = sck XOR the CRC16
                                                polynomial

                    }
              else

                    {
                    sck.i>>=1;              // shift the remainder right 1 bit
(divide by 2)

                    sck.b.lsb=0;            // and set the MSB to 0
                    }
              }
        }
    d->crc = (sck.i)<<8;    // update the CRC in the data structure
    d->crc |= (sck.i&0xFF00)>>8;
    free (msg);                   // free the temporary storage area
    }
```

Appendix C

Serial link design

C.1   Strategies in writing protocol software

Bearing in mind the complexity of implementing a protocol program, the anecdotal stories of cost over-runs in writing protocols are not surprising. This section will examine the software implementation of a protocol.

Protocol software has to support bi-directional communications between two devices. This requires the appropriate data (of interest to the user) being packaged in a 'system level envelope' by the transmitter, and decoded by the receiver. The system level is usually of fixed length and describes fully how the protocol works. The data level, on the other hand, can often be of a variable length.

The process of writing a protocol involves various levels of sophistication and the following factors have to be considered in implementing an appropriate protocol:

Cost	The budget size has to be carefully assessed against the required level of protocol implementation.
Level of performance	If you only require a low level of performance there is no point in implementing the full protocol at an increased cost.
Future requirements	Future requirements may encourage the programmer to include additional features now, which may only be used in the future.
Risk and security	Where possible failure of the serial link has the potential for catastrophic results, it may be prudent to put significant effort into protocol development.
Access to information, technical support	Many vendors are very reluctant to release all the information about their particular protocol for fear of compromising their market position, or do not have the local organization to provide adequate technical support.

The three levels of writing protocol software are:
- Simple one-way asynchronous
- Simple one-way synchronous
- Bi-directional asynchronous

Simple one-way asynchronous

This allows the programmer to drive the software development with a protocol that constructs a message and transmits it from device A to device B. The response from device B would then be crudely received and displayed by device A. Any messages generated independently by device B (without initiation from device A) would not be read by device A.

This would apply to both read and write type messages from device A. In the case of read messages, the user can verify that the correct data status was received in the return string. In the case of write messages, the user can confirm that the address specified in the request message has been updated correctly by the response message.

Simple one-way synchronous

This implementation would be to extend the first option to make the program in device A respond to synchronous messages from device B. By synchronous we mean that the program in device A would enter a mode where it would wait for a command to be issued from device B and respond accordingly. During this time device A would not be able to send a command to device B.

Bi-directional asynchronous

This option would provide asynchronous bi-directional communications between the two devices. That is, device A could send commands to device B, and simultaneously service any requests received from device A.

This option would require an interrupt service routine to initiate the response to a device 'A' command, as it arrives. This is important because the speed of response is also a constraint imposed by the protocol. A response that is too slow may have to be ignored.

The disadvantages of this option are the complexity of the protocol software required. In addition, the device used for such a task would have to be sufficiently fast to allow the interrupts to be serviced, and to handle its own processing requirements.

C.2   A typical program structure

A block diagram of a typical program structure is given in Figure C.1. The structure has been used successfully in implementing a number of different protocol structures, but is included here only as a guide.

Typically, most industrial protocols are involved in the following operations:
- Read digital data
- Read analog data
- Write digital data
- Write analog data

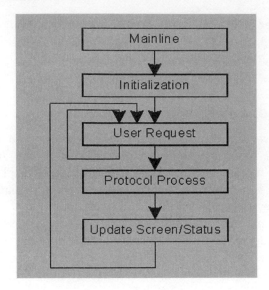

Figure C.1
Block diagram of typical protocol program structure

The following describes each part of the typical protocol program structure.

Initialization

This routine runs at the start of the program only and comprises a number of routines (or tasks).

Variable initialization

The program should declare adequate storage for the data area, to be used for transfer of the data points, across the serial link by the protocol. The data areas used in each communicating device do not necessarily have to be identical in size or even structure. The easiest way to use storage is to define an array of variables representing a block of addresses. All data should be initialized before the other program steps are commenced.

Device status and display initialization

The display and status variables should be initialized. This includes putting the display into the correct video mode and writing all static text to the screen. The communication port should also be initialized.

User request

This function processes incoming commands from the user. The use of interrupts here would be ideal.

C.3 Protocol process

This is the main body of the program; the steps are as follows:

- Get any input required from the user and write to or read from the other device.
- Compute any parts of the protocol that require calculation, e.g. CRC check word, byte count etc.

- Send a command message and wait for the response.
- Update the display showing the response message, system and data fields.
- Read any input data from the other device and decode as per the protocol structure.
- Identify, define any error messages that occur and pass onto the next step.

Update device status

This is where any clean-up operations are performed. This includes such items as communications port status, or a message indicating the success or failure of the preceding operation. The 'protocol process' routine could return an error code to the main part of the program.

C.4 Typical problems encountered in protocol software

Some typical practical problems, which a programmer may encounter when implementing protocol software, are listed below. A lot of these may sound like commonsense but it is surprising how many times they are ignored.

Typical problems are:

- The list of data points transferred over the communications link is normally dynamic. Appropriate data structures to handle the variation in the points transferred (and their addresses) are not implemented.
- Speed of transfer of data across the communications link is too slow because of the quantity of points transferred across. A prioritized or exception reporting scheme of transfer of data points can solve this problem.
- Buffer size of receiver is inadequate. This causes loss of data or delays in transferring information across the link.
- Receiving station, CPU cannot react fast enough to service data coming in, because of overload generated by other activities.
- Physical disruption to communications channel (by a link breakage) causes a catastrophic collapse in re-establishing transfer of the latest data because of inadequate error handling and the build up of old messages.
- The use of interrupt handling of messages is avoided because of the complexity associated with this feature. This results in a significant degradation in performance because of the overload of the CPU.
- Scaling of data at transmitter and receiver is often not done correctly.

C.5 Fragments of protocol programs written in BASIC

A program is included at the end of this chapter, which demonstrates the use of a protocol with a PC (refer to section C.11). This is to illustrate the basics of constructing a protocol and the various functions used.

Microsoft QuickBASIC has been chosen for illustrating the basics of protocol construction, for the following reasons:

- It is a simple and straightforward computer language to use.
- It is relatively powerful and efficient (especially in the later versions).
- Many people are familiar with it.
- It can be compiled (instead of merely interpreted).
- It is low cost and freely available.

It should be emphasized that QuickBASIC is not the most efficient computer language for this sort of work. Most programmers elect to use C language because of its portability, power and efficiency. C is, however, a fairly cryptic language, difficult to remember, and learning it is time-consuming. Hence, it has not been used for this example.

Microsoft QuickBASIC is a programming environment that includes all the tools needed for writing, editing, running, and de-bugging programs. A full help facility is available online to assist in writing the programs. The software has to run on an IBM PC or IBM PC compatible, which uses MS-DOS.

BASIC program implementation

A few elementary 'housekeeping' rules are necessary when using QuickBASIC language. Comments have been added to all QuickBASIC statements, used in the example program, to help you understand the programming process.

The following points should be remembered when writing a program. Although obviously directly related to QuickBASIC, the concepts will be applicable to implementations in other languages.

In this discussion, COM1 and COM2 refer to the two serial ports on the IBM or compatible PC. The following opening statement, which makes BASIC as tolerant as possible of hardware-related problems, should always be used when uncertain of the hardware and software configuration:

OPEN 'COM1:300,N,8,1,BIN,CD0,CS0,DS0,OP0,RS,TB2048, RB2048' AS #1

(This OPEN is FOR RANDOM access). The following is an explanation of each recommended parameter used in the OPEN statement:

- The higher the baud rate, the greater the chances of problems; thus, 300 baud is unlikely to give any problems. 56 K baud is the highest speed possible over most telephone lines, due to their limited high-frequency capability. 19 200 baud, which requires a direct wire connection, is most likely to cause problems. (Possible baud rates for QuickBASIC are: 75, 110, 150, 300, 600, 1200, 1800, 2400, 4800, 9600 and 19 200).
- Parity usually does not help significantly. Because of this, no parity (*N*) is recommended. As discussed in an earlier chapter, parity error detection is not a very efficient way of identifying errors.

For those devices that require parity, the parity enable (PE) option should be used in the OPEN COM statement, which is required to turn on parity checking. When the PE option turns on parity checking, a 'Device I/O error' occurs if the two communication programs have two different parities (parity can be Even, Odd, None, Space or Mark). For example, a 'Device I/O error' occurs when two programs try to talk to each other across a serial line using the following two different OPEN COM statements.

OPEN 'COM,1200,O,7,2,PE'FOR RANDOM AS #1

and

OPEN "COM2:1200,E,7,2,PE"FOR RANDOM AS#2

If the PE option is removed from the OPEN COM statements above, no error message is displayed.

- The above example uses 8 data bits and 1 stop bit. 8 data bits requires no parity (*N*), because of the size limit for BASIC's communications data frame (10 bits).
- The BIN (binary mode) is the default. Note: The ASC option does NOT support XON/XOFF protocol, and the XON and XOFF characters are passed without special handling.
- Ignoring hardware handshaking often corrects many problems. Thus, if the application does not require handshaking, try turning off the following hardware line-checking:
  - CD0   Turns off time out for data carrier detect (DCD) line
  - CS0   Turns off time out for clear to send (CTS) line
  - DS0   Turns off time out for data set ready (DSR) line
  - OP0   Turns off time out for a successful OPEN
- RS suppresses detection of request to send (RTS).
- For buffer-related problems, increase the transmit and receiver buffer sizes above the 512-byte default:
  - TB2048 = Increases the transmit buffer size to 2048 bytes
  - RB2048 = Increases the receive buffer size to 2048 bytes

A large receive buffer can work around BASIC delays caused by statements, like the graphics function, PAINT, which use the processor intensively.

The following are additional important hints for troubleshooting communications problems:

- Use the INPUT$(x) function in conjunction with the LOC (n) function to receive all input from the communications device (where 'x' is the number of characters returned by LOC(n)), which is the number of characters in the input queue waiting to be read. 'n' is the file number that is OPENed for 'COM1:' or 'COM2:'.
- Avoid using the INPUT#n statement to input from the communications port because INPUT#n waits for a carriage return (ASCII 13) character.
- Avoid using the GET#n statement for communications because GET#n waits for the buffer to fill (and buffer overrun could then occur).
- Avoid using the PUT#n statement for communications and use the PRINT#n statement instead. For example, in QuickBASIC 4.00b and 4.50, in BASIC Compiler 6.00 and 6.00b, and in BASIC PDS 7.00 and 7.10 using the PUT#n,,x$ syntax for sending a variable length string variable as the third argument of the PUT#n statement sends an extra 2 bytes containing the string length before the actual string. These 2 length bytes sent to the communications port may confuse the receiving program, if it is not designed to handle them. No length bytes are sent with PUT#n,,x$ in QuickBASIC 4.00 (QuickBASIC versions earlier than 4.00 don't offer the feature to use a variable as the third argument of the PUT#n statement).
- Many communications problems can only be shown on certain hardware configurations and are difficult to resolve or duplicate on other computers. Experimenting with a direct connection (with a short null modem cable) is recommended instead of with a phone/modem link, between sender and receiver, to isolate problems on a given configuration.

- The wiring scheme for cables varies widely. Check the pin wiring on the cable connectors. For direct cable connections, a long or high-resistance cable is more likely to give problems than a short, low-resistance one.
- If both 'COM1:' and 'COM 2:' are open, 'COM2:' will be serviced first. At high baud rates, 'COM1:' can lose characters when competing for processor time with 'COM2:'.
- Using the ON COM GOSUB statement, instead of polling the LOC(n) function to detect communications input, can sometimes work around timing or buffering problems caused by delays in BASIC. Delays in BASIC can be caused by string space garbage collection, PAINT statements, or other operations that heavily use the processor.

C.6 Management of data points over the serial link

Although possibly considered a trivial subject by most engineers who are more concerned with the development and commissioning of the data communications system, the management of the data coming over the link can be a challenging issue.

The main reasons why this requires attention are:

- Data can vary from a few points to a few thousand.
- Data points coming across a serial link can vary in terms of addressing as the design proceeds.
- Update times of serial data points can vary from point to point.
- Tag names of serial points can vary as the design and implementation proceeds.

Typical parameters that need to be recorded (preferably in a database program such as dBase to allow easy manipulation of the data) are:

- Tag name of the data point
- Description of the data point
- Address of point (e.g. Modbus address)
- Update time for point
- Scaling factor
- Maximum and minimum and ranges (including the possibility of the data becoming negative)
- Data format (maximum number of digits)
- Type of data for point (e.g. ASCII, integer, floating point)
- Destination of point
- Source of point

An area, which always causes problems, is scaling of the data on the link. For example, if the protocol restricts the range of values over the link to 0 to 4095 (i.e. a 12 bit quantity) and the actual engineering (or 'real world' quantities) are −10 kPa to 20 000 kPa, some delicate footwork has to be done. This is to ensure that there are no problems with the scaling at the transmitting end, transfer of the data across the link and rescaling at the receiving end. In addition, there should be a careful analysis of the loss in resolution caused by scaling.

C.7 Suggested testing philosophy for a communications system

Data communication is a strategic part of a control system. Failure of a communications link could be the cause of information loss from thousands of data points. It is imperative, therefore, that a communications system is thoroughly tested within the framework of a rigorous standard.

There are numerous reasons for the test requirements to be more demanding than those for a standard control system; some of the main reasons are:

- Information losses resulting from a failure of a communications link can be catastrophic.
- Loading factors of a data communications system can vary considerably and may lead to failure if the system is pushed to the limit.
- The communications interface can be complex consisting of hardware/firmware and software components.
- The communication link normally serves as an interface between two dissimilar systems (sometimes consisting of different design personnel, different hardware), which raises the level of technical risk and the need for common understanding of the overall requirements of the system.

A good framework in which to do the testing is ANSI/IEEE standard 829-1983 for software test documentation. While some engineers may be less than enthusiastic about formal testing procedures of this nature, the investment in time and effort is worthwhile in creating a high quality final product, engineered with proven standards of performance. The authors can testify from bitter experience that this approach pays off.

A typical test procedure (or master test plan) for the link between a PC and the Modbus port of a new PLC is sketched out below.

C.8 Typical test procedure example

The test specification procedures and recording practices have been prepared in accordance with ANSI/IEEE Std 829-1983. All software written or modified for this installation will be tested according to these guidelines.

This test is to confirm that the link from the PC to the Modbus port of the new PLC operates correctly as per the specifications.

Features to be tested

The functions to be tested will be derived from the serial link requirements specification PC-MOD1 and functional specification PC-MOD2. These functions will be grouped under the following headings:

- Hardware/firmware checks
- Interface protocol
- Interface I/O points (and addresses)
- Control strategy
- Program structure

Test environment

A PC/AT (the monitor) will interface via a second PC/AT (or protocol analyzer) to the Modbus port of the serial hardware being tested. Serial port (COM 1) of the second

PC/AT will connect to the PC/AT monitor. Serial port (COM 2) will connect to the Modbus port of the serial hardware being tested.

The relevant 'C' language compiled software modules for the serial link will be downloaded into the monitor PC/AT.

The appropriate EPROM (Revision C, 10 Nov.'91) for the PLC will be inserted into the PLC communication board. Test data will be downloaded into the monitor PC controller and the proprietary unit control system.

Testing approach

There are no risks and contingencies at the initial phase of testing, as this is performed offline and merely tests the serial interface system. The second stage of testing is envisaged to directly interface with the operational hardware, but will be done in a manual mode at the specific construction yard. The third phase of testing will be the commissioning phase and will be carried out at the plant.

The second and third phases of the testing which do have risks will not form part of this test procedure, but will be incorporated into an overall test program covering all aspects of testing.

General strategy

The test hardware will be connected to the proprietary hardware under test. The appropriate version of compiled C code and data will be downloaded to the monitor PC. The vendor will download certain specified data structures in the unit controller. Each item of data will be transferred between the two nodes of the link in the appropriate form.

Specifically the following characteristics will be checked:

- Accuracy of data transfer
- Average speed of data transfer
- Loopback and diagnostic tests where applicable
- Loading of link with high traffic loads
- Interaction of multiple nodes on link (e.g. multiplexers, dual controllers)
- Transmission of incorrect function requests
- Interface between 'C' code and the EPROM on the proprietary hardware
- Interface 'C' code and the user software in PLC
- Interface between 'C' code and the serial port hardware/firmware
- Handling of failure of:
  - Serial link
  - Proprietary hardware/firmware/software
- Monitor hardware/firmware/software (with reference to fallback strategies, recovery times, diagnostics, error messages)
- High levels of electrostatic/electromagnetic noise on the link
- Adequacy of earthing systems for each mode of line (including earth potential rise)
- Performance of error checking features of the link (e.g. using CRC-16)

Note: A PC-based protocol analyzer will be used to confirm that the data structures being transmitted down the link and the appropriate responses are correct. This will be in addition to the diagnostic messages generated by both the monitor PC and the PLC serial hardware being tested.

Acceptance tests on the various portions of the system will occur at different stages of the testing. Tests will be jointly performed by the client and the contractor. No modified software will be available for use by the client until it has been fully tested and accepted by the client.

Associated test documentation

Full test documentation should be filled out correctly and stored in a central safe location.

C.9   An example serial data communication link

This section contains an example of a serial data communication link for the control of a variable speed drive using the EIA-485 interface and ANSI-X3.28 protocol.

'Smart' instrumentation and other digital sensors and actuators are increasingly being used in factory automation and industrial process control systems. A 'sensor' is a general term that refers to instruments, monitors, etc. that measure field variables such as temperature, pressure, levels, flow and power in a process control system. An 'actuator' is a general term that refers to devices located in the field, such as valves, variable speed drives, positioners, servos, etc., that implement instructions from the control system.

Making effective use of these devices depends on their ability to transfer data reliably and quickly to and from other controlling devices, such as computers, PLCs, DCSs, etc. via a common data communications network. Data communications at this level is usually referred to as the 'field level' communications and the type of networks used are often called the 'field bus'.

Data communications at the field level is usually reliable when all the equipment comes from one manufacturer. When several different types of equipment from various manufacturers are required to communicate on the same network, difficulties always seem to appear. One major reason is that no clear and universally acceptable data communication network standard has yet emerged for systems for the field level.

The process of developing and implementing acceptable international standards is a painfully slow process and a solution to this problem is still a long way off. In the meantime, manufacturers have created their own standards or have used a combination of available standards that may have been developed for other similar applications.

Therefore, the practical problem of controlling a field actuator device, such as a variable speed drive (VSD), from an intelligent control device, such as a PLC or a PC, needs to be addressed on an application by application basis. This section describes the process of designing and implementing a simple data communications system for transferring data between an IBM compatible PC and an AC VSD to achieve the following:

- To read data from the list of parameter registers in the VSD, transfer them to the PC and display variables such as speed, current etc., on the PC monitor.
- To write data to the VSD parameter registers for starting, stopping, changing settings (e.g. speed reference), or adjusting any other variable in the VSD, as would be required in an industrial application.

Most modern VSDs have some form of communications capability, usually based on a well known physical standard, such as EIA-232 or EIA-485. The transfer of data can be controlled by a suitable program (written by the user) based on one of the ASCII character protocols. The program should be able to address multiple VSDs on the network without having a problem with data collision on the network. Typically, the programs use the poll/response method with one 'master' (PC or PLC) in control of the network and several 'slaves' (VSDs). The slaves respond only when they are polled by the master. Although this approach works quite well, it has some limitations that affect the overall performance of the system:

- This type of system is slow because it uses low baud rates, with inefficient ASCII coding. The master also needs to address each slave individually. This type of data communications solution is not suitable for controlling several drives in applications with fast speed and torque dynamics.
- There is no available software standardization, so a special program has to be written for each application.

The physical interface

The physical connections between the PC and VSDs are according to the EIA-232 and EIA-485 interface standards, both of which are covered in detail in Chapter 3 The standard PC is fitted with an EIA-232 port. The standard VSD port is EIA-485, suitable for multidropping up to 32 units.

From Chapter 3, it is clear that EIA-232 and EIA-485 are not directly compatible and the two devices cannot be directly connected and expected to work.

This apparent mismatch at the physical level can be overcome by one of the following methods:

- Interface Converter – an EIA-232/EIA-485 interface converter can be connected between the two devices to convert the voltage levels and the connection configuration from one to the other (unbalanced to differential). An interface converter should be physically located close to the EIA-232 port (i.e. at the PC end) to take advantage of the better performance characteristics of the EIA-485 interface for the longer distance to the VSD in the field. The interface between the PC and the interface converter is one-to-one, while the EIA-485 side may have several drives (up to 32) connected in a multidrop configuration. The internal connection details, of an EIA-232/EIA-485 converter, are shown in figure below.

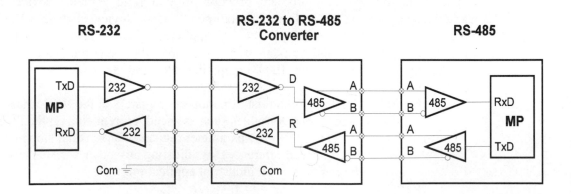

Figure C.2
Block diagram of an EIA-232/EIA-485 converter

- EIA-485 PC interface card – plug-in EIA-485 interface cards are available for IBM compatible PCs for mounting directly onto the motherboard. Similar cards are also available for PLCs. This card must be configured as a separate port in the PC. The PC can then be connected directly to the EIA-485 network.

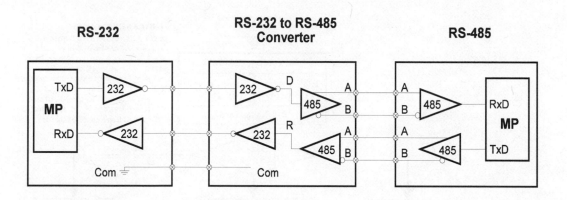

Figure C.3
Block diagram of an EIA-232/EIA-485 converter

The software interface

Once the physical interface problems have been solved, the flow of data between the PC and the VSD must be controlled by software located in the master device. In our example, the program is based on ANSI-X3.28-2.5-A4, which is an ASCII based protocol that defines the format, order, and syntax of the characters. There is no standard format, or content, for this type of program and it is usually written by the user to suit the application. The program below is an example of a simple program written in QuickBASIC for demonstration purposes only.

In accordance with ANSI-X3.28, the 10 bit character format is as follows:

- 1 Start Bit   : Logic 0
- 7 Data Bits   : ASCII Code for each character
- 1 Parity Bit   : Even or None
- 1 Stop Bit   : Logic 1

There are two styles of message order and syntax:

- Read message
- Write message

The read message comprises of a maximum 9 characters in the order shown in the following flow chart. The read message is used to transfer data from the VSDs to the master. This data is usually the field data, such as speed, current, etc. or the VSD's setting parameters.

The block checksum character (BCC) is a single character generated from all the data in the message and is used to detect errors in the transmitted data.

The write message comprises of a maximum 17 characters in the order shown in the following flowchart. The write message is used to transfer data from the master to the VSDs. This data is used to issue commands to the field device or change parameters (e.g. start, stop, change speed).

The baud rate can be set to any one of the 'standard' values between 300 to 19,200 bps.

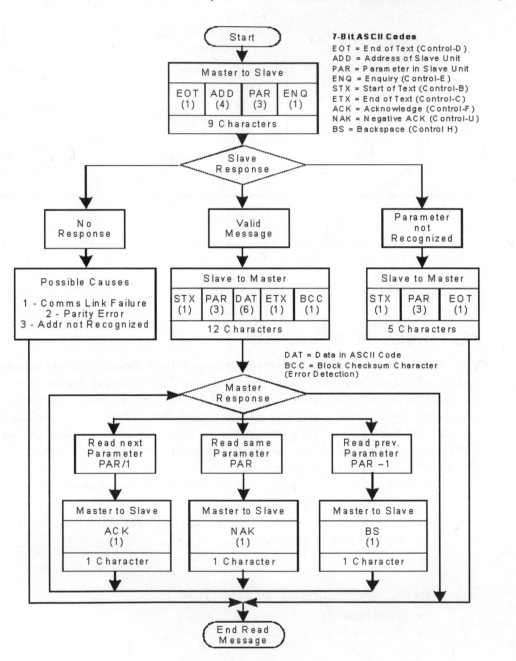

Figure C.4
Flow chart showing the order and syntax of the read message

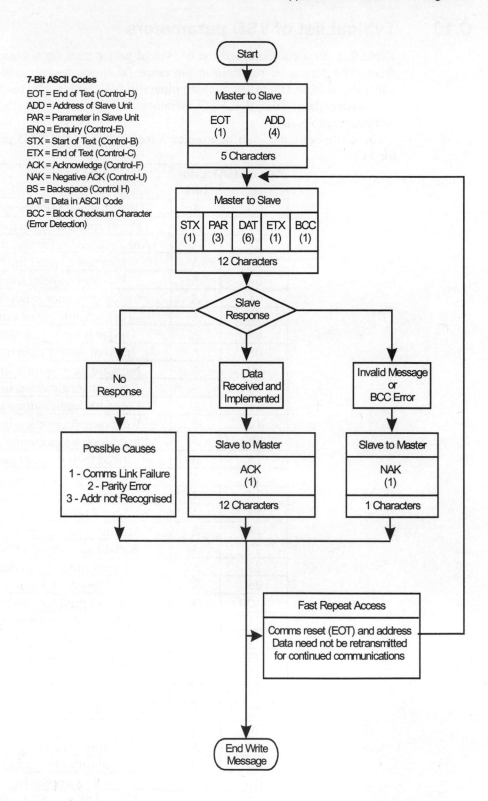

7-Bit ASCII Codes
EOT = End of Text (Control-D)
ADD = Address of Slave Unit
PAR = Parameter in Slave Unit
ENQ = Enquiry (Control-E)
STX = Start of Text (Control-B)
ETX = End of Text (Control-C)
ACK = Acknowledge (Control-F)
NAK = Negative ACK (Control-U)
BS = Backspace (Control H)
DAT = Data in ASCII Code
BCC = Block Checksum Character
(Error Detection)

Figure C.5
Flow chart showing the order and syntax for the write message

C.10   Typical list of VSD parameters

Table C.1 shows a shortened list of typical parameters for a ontrol Techniques 'Vector' drive. The parameter registers in the range 00 to 99 contain analog (numerical) values, while those from 100 to 199 contain binary digital values. The read command can be used to transfer data from the VSD parameter registers to the PC over the serial data communications link.

The write command can be used to transfer data to the VSD parameter registers from the PC.

Parameter no.	Analog/ bits	Description	Default Value
00	A	Digital speed reference, run	100
01	A	Digital speed reference, inch	– 100
02	A	Analog speed reference offset	0
03	A	Minimum speed limit	– 1500
04	A	Maximum speed limit	+ 1500
05	A	Analog reference input filter	32
06	A	Torque limit – motoring	150
07	A	Torque limit – generating	150
08	A	Internal torque reference	0
09	A	Forward acceleration limit	0.01
10	A	Reverse acceleration limit	0.01
11	A	Forward deceleration limit	0.01
12	A	Reverse deceleration limit	0.01
13	A	Speed loop proportional gain	1.5
14	A	Speed loop internal gain	1.5
15	A	Speed loop derivative gain	0
16	A	Analog output scaling	1.67
17	A	Speed reference select	3
18	A	Analog speed input scaling	600
22	A	Serial link – drive address	01
23	A	Serial link – baud rate	9600
25	A	Security key	0
40	A	Drive model number	n/a
41	A	Motor full load current	n/a
42	A	Motor magnetizing current	n/a
43	A	Motor base frequency	n/a
70	A	Motor speed	n/a
71	A	Motor frequency	n/a
75	A	Active current	n/a
78	A	DC bus voltage	n/a
82	A	Motor line current	n/a
100	B	Security key enable	n/a
101	B	Miscellaneous trip	1
102	B	Drive enable	1
103	B	Drive reset	n/a
104	B	Zero torque demand	1
105	B	Torque/speed control mode	1

106	B	Torque reference select	1
120	B	Serial link – parity enable	1
121	B	Serial link – block checksum enable	1
133	B	Drive healthy	n/a

Table C.1
Example list of VSD parameters

C.11   Example

```
'* INSTRUMENT DATA COMMUNICATIONS
'*
'* Example of PC program written in Microsoft QuickBASIC Version 4.5
'* for IBM compatible PC to a Control Techniques CD Variable Speed Drive.
'*
'* Creation date : 02-12-91
'*
'*
'* SERIAL INTERFACE CONNECTION
'* This program provides for communication to a Commander CD Variable
'* Speed Drive through one of the PC's EIA-232 serial interface ports.
'* An EIA-485 interface would be preferable because of its noise immunity.
'*
'*
'* THE PROTOCOL
'* The protocol implemented is ANSI-X3.28-2.5-A4. It defines the
'* format and order of characters and the syntax of the commands.
'* A character consists 10 bits, comprising 1 start bit (logic-0),
'* followed by 7 data bits (ASCII Code), 1 parity bit (even or none)
'* and the final bit is 1 stop bit (logic-1).
'*
'* There are two styles of commands.
'* The READ command allows reading of all the drives parameters.
'* The WRITE command allows the read/write to be changed.
ON ERROR GOTO FAULT    'Activate error trapping
CLS                    'Clears the screen

OPEN "COM1:9600,E,7,1,CD0,CS0,DS0,OP0,RS" FOR RANDOM AS #1
'* Opens the PC's serial port 1 for serial bit-by-bit communication
'* with the peripheral device.  (The OPEN is FOR RANDOM access)
'* An explanation of each parameter used in the OPEN statement follows:

'* 1. 9600 specifies the Baud Rate (bits transmitted each second). The
'*    baud rate setting for the drive and the host (PC) must be set
'*    equal to allow communication to take place.

'* 2. E sets the parity of the drive to EVEN.  With even parity selected,
'*    the parity bit is set to logic 1 when the data segment of the
'*    character consists of an odd number of logic 1's.
```

```
'* 3. 7 bits in each byte of data transmitted or received constitute
'*    actual data.
'* 4. Since this application does not require handshaking, setting CD,
'*    CS, DS and OP to 0 tells the hardware to ignore handshaking
'*        CD0 = Turns off time-out for Data Carrier Detect (DCD) line
'*        CS0 = Turns off time-out for Clear To Send (CTS) line
'*        DS0 = Turns off time-out for Data Set Ready (DSR) line
'*        OP0 = Turns off time-out for a successful OPEN
'* 5. RS suppresses detection of Request To Send (RTS)

MENU: VIEW PRINT 1 TO 24     'set text viewport
CLS
LOCATE 4, 5: PRINT "VECTOR Drive communications program"
LOCATE 5, 5: PRINT "~~~~~~~~~~~~~~~~~~~"
LOCATE 7, 5: PRINT " 1.  List Parameter Values"
LOCATE 8, 5: PRINT " 2.  Get a specific particular parameter from VECTOR"
LOCATE 9, 5: PRINT " 3.  Write data to VECTOR"
LOCATE 10, 5: PRINT " 4.  Quit"
LOCATE 12, 5: PRINT " Select an option (1-4)"

OPT: '* Wait for key to be pressed
    KY$ = INKEY$: IF KY$ = "" THEN GOTO OPT
    A = VAL(KY$)
    IF A >= 4 THEN GOSUB QUIT
    ON A GOSUB PARAMVALUES, GETPARAM, WRITEDATA
    GOTO MENU

QUIT: CLOSE #1   ' end communication with the Flux Vector Drive
    END         ' end the basic program


'*************************************************************


' Read all the drive parameters
' ~~~~~~~~~~~~~~~~~~~~~~~~~~~~~~~
' The message format is:   <EOT>    ADD     PAR     <ENQ>
'                          1 char  4 chars  3 chars  1 char

' When the ASCII control character <EOT> is sent it initializes the drive
' connected to the serial link.   (<EOT> - ASCII code 04 HEX, CONTROL D)

' ADD represents the drive address. The drive address identifies which
' device connected to the serial link is to be communicated with. For
' data integrity the digits are sent twice. For example if addressing
' drive 31 send 3311.

' PAR - this is the parameter which is required to be read. It is a maximum
' of 3 characters. e.g. For the Minimum Speed Limit, Pr-3 : just the number
' 003 is sent

' The message is terminated by the ASCII control character <ENQ>.
```

```
' (<ENQ> - ASCII CODE 05 HEX, CONTROL E)
'*************************************************************

' Response from the drive: <STX>    PAR    DATA    <ETX>    BCC
' ~~~~~~~~~~~~~~~~~~~~~~~~    1 char  3 chars  6 chars  1 char  1 char

' <STX> is an ASCII control character. This indicates to the host the start
'  of the reply.            (<STX> - ASCII code 02 HEX, CONTROL B)

' PAR - The drive parameter. (See above for description)

' DATA - This symbol represents a parameter's numerical value. It is 1-6
' numeric characters in length, with optional decimal point and sign
' character, and is accurate to 1 decimal point.
' <ETX> is an ASCII control character. It indicates to the host that the
' data is finished.
' BCC, Block checksum - The final character is generated by the drive to
' allow the host which is receiving data to perform an error check on the
' data received. This character is not sent if the drive is set for BCC
' disabled, bit parameter b-21. BCC is the exclusive-OR of all the characters
' after the <STX> character up to and including the <ETX> character.
' If the BCC is disabled the ASCII control character <CR> is sent.
' NOTE: The same parameter can be re-read simply by sending the ASCII
' control character <NAK> (ASCII code 15 HEX, CONTROL U).
' Alternatively it is possible to step forward or backwards through the
' parameters sequentially by sending the ASCII control characters <ACK>
' (ASCII code 6 HEX, CONTROL F) OR <BS> (ASCII code 8 HEX, CONTROL H)
'*************************************************************


PARAMVALUES:
 CLS
 txd$ = CHR$(4) + "0011000" + CHR$(5)     'String transmitted
                      '<EOT> - CHR$(4)
                      '<ENQ> - CHR$(5)
 rxd$ = ""
 LOCATE 1, 5: PRINT "STRING TRANSMITTED IS :- "; txd$
 LOCATE 3, 1: PRINT "PARAMETER"
 LOCATE 3, 14: PRINT "VALUE"
 LOCATE 3, 40: PRINT "PARAMETER"
 LOCATE 3, 54: PRINT "VALUE"
 VIEW PRINT 4 TO 25        'set text viewport
 x = 4              'x-coord cursor position
 i = 0              'variable to keep track of y-coord position
 parameter$ = "00"
 PRINT #1, txd$          'Writes data to Coms port 1
 GOSUB delay

DO UNTIL parameter$ = "83"     'display up to parameter 83
   rxd$ = INPUT$(LOC(1), #1)    'receive input from the comms device
   STXloc = INSTR(rxd$, CHR$(2))  'find position of control character
```

```
                         '<STX> to obtain start of message

        rxd$ = MID$(rxd$, STXloc, 11)   'remove unwanted noise from response OR
                            'obtain required info from response.
                            'Ignore BCC - ie BCC disabled
        parameter$ = MID$(rxd$, 2, 3)
        IF VAL(parameter$) > 83 THEN RETURN
        parameter$ = MID$(rxd$, 3, 2)   'extract parameter from response
        dat$ = MID$(rxd$, 5, 6)         'extract data from driver's response
        GOSUB CURSORPOSI
        LOCATE y, x
        PRINT parameter$; "        "; dat$
        IF VAL(parameter$) >= 83 THEN GOSUB MENURETURN
        prevrxd$ = rxd$            'keep record of the previous received
                            'message from the drive in case of error
        PRINT #1, CHR$(6)          'Send <ACK> to the drive to step forward
                         'through the parameters sequentially
      GOSUB delay
    LOOP
    RETURN
GETPARAM: '******************************
      'Get a specific parameter from VECTOR
      CLS
      LOCATE 3, 3: PRINT "1.   Minimum Speed Limit"
      LOCATE 4, 3: PRINT "2.   Maximum Speed Limit"
      LOCATE 5, 3: PRINT "3.   Drive Address"
      LOCATE 6, 3: PRINT "4.   Baud Rate"
      LOCATE 7, 3: PRINT "5.   Motor Speed"
      LOCATE 8, 3: PRINT "6.   Motor Frequency"
      LOCATE 9, 3: PRINT "7.   Active Current"
      LOCATE 10, 3: PRINT "8.   Drive Status"
      LOCATE 11, 3: PRINT "9.   Parity Status"
      LOCATE 12, 3: PRINT "10.  Block CheckSum Status "
      LOCATE 13, 3: PRINT "11.  Hardware Status"
      LOCATE 14, 3: PRINT "12.  Return to main menu"
      LOCATE 16, 1: PRINT "Choose a parameter and press 'return':"
      INPUT PR$
      IF LEN(PR$) = 0 OR LEN(PR$) > 2 OR PR$ = "12" THEN RETURN
      IF PR$ = "1" THEN GOSUB GP1
      IF PR$ = "2" THEN GOSUB GP2
      IF PR$ = "3" THEN GOSUB GP3
      IF PR$ = "4" THEN GOSUB GP4
      IF PR$ = "5" THEN GOSUB GP5
      IF PR$ = "6" THEN GOSUB GP6
      IF PR$ = "7" THEN GOSUB GP7
      IF PR$ = "8" THEN GOSUB GP8
      IF PR$ = "9" THEN GOSUB GP9
      IF PR$ = "10" THEN GOSUB GP10
      IF PR$ = "11" THEN GOSUB GP11
```

```
TRANSMIT: txd$ = CHR$(4) + "0011" + DP$ + CHR$(5)
        rxd$ = ""
        PRINT #1, txd$
        GOSUB delay

        rxd$ = INPUT$(LOC(1), #1)
        IF rxd$ = "" THEN GOTO TRANSMIT
        STXloc = INSTR(rxd$, CHR$(2))  'Filter out any noise if it exists
        rxd$ = MID$(rxd$, STXloc, 11)
        IF MID$(rxd$, 2, 3) <> DP$ THEN GOTO TRANSMIT
        dat$ = MID$(rxd$, 5, 6)
        LOCATE 18, 4: PRINT MSG$; " "; dat$; " "; UNIT$
        LOCATE 20, 1: PRINT "Would you like to read another parameter (Y/N)"

RAGAIN: KY$ = INKEY$: IF KY$ = "" THEN GOTO RAGAIN
        IF KY$ = "y" OR KY$ = "Y" THEN GOTO GETPARAM
        RETURN

GP1: DP$ = "003"
     MSG$ = "The minimum speed limit is: "
     UNIT$ = "r.p.m."
     RETURN
GP2: DP$ = "004"
     MSG$ = "The maximum speed limit is: "
     UNIT$ = "r.p.m."
     RETURN
GP3: DP$ = "022"
     MSG$ = "The drive address is:"
     UNIT$ = ""
     RETURN
GP4: DP$ = "023"
     MSG$ = "The baud rate is:"
     UNIT$ = "baud"
     RETURN
GP5: DP$ = "070"
     MSG$ = "The motor speed is:"
     UNIT$ = "r.p.m."
     RETURN
GP6: DP$ = "071"
     MSG$ = "The motor frequency is:"
     UNIT$ = "Hz"
     RETURN
GP7: DP$ = "075"
     MSG$ = "The active current is:"
     UNIT$ = "%"
     RETURN
GP8: DP$ = "102"
     MSG$ = "The drive is:"
     UNIT$ = "  0-drive enabled   1-drive disabled"
     RETURN
```

```
GP9: DP$ = "120"
    MSG$ = "The parity is:"
    UNIT$ = "  0-parity disabled   1-parity enabled"
    RETURN
GP10: DP$ = "121"
    MSG$ = "The block checksum is:"
    UNIT$ = "  0-BCC disabled   1-BCC enabled"
    RETURN
GP11: DP$ = "190"
    MSG$ = "The hardware status is:"
    UNIT$ = "  0-Inactive   1-Active"
    RETURN
'****************************************************************
********
'The Write Command.  The message sent to the drive is :-

'  <EOT>   ADD   <STX>   PAR   DATA   <ETX>   BCC
'  1 char  4 chars  1 char  3 chars  6 chars  1 char  1 char

' If the drive parameter, data or the BCC is in error then the control
' character <NAK> is sent.
' If it is required to write further data to the drive, it is not necessary
' to re-send the initialization character or the drive address. The drive
' parameter characters are the first characters sent.
'****************************************************************
********
WRITEDATA: '* Send data to VECTOR
    CLS
    COUNT = 0
    LOCATE 3, 3: PRINT "1.   Minimum Speed Limit"
    LOCATE 4, 3: PRINT "2.   Maximum Speed Limit"
    LOCATE 5, 3: PRINT "3.   Drive Status"
    LOCATE 6, 3: PRINT "4.   Return to main menu"
    LOCATE 8, 1: PRINT "Choose a parameter: (1,2,3 or 4)";

WDMENU: KY$ = INKEY$: IF KY$ = "" THEN GOTO WDMENU
    A = VAL(KY$)
    IF A >= 4 THEN GOSUB MENU
    ON A GOSUB WD1, WD2, WD3
    COUNT = COUNT + 1

WDSEND: IF COUNT <= 1 THEN
    txd$ = CHR$(4) + "0011" + CHR$(2) + DP$ + DATA$ + CHR$(3) + " "
    ELSEIF COUNT > 1 THEN
    txd$ = DP$ + DATA$ + CHR$(3) + " "
    END IF
    rxd$ = ""
    PRINT #1, txd$
    FOR j = 1 TO 3
     GOSUB delay
```

```
        NEXT
        rxd$ = INPUT$(LOC(1), #1)
        IF rxd$ = "" THEN GOTO WDSEND
        '*********** First locate the beginning of the received string
        FOR k = 1 TO LEN(rxd$)
          ch$ = MID$(rxd$, k, 1)
          IF ch$ = CHR$(6) OR ch$ = CHR$(15) THEN messageposi = k
        NEXT

        rxd$ = MID$(rxd$, messageposi, 1)

        IF rxd$ = CHR$(15) THEN
          PRINT " I/O error"; rxd$
        ELSEIF rxd$ = CHR$(6) THEN
          PRINT "successful write"; rxd$
        ELSEIF rxd$ <> CHR$(15) OR rxd$ <> CHR$(6) THEN
          GOTO WDSEND
        END IF
        LOCATE 20, 1: PRINT "Would you like to write to another parameter (Y/N)"

WAGAIN: KY$ = INKEY$: IF KY$ = "" THEN GOTO WAGAIN
        IF KY$ = "y" OR KY$ = "Y" THEN GOTO WRITEDATA
        RETURN

WD1: DP$ = "003"
        GOSUB INPUTDATA
        RETURN
WD2: DP$ = "004"
        GOSUB INPUTDATA
        RETURN
WD3: DP$ = "102"
        LOCATE 15, 1: INPUT "Do you wish to (E)nable or (D)isable the drive"; ABLE$
        IF ABLE$ = "E" THEN
          DATA$ = " 0000."
        ELSEIF ABLE$ = "D" THEN
          DATA$ = " 0001."
        ELSEIF ABLE$ <> "E" OR ABLE$ <> "D" THEN
          GOTO WD3
        END IF
        RETURN


    '*************************************************************
    *********
    'Error handling routine

    ' This is the first statement the program branches to after an error.
    ' The error handler is placed where it cannot be executed during the normal
    ' flow of program execution.
    '*************************************************************
    **********
```

```
FAULT: LOCATE 23, 1
    PRINT "ERL="; ERL, "ERR="; ERR;
    GOSUB PAUSE
    rxd$ = prevrxd$
    RESUME NEXT

PAUSE: LOCATE 24, 1: PRINT "Press any key to continue ... ";
KEYPRESSED: KY$ = INKEY$
    IF KY$ = "" THEN GOTO KEYPRESSED
    LOCATE 23, 1: PRINT "                      "
    LOCATE 24, 1: PRINT "                      "
    RETURN

NEXTSCR: LOCATE 24, 1: PRINT "Press any key for next screen of Parameters  ... ";
KYPRESSED: KY$ = INKEY$
    IF KY$ = "" THEN GOTO KYPRESSED
    RETURN
MENURETURN: LOCATE 24, 1: PRINT "Press any key to return to the main menu
..."
KPRESSED: KY$ = INKEY$
    IF KY$ = "" THEN GOTO KYPRESSED
    RETURN

CURSORPOSI:
    i = i + 1
    IF i <= 19 THEN
      y = i + 3
      x = 4
    ELSEIF i > 19 AND i <= 38 THEN
      y = i - 16
      x = 44
    ELSEIF i > 38 AND i <= 57 THEN
      y = i - 35
      x = 4
    ELSEIF i > 57 THEN
      y = i - 54
      x = 44
    END IF
    IF i = 39 THEN GOSUB NEXTSCR: CLS
    RETURN
delay: 'Delay
  FOR k = 1 TO 10
  z = SIN(z * 3.4) + COS(z * 3.45)
  NEXT k
  RETURN

INPUTDATA:
    DECPOSI = 0
```

```
      LOCATE 12, 1: INPUT "Enter data to be sent (-1500 to 1500)  (No decimals) ",
dat$
    IF ABS(VAL(dat$)) > 1500 OR LEN(dat$) > 5 THEN GOTO WD1

    FOR CNT = 1 TO LEN(dat$)
      DUM$ = MID$(dat$, CNT, 1)
      IF DUM$ = "." THEN DECPOSI = CNT
    NEXT

    IF LEFT$(dat$, 1) = "-" THEN
      IF DECPOSI = 0 THEN    'No decimal
       IF LEN(dat$) = 3 THEN dat$ = "-" + "00" + RIGHT$(dat$, 2) + "."
       IF LEN(dat$) = 4 THEN dat$ = "-" + "0" + RIGHT$(dat$, 3) + "."
       IF LEN(dat$) = 5 THEN dat$ = dat$ + "."
      ELSEIF DECPOSI <> 0 THEN
       LOCATE 15, 1: INPUT "Please enter a whole number:"; b
       GOTO WD1
      END IF
  ELSEIF LEFT$(dat$, 1) <> "-" THEN
      IF DECPOSI = 0 THEN    'No decimal
       IF LEN(dat$) = 2 THEN dat$ = " " + "00" + RIGHT$(dat$, 2) + "."
       IF LEN(dat$) = 3 THEN dat$ = " " + "0" + RIGHT$(dat$, 3) + "."
       IF LEN(dat$) = 4 THEN dat$ = " " + RIGHT$(dat$, 4) + "."
       IF LEN(dat$) = 5 THEN dat$ = dat$ + "."
      ELSEIF DECPOSI <> 0 THEN
       LOCATE 15, 1: INPUT "Please enter a whole number:"; b
       GOTO WD1
      END IF
    END IF
    DATA$ = dat$
    RETURN
```

Appendix D

Glossary

ABM	Asynchronous balanced mode
ACE	Association control element
ACE	Asynchronous communications element. Similar to UART
ACK	Acknowledge (ASCII - control F)
Active filter	Active circuit devices (usually amplifiers), with passive circuit elements (resistors and capacitors) and which have characteristics that more closely match ideal filters than do passive filters.
Active passive device	Device capable of supplying the current for the loop (active) or one that must draw its power from connected equipment (passive).
ADCCP	Advanced data communication control procedure
ADDR	Address field
Address	A normally unique designator for location of data or the identity of a peripheral device that allows each device on a single communications line to respond to its own message.
Algorithm	Normally used as a basis for writing a computer program. This is a set of rules with a finite number of steps for solving a problem.
Alias frequency	A false lower frequency component that appears in data reconstructed from original data acquired at an insufficient sampling rate (which is less than two (2) times the maximum frequency of the original data).
ALU	Arithmetic logic unit

Amplitude flatness	A measure of how close to constant the gain of a circuit remains over a range of frequencies.
Amplitude modulation	A modulation technique (also referred to as AM or ASK) used to allow data to be transmitted across an analog network, such as a switched telephone network. The amplitude of a single (carrier) frequency is varied or modulated between two levels – one for binary 0 and one for binary 1.
Analog	A continuous real time phenomena where the information values are represented in a variable and continuous waveform.
ANSI	American National Standards Institute – the principal standards development body in the USA.
APM	Alternating pulse modulation
Appletalk	A proprietary computer networking standard initiated by the Apple Computer for use in connecting the Macintosh range of computers and peripherals (including laser writer printers). This standard operates at 230 kbps.
Application layer	The highest layer of the seven layer ISO/OSI reference model structure, which contains all user or application programs.
Arithmetic logic unit	The element(s) in a processing system that perform(s) the mathematical functions such as addition, subtraction, multiplication, division, inversion, AND, OR, NAND and NOR.
ARP	Address resolution protocol A transmission control protocol/ Internet protocol (TCP/IP) process that maps an IP address to Ethernet address, required by TCP/IP for use with Ethernet.
ARQ	Automatic request for transmission A request by the receiver for the transmitter to retransmit a block or frame because of errors detected in the originally received message.
AS	Australian standard
ASCII	American standard code for information interchange. A universal standard for encoding alphanumeric characters into 7 or 8 binary bits. Drawn up by ANSI to ensure compatibility between different computer systems.
AS-i	Actuator sensor interface
ASIC	Application specific integrated circuit
ASK	Amplitude shift keying – see Amplitude modulation

ASN.1	Abstract syntax notation 1 – an abstract syntax used to define the structure of the protocol data units associated with a particular protocol entity.
Asynchronous	Communications where characters can be transmitted at an arbitrary unsynchronized point in time and where the time intervals between transmitted characters may be of varying lengths. Communication is controlled by start and stop bits at the beginning and end of each character.
Attenuation	The decrease in the magnitude of strength (or power) of a signal. In cables, generally expressed in dB per unit length.
AWG	American wire gauge

B

Balanced circuit	A circuit so arranged that the impressed voltages on each conductor of the pair are equal in magnitude but opposite in polarity with respect to ground.
Bandpass filter	A filter that allows only a fixed range of frequencies to pass through. All other frequencies outside this range (or band) are sharply reduced in magnitude.
Bandwidth	The range of frequencies available expressed as the difference between the highest and lowest frequencies is expressed in Hertz (or cycles per second).
Base address	A memory address that serves as the reference point. All other points are located by offsetting in relation to the base address.
Baseband	Baseband operation is the direct transmission of data over a transmission medium without the prior modulation on a high frequency carrier band.
Baud	Unit of signaling speed derived from the number of events per second (normally bits per second). However if each event has more than one bit associated with it the baud rate and bits per second are not equal.
Baudot	Data transmission code in which five bits represent one character. 64 alphanumeric characters can be represented. This code is used in many teleprinter systems with one start bit and 1.42 stop bits added.
BCC	Block check calculation
BCC	Block check character – error checking scheme with one check character; a good example being block sum check.
BCD	Binary coded decimal A code used for representing decimal digits in a binary code.
BEL	Bell (ASCII for control-G)
Bell 212	An AT&T specification of full duplex, asynchronous or synchronous 1200 baud data transmission for use on the public telephone networks.
BER	Bit error rate

BERT/BLERT	Bit error rate/block error rate testing – an error checking technique that compares a received data pattern with a known transmitted data pattern to determine transmission line quality.
BIN	Binary digits
BIOS	Basic input/output system
Bipolar	A signal range that includes both positive and negative values.
BISYNC	Binary synchronous communications protocol
Bit	Derived from "**BI**nary Digi**T**", a one or zero condition in the binary system.
Bit stuffing with zero bit insertion	A technique used to allow pure binary data to be transmitted on a synchronous transmission line. Each message block (frame) is encapsulated between two flags that are special bit sequences. Then if the message data contains a possibly similar sequence, an additional (zero) bit is inserted into the data stream by the sender, and is subsequently removed by the receiving device. The transmission method is then said to be data transparent.
Bits per second (bps)	Unit of data transmission rate.
Block sum check	This is used for the detection of errors when data is being transmitted. It comprises a set of binary digits (bits) which are the modulo 2 sum of the individual characters or octets in a frame (block) or message.
Bridge	A device to connect similar subnetworks without its own network address. Used mostly to reduce the network load.
Broadband	A communications channel that has greater bandwidth than a voice grade line and is potentially capable of greater transmission rates. Opposite of baseband. In wideband operation the data to be transmitted are first modulated on a high frequency carrier signal. They can then be simultaneously transmitted with other data modulated on a different carrier signal on the same transmission medium.
Broadcast	A message on a bus intended for all devices that requires no reply.
BS	Backspace (ASCII Control-H)
BS	British standard
BSC	Bisynchronous transmission A byte or character oriented communication protocol that has become the industry standard (created by IBM). It uses a defined set of control characters for synchronized transmission of binary coded data between stations in a data communications system.
BSP	Binary synchronous protocol

Buffer	An intermediate temporary storage device used to compensate for a difference in data rate and data flow between two devices (also called a spooler for interfacing a computer and a printer).
Burst mode	A high-speed data transfer in which the address of the data is sent followed by back to back data words while a physical signal is asserted.
Bus	A data path shared by many devices with one or more conductors for transmitting signals, data or power.
Byte	A term referring to eight associated bits of information; sometimes called a 'character'.

C

CAN	Controller area network
Capacitance	Storage of electrically separated charges between two plates having different potentials. The value is proportional to the surface area of the plates and inversely proportional to the distance between them.
Capacitance (mutual)	The capacitance between two conductors with all other conductors, including shield, short-circuited to the ground.
CATV	Community Antenna Television
CCITT	See ITU
Cellular polyethylene	Expanded or 'foam' polyethylene consisting of individual closed cells suspended in a polyethylene medium.
Character	Letter, numeral, punctuation, control figure or any other symbol contained in a message.
Characteristic impedance	The impedance that, when connected to the output terminals of a transmission line of any length, makes the line appear infinitely long. The ratio of voltage to current at every point along a transmission line on which there are no standing waves.
CIC	Controller in charge
Clock	The source(s) of timing signals for sequencing electronic events e.g. synchronous data transfer.
CMD	Command byte
CMR	Common mode rejection
CMRR	Common mode rejection ratio
CMV	Common mode voltage
Common carrier	A private data communications utility company that furnishes communications services to the general public.

Composite link	The line or circuit connecting a pair of multiplexers or concentrators; the circuit carrying multiplexed data.
Contention	The facility provided by the dial network or a data PABX which allows multiple terminals to compete on a first come, first served basis for a smaller number of computer posts.
CPU	Central processing unit
CR	Carriage return (ASCII control-M)
CRC	Cyclic redundancy check – an error-checking mechanism using a polynomial algorithm based on the content of a message frame at the transmitter and included in a field appended to the frame. At the receiver, it is then compared with the result of the calculation that is performed by the receiver. Also referred to as CRC-16.
CRL	Communication relationship list
Cross talk	A situation where a signal from a communications channel interferes with an associated channel's signals.
Crossed planning	Wiring configuration that allows two DTE or DCE devices to communicate. Essentially it involves connecting pin 2 to pin 3 of the two devices.
Crossover	In communications, a conductor that runs through the cable and connects to a different pin number at each end.
CSMA/CD	Carrier sense multiple access/collision detection – when two senders transmit at the same time on a local area network; they both cease transmission and signal that a collision has occurred. Each then tries again after waiting for a predetermined time period.
CTS	Clear to send
Current Loop	Communication method that allows data to be transmitted over a longer distance with a higher noise immunity level than with the standard EIS-232-C voltage method. A mark (a binary 1) is represented by current of 20 mA and a space (or binary 0) is represented by the absence of current.

D

DAQ	Data acquisition
Data integrity	A performance measure based on the rate of undetected errors.
Data link layer	This corresponds to layer 2 of the ISO reference model for open systems interconnection. It is concerned with the reliable transfer of data (no residual transmission errors) across the data link being used.
Data reduction	The process of analyzing large quantities of data in order to extract some statistical summary of the underlying parameters.

Datagram	A type of service offered on a packet-switched data network. A datagram is a self contained packet of information that is sent through the network with minimum protocol overheads.
DCD	Data carrier detect
DCE	Data communications equipment or data circuit-terminating equipment – devices that provide the functions required to establish, maintain, and terminate a data transmission connection. Normally it refers to a modem.
DCS	Distributed control systems
Decibel (dB)	A logarithmic measure of the ratio of two signal levels where $dB = 20\log_{10} V1/V2$ or where $dB = 10\log_{10} P1/P2$ and where V refers to voltage or P refers to power. Note that it has no units of measurement.
Default	A value or setup condition assigned, which is automatically assumed for the system unless otherwise explicitly specified.
Delay distortion	Distortion of a signal caused by the frequency components making up the signal having different propagation velocities across a transmission medium.
DES	Data encryption standard
DFM	Direct frequency modulation
Dielectric constant (E)	The ratio of the capacitance using the material in question as the dielectric, to the capacitance resulting when the material is replaced by air.
Digital	A signal which has definite states (normally two).
DIN	Deutsches Institut Fur Normierung
DIP	Dual in line package, referring to integrated circuits and switches.
Direct memory access	A technique of transferring data between the computer memory and a device on the computer bus without the intervention of the microprocessor. Also abbreviated to DMA.
DISC	Disconnect
DLE	Data link escape (ASCII character)
DNA	Distributed network architecture
DPI	Dots per inch
DPLL	Digital phase locked loop
DR	Dynamic range The ratio of the full-scale range (FSR) of a data converter to the smallest difference it can resolve. DR = 2n where n is the resolution in bits.

Driver software	A program that acts as the interface between a higher-level coding structure and the lower level hardware/firmware component of a computer.
DSP	Digital signal processing
DSR	Data set ready or DCE ready in EIA-232D/E – A EIA-232 modem interface control signal which indicates that the terminal is ready for transmission.
DTE	Data terminal equipment – devices acting as data source or data sink, or both.
DTR	Data terminal ready or DTE ready in EIA-232D/E
Duplex	The ability to send and receive data simultaneously over the same communications line.

E

EBCDIC	Extended binary coded decimal interchange code An eight bit character code used primarily in IBM equipment. The code allows for 256 different bit patterns.
EDAC	Error detection and correction
EFTPOS	Electronic funds transfer at the point of sale
EIA	Electronic Industries Association – a standards organization in the USA specializing in the electrical and functional characteristics of interface equipment.
EISA	Enhanced industry standard architecture
EMI/RFI	Electromagnetic interference/radio frequency interference 'background noise' that could modify or destroy data transmission.
EMS	Expanded memory specification
Emulation	The imitation of a computer system performed by a combination of hardware and software that allows programs to run between incompatible systems.
ENQ	Enquiry (ASCII Control-E)
EOT	End of transmission (ASCII Control-D)
EPA	Enhanced performance architecture
EPR	Earth potential rise
EPROM	Erasable programmable read only memory – non-volatile semiconductor memory that is erasable in an ultra violet light and reprogrammable.
Error rate	The ratio of the average number of bits that will be corrupted to the total number of bits that are transmitted for a data link or system.
ESC	Escape (ASCII character)
ESD	Electrostatic discharge
ETB	End of transmission block

Ethernet	Name of a widely used LAN, based on the CSMA/CD bus access method (IEEE 802.3). Ethernet is the basis of the TOP bus topology.
ETX	End of text (ASCII control-C)
Even parity	A data verification method normally implemented in hardware in which each character must have an even number of 'ON' bits.

F

Farad	Unit of capacitance whereby a charge of one coulomb produces a one volt potential difference.
FAS	Fieldbus access sublayer
FCC	Federal communications commission
FCS	Frame check sequence – a general term given to the additional bits appended to a transmitted frame or message by the source to enable the receiver to detect possible transmission errors.
FDM	Frequency division multiplexer– a device that divides the available transmission frequency range in narrower bands, each of which is used for a separate channel.
FIB	Factory information bus
FIFO	First in, first out
Filled cable	A telephone cable construction in which the cable core is filled with a material that will prevent moisture from entering or passing along the cable.
FIP	Factory instrumentation protocol
Firmware	A computer program or software stored permanently in PROM or ROM or semi-permanently in EPROM.
Flame retardancy	The ability of a material not to propagate flame once the flame source is removed.
Flow control	The procedure for regulating the flow of data between two devices preventing the loss of data once a device's buffer has reached its capacity.
FMS	Fieldbus message specification
FNC	Function byte
Frame	The unit of information transferred across a data link. Typically, there are control frames for link management and information frames for the transfer of message data.
Frequency modulation	A modulation technique (abbreviated to FM) used to allow data to be transmitted across an analog network where the frequency is varied between two levels – one for binary '0'

and one for binary '1'. Also known as frequency shift keying (or FSK).

Frequency	Refers to the number of cycles per second.
FRMR	Frame reject
FSK	Frequency shift keying, see frequency modulation
Full duplex	Simultaneous two-way independent transmission in both directions (4 wire). See Duplex.

G

G	Giga (metric system prefix – 10^9)
Gateway	A device to connect two different networks which translates the different protocols.
GMSK	Gaussian minimum shift keying
GPIB	General purpose interface bus – an interface standard used for parallel data communication, usually used for controlling electronic instruments from a computer. Also known as IEEE 488 standard.
Ground	An electrically neutral circuit that has the same potential as the earth. A reference point for an electrical system also intended for safety purposes.

H

Half duplex	Transmissions in either direction, but not simultaneously.
Hamming Distance	A measure of the effectiveness of error checking. The higher the Hamming distance (HD) index, the safer is the data transmission.
Handshaking	Exchange of predetermined signals between two devices establishing a connection.
Hardware	Refers to the physical components of a device, such as a computer, sensor, controller or data communications system. These are the physical items that one can see.
HART	Highway addressable remote transducers
HDLC	High level data link control. The international standard communication protocol defined by ISO to control the exchange of data across either a point-to-point data link or a multidrop data link.
Hertz (Hz)	A term replacing cycles per second as a unit of frequency.
Hex	Hexadecimal
HF	High frequency
Host	This is normally a computer belonging to a user that contains (hosts) the communication hardware and software necessary to connect the computer to a data communications network.

HSE	High speed Ethernet

I

I/O address	A method that allows the CPU to distinguish between different boards in a system. All boards must have different addresses.
IA5	International alphabet number 5
IC	Integrated circuit
ICS	Instrumentation and control system
IDF	Intermediate distribution frame
IEC	International Electrotechnical Commission
IEE	Institution of Electrical Engineers – an American based international professional society that issues its own standards and is a member of ANSI and ISO.
IEEE	Institute of Electrical and Electronic Engineers
IFC	International Fieldbus Consortium
ILD	Injection laser diode
Impedance	The total opposition that a circuit offers to the flow of alternating current or any other varying current at a particular frequency. It is a combination of resistance R and reactance X, measured in Ohms.
Inductance	The property of a circuit or circuit element that opposes a change in current flow, thus causing current changes to lag behind voltage changes. It is measured in henrys.
Insulation resistance (IR)	That resistance offered by insulation to an impressed dc voltage, tending to produce a leakage current though the insulation.
Interface	A shared boundary defined by common physical interconnection characteristics, signal characteristics and measurement of interchanged signals.
Interrupt handler	The section of the program that performs the necessary operation to service an interrupt when it occurs.
Interrupt	An external event indicating that the CPU should suspend its current task to service a designated activity.
IP	Internet protocol
IRQ	Interrupt request line
ISA	Industry Standard Architecture (for IBM Personal Computers)
ISB	Intrinsically safe barrier
ISDN	Integrated services digital network – the new generation of worldwide telecommunications network that utilizes digital

techniques for both transmission and switching. It supports both voice and data communications.

ISO	International Standards Organization
ISP	Interoperable systems project
ISR	Interrupt service routine, see interrupt handler
ITB	End of intermediate block
ITS	Interface terminal strip
ITU	International Telecommunications Union – formerly CCITT (Consultative Committee International Telegraph and Telephone). An international association that sets worldwide standards (e.g. V.21, V.22, V.22bis).

J

Jumper	A wire connecting one or more pins on the one end of a cable only.

K

k (kilo)	This is 2^{10} or 1024 in computer terminology, e.g. 1 kB = 1024 bytes.

L

LAN	Local area network – a data communications system confined to a limited geographic area typically about 10 km with moderate to high data rates (100 kbps to 50 Mbps). Some type of switching technology is used, but common carrier circuits are not used.
LAN	Local area network. See Local area network.
LAP-M	Link access protocol modem
LAS	Link active scheduler
LCD	Liquid crystal display – a low-power display system used on many laptops and other digital equipment.
LDM	Limited distance modem – a signal converter which conditions and boosts a digital signal so that it may be transmitted further than a standard EIA-232 signal.
Leased (or Private) line	A private telephone line without inter-exchange switching arrangements.
LED	Light emitting diode. A semiconductor light source that emits visible light or infra red radiation.
LF	Line feed (ASCII Control-J)
Line driver	A signal converter that conditions a signal to ensure reliable transmission over an extended distance.

Line turnaround	The reversing of transmission direction from transmitter to receiver or vice versa when a half duplex circuit is used.
Linearity	A relationship where the output is directly proportional to the input.
Link layer	Layer 2 of the ISO/OSI reference model. Also known as the data link layer.
Listener	A device on the GPIB bus that receives information from the bus.
LLC	Logical link control (IEEE 802)
LLI	Lower layer interface
Loaded line	A telephone line equipped with loading coils to add inductance in order to minimize amplitude distortion.
Loop resistance	The measured resistance of two conductors forming a circuit.
Loopback	Type of diagnostic test in which the transmitted signal is returned on the sending device after passing through all, or a portion of, a data communication link or network. A loopback test permits the comparison of a returned signal with the transmitted signal.
LRC	Longitudinal redundancy check
LSB	Least significant bits – the digits on the right hand side of the written HEX or BIN codes.
LSD	Least significant digit

M

M	Mega. Metric system prefix for 10^6.
m	Meter. Metric system unit for length.
MAC	Media Access Control (IEEE 802).
MAN	Metropolitan Area Network
Manchester encoding	Digital technique (specified for the IEEE 802.3 Ethernet baseband network standard) in which each bit period is divided into two complementary halves; a negative to positive voltage transition in the middle of the bit period designates a binary '1', whilst a positive to negative transition represents a '0'. The encoding technique also allows the receiving device to recover the transmitted clock from the incoming data stream (self clocking).
MAP 3.0	Standard profile for manufacturing developed by MAP.
MAP	Manufacturing automation protocol – a suite of network protocols originated by General Motors, which follow the seven layers of the OSI model. A reduced implementation is referred to as a mini-MAP.
Mark	This is equivalent to a binary 1.

Master/slave	Bus access method whereby the right to transmit is assigned to one device only, the **master,** and all the other devices, the **slaves** may only transmit when requested.
MDF	Main distribution frame
MIPS	Million instructions per second
MMI	Man-machine-interface
MMS	Manufacturing message services – a protocol entity forming part of the application layer. It is intended for use specifically in the manufacturing or process control industry. It enables a supervisory computer to control the operation of a distributed community of computer-based devices.
MNP	Microcom networking protocol
Modem eliminator	A device used to connect a local terminal and a computer port in lieu of the pair of modems to which they would ordinarily connect, allow DTE to DTE data and control signal connections otherwise not easily achieved by standard cables or connections.
Modem	MODulator/DEModulator – a device used to convert serial digital data from a transmitting terminal to a signal suitable for transmission over a telephone channel or to reconvert the transmitted signal to serial digital data for the receiving terminal.
MOS	Metal oxide semiconductor
MOV	Metal oxide varistor
MSB	Most significant bits – the digits on the left hand side of the written HEX or BIN codes.
MSD	Most significant digit
MTBF	Mean time between failures
MTTR	Mean time to repair
Multidrop	A single communication line or bus used to connect three or more points.
Multiplexer (MUX)	A device used for division of a communication link into two or more channels either by using frequency division or time division.

N

NAK	Negative acknowledge (ASCII Control-U)
Network architecture	A set of design principles including the organization of functions and the description of data formats and procedures used as the basis for the design and implementation of a network (ISO).

Network layer	Layer 3 in the ISO/OSI reference model, the logical network entity that services the transport layer responsible for ensuring that data passed to it from the transport layer is routed and delivered throughout the network.
Network topology	The physical and logical relationship of nodes in a network; the schematic arrangement of the links and nodes of a network typically in the form of a star, ring, tree or bus topology.
Network	An interconnected group of nodes or stations.
NMRR	Normal mode rejection ratio
Node	A point of interconnection to a network.
Noise	A name given to the extraneous electrical signals that may be generated or picked up in a transmission line. If the noise signal is large compared with the data carrying signal, the latter may be corrupted resulting in transmission errors.
NOS	Network operating system
NRM	Unbalanced normal response mode
NRZ	Non return to zero – pulses in alternating directions for successive 1 bits but no change from existing signal voltage for 0 bits.
NRZI	Non return to zero inverted
Null modem	A device that connects two DTE devices directly by emulating the physical connections of a DCE device.
Nyquist sampling theorem	In order to recover all the information about a specified signal it must be sampled at least at twice the maximum frequency component of the specified signal.

O

OD	Object dictionary
Ohm (Ω)	Unit of resistance such that a constant current of one ampere produces a potential difference of one Volt across a conductor.
Optical isolation	Two networks with no electrical continuity in their connection because an optoelectronic transmitter and receiver have been used.
OSI	Open systems interconnection

P

Packet	A group of bits (including data and call control signals) transmitted as a whole on a packet switching network. Usually smaller than a transmission block.
PAD	Packet access device – an interface between a terminal or computer and a packet switching network.

Parallel transmission	The transmission model where a number of bits is sent simultaneously over separate parallel lines. Usually unidirectional such as the Centronics interface for a printer.
Parity bit	A bit that is set to a '0' or '1' to ensure that the total number of 1 bits in the data field is even or odd.
Parity check	The addition of non-information bits that make up a transmission block to ensure that the total number of bits is always even (even parity) or odd (odd parity). Used to detect transmission errors but rapidly losing popularity because of its weakness in detecting errors.
Passive filter	A circuit using only passive electronic components such as resistors, capacitors and inductors.
PBX	Private branch exchange
PCIP	Personal computer instrument products
PDU	Protocol data unit
Peripherals	The input/output and data storage devices attached to a computer e.g. disk drives, printers, keyboards, display, communication boards, etc.
Phase modulation	The sine wave or carrier changes phase in accordance with the information to be transmitted.
Phase shift keying	A modulation technique (also referred to as PSK) used to convert binary data into an analog form comprising a single sinusoidal frequency signal whose phase varies according to the data being transmitted.
Physical layer	Layer 1 of the ISO/OSI reference model, concerned with the electrical and mechanical specifications of the network termination equipment.
PID	Proportional integral derivative – a form of closed loop control.
PLC	Programmable logic controller
Point to point	A connection between only two items of equipment.
Polling	A means of controlling devices on a multipoint line. A controller queries devices for a response.
Polyethylene	A family of insulators derived from the polymerization of ethylene gas and characterized by outstanding electrical properties, including high IR, low dielectric constant, and low dielectric loss across the frequency spectrum.
Polyvinyl chloride (PVC)	A general-purpose family of insulations whose basic constituent is polyvinyl chloride or its copolymer with vinyl acetate. Plasticizers, stabilisers, pigments and fillers are

	added to improve mechanical and/or electrical properties of this material.
Port	A place of access to a device or network, used for input/output of digital and analog signals.
Presentation layer	Layer 6 of the ISO/OSI reference model, concerned with negotiating suitable transfer syntax for use during an application. If this is different from the local syntax, the translation to/from this syntax.
Profibus	Process field bus developed by a consortium of mainly German companies with the aim of standardization.
Protocol entity	The code that controls the operation of a protocol layer.
Protocol	A formal set of conventions governing the formatting, control procedures and relative timing of message exchange between two communicating systems.
PSDN	Public switched data network Any switching data communications system, such as telex and public telephone networks, which provides circuit switching to many customers.
PSK	See Phase shift keying
PSTN	Public switched telephone network – this is the term used to describe the (analog) public telephone network.
PTT	Post, Telephone and Telecommunications Authority or: push to talk signal
PV	Primary variable

Q

QAM	Quadrature amplitude modulation
QPSK	Quadrature phase shift keying

R

R/W	Read/write
RAM	Random access memory – semiconductor read/write volatile memory. Data is lost if the power is turned off.
Reactance	The opposition offered to the flow of alternating current by inductance or capacitance of a component or circuit.
REJ	Reject
Repeater	An amplifier that regenerates the signal and thus expands the network.
Resistance	The ratio of voltage to electrical current for a given circuit measured in Ohms.
Response time	The elapsed time between the generation of the last character of a message at a terminal and the receipt of the first character of the reply. It includes terminal delay and network delay.

RF	Radio frequency
RFI	Radio frequency interference
Ring	Network topology commonly used for interconnection of communities of digital devices distributed over a localized area, e.g. a factory or office block. Each device is connected to its nearest neighbors until all the devices are connected in a closed loop or ring. Data is transmitted in one direction only. As each message circulates around the ring, it is read by each device connected in the ring.
RMS	Root mean square
RNR	Receiver not ready
ROM	Read only memory – computer memory in which data can be routinely read but written to only once using special means when the ROM is manufactured. A ROM is used for storing data or programs on a permanent basis.
Router	A linking device between network segments which may differ in layers 1, 2a and 2b of the ISO/OSI reference model.
RR	Receiver ready
RS	Recommended standard (e.g. RS-232C) – newer designations use the prefix EIA (e.g. EIA-RS-232C or just EIA-232C).
RS-232-C	Interface between DTE and DCE, employing serial binary data exchange. Typical maximum specifications are 15 m (50 feet) at 19200 Baud.
RS-422	Interface between DTE and DCE employing the electrical characteristics of balanced voltage interface circuits.
RS-423	Interface between DTE and DCE, employing the electrical characteristics of unbalanced voltage digital interface circuits.
RS-449	General purpose 37 pin and 9 pin interface for DCE and DTE employing serial binary interchange.
RS-485	The recommended standard of the EIA that specifies the electrical characteristics of drivers and receivers for use in balanced digital multipoint systems.
RSSI	Receiver signal strength indicator
RTS	Request to send
RTU	Remote terminal unit – terminal unit situated remotely from the main control system.
RxRDY	Receiver ready

S

S/N	Signal to noise (ratio)
SAA	Standards Association of Australia
SAP	Service access point

SDLC	Synchronous data link control – IBM standard protocol superseding the bisynchronous standard.
SDM	Space division multiplexing
SDS	Smart distributed system
Serial transmission	The most common transmission mode in which information bits are sent sequentially on a single data channel.
Session layer	Layer 5 of the ISO/OSI reference model, concerned with the establishment of a logical connection between two application entities and with controlling the dialogue (message exchange) between them.
SFD	The start of frame delimiter
Short haul modem	A signal converter that conditions a digital signal for transmission over dc continuous private line metallic circuits, without interfering with adjacent pairs of wires in the same telephone cables.
Signal to noise ratio	The ratio of signal strength to the level of noise.
Simplex transmissions	Data transmission in one direction only.
Slew rate	This is defined as the rate at which the voltage changes from one value to another.
SNA	Subnetwork access, or systems network architecture
SNDC	Subnetwork dependent convergence
SNIC	Subnetwork independent convergence
SNR	Signal to noise ratio
Software	Refers to the programs that are written by a user to control the actions of a microprocessor or a computer. These may be written in one of many different programming languages and may be changed by the user from time to time.
SOH	Start of header (ASCII Control-A)
Space	Absence of signal. This is equivalent to a binary 0.
Spark test	A test designed to locate imperfections (usually pinholes) in the insulation of a wire or cable by application of a voltage for a very short period of time while the wire is being drawn through the electrode field.
SRC	Source node of a message
SREJ	Selective reject
Star	A type of network topology in which there is a central node that performs all switching (and hence routing) functions.
Statistical multiplexer	A device used to enable a number of lower bit rate devices, normally situated in the same location, to share a single,

higher bit rate transmission line. The devices usually have human operators and hence data is transmitted on the shared line on a statistical basis rather than, as is the case with a basic multiplexer, on a pre-allocated basis. It endeavors to exploit the fact that each device operates at a much lower mean rate than its maximum rate.

STP	Shielded twisted pair
Straight through pinning	RS-232 and RS-422 configuration that match DTE to DCE, pin for pin (pin 1 with pin 1, pin 2 with pin 2, etc.).
STX	Start of text (ASCII Control-B).
Switched line	A communication link for which the physical path may vary with each usage, such as the public telephone network.
SYN	Synchronous Idle
Synchronization	The coordination of the activities of several circuit elements.
Synchronous transmission	Transmission in which data bits are sent at a fixed rate, with the transmitter and receiver synchronized. Synchronized transmission eliminates the need for start and stop bits.

T

Talker	A device on the GPIB bus that simply sends information on to the bus without actually controlling the bus.
TCP	Transmission control protocol
TCU	Trunk coupling unit
TDM	Time division multiplexer A device that accepts multiple channels on a single transmission line by connecting terminals, one at a time, at regular intervals, interleaving bits (bit TDM) or characters (character TDM) from each terminal.
Telegram	In general a data block which is transmitted on the network. Usually comprises address, information and check characters.
Temperature rating	the maximum and minimum temperature at which an insulating material may be used in continuous operation without loss of its basic properties.
TIA	Telecommunications Industry Association
Time sharing	A method of computer operation that allows several interactive terminals to use one computer.
TNS	Transaction bytes
Token ring	Collision free, deterministic bus access method as per IEEE 802.2 ring topology.

TOP	Technical Office Protocol – a user association in USA which is primarily concerned with open communications in offices.
Topology	Physical configuration of network nodes, e.g. bus, ring, star, tree.
Transceiver	Transmitter/receiver – network access point for IEEE 803.2 networks.
Transient	An abrupt change in voltage of short duration.
Transport layer	Layer 4 of the ISO/OSI reference model, concerned with providing a network independent reliable message interchange service to the application oriented layers (Layers 5 through 7).
Trunk	A single circuit between two points, both of which are switching centers or individual distribution points. A trunk usually handles many channels simultaneously.
TTL	Transistor-transistor logic
Twisted pair	A data transmission medium, consisting of two insulated copper wires twisted together. This improves its immunity to interference from nearby electrical sources that may corrupt the transmitted signal.

U

UART	Universal asynchronous receiver/transmitter – an electronic circuit that translates the data format between a parallel representation, within a computer, and the serial method of transmitting data over a communications line.
UHF	Ultra high frequency
Unbalanced circuit	A transmission line in which voltages on the two conductors are unequal with respect to ground e.g. a coaxial cable.
Unloaded line	A line with no loaded coils that reduce line loss at audio frequencies.
UP	Unnumbered poll
USB	Universal serial bus
USRT	Universal synchronous receiver/transmitter. See UART.
UTP	Unshielded twisted pair

V

V.35	ITU standard governing the transmission at 48 kbps over 60 to 108 kHz group band circuits.
Velocity of propagation	The speed of an electrical signal down a length of cable compared to speed in free space expressed as a percentage.
VFD	Virtual field device – a software image of a field device describing the objects supplied by it e.g. measured data,

events, status etc. which can be accessed by another network.

VHF	Very high frequency
VLAN	Virtual LAN
Volatile memory	An electronic storage medium that loses all data when power is removed.
Voltage rating	The highest voltage that may be continuously applied to a wire in conformance with standards of specifications.
VRC	Vertical redundancy check
VSD	Variable speed drive
VT	Virtual terminal

W

WAN	Wide area network
Word	The standard number of bits that a processor or memory manipulates at one time. Typically, a word has 16 bits.

X

X.21	ITU standard governing interface between DTE and DCE devices for synchronous operation on public data networks.
X.25	ITU standard governing interface between DTE and DCE device for terminals operating in the packet mode on public data networks.
X.25 Pad	A device that permits communication between non X.25 devices and the devices in an X.25 network.
X.3/X.28/X.29	A set of internationally agreed standard protocols defined to allow a character oriented device, such as a visual display terminal, to be connected to a packet switched data network.
X-ON/X-OFF	Transmitter on/transmitter off – control characters used for flow control, instructing a terminal to start transmission (X-ON or control-S) and end transmission (X-OFF or control-Q).
XOR	Exclusive-OR

Index